Costs and Returns for
Agricultural Commodities

Costs and Returns for Agricultural Commodities

Advances in Concepts and Measurement

EDITED BY

Mary C. Ahearn
and Utpal Vasavada

Routledge
Taylor & Francis Group

NEW YORK AND LONDON

First published 1992 by Westview Press, Inc.

Published 2021 by Routledge
605 Third Avenue, New York, NY 10017
2 Park Square, Milton Park, Abingdon, Oxon OX14 4RN

Routledge is an imprint of the Taylor & Francis Group, an informa business

Library of Congress Cataloging-in-Publication Data
Costs and returns for agricultural commodities : advances in concepts
  and measurement / edited by Mary C. Ahearn and Utpal Vasavada.
    p. cm.
  Includes bibliographical references.
  ISBN 0-8133-0369-9
  1. Agriculture—Economic aspects—Mathematical models. 2. Produce
trade—Mathematical models. I. Ahearn, Mary. II. Vasavada, Utpal.
HD1433.C67 1992
338.1'3—dc20   92-3840
                                                                         CIP

ISBN 13: 978-0-3670-0652-5 (hbk)
ISBN 13: 978-0-3671-5639-8 (pbk)

Contents

Foreword, John E. Lee, Jr.	*ix*
Preface	*xi*
Acknowledgments	*xiii*
About the Contributors	*xv*

Introduction
Mary C. Ahearn — 1

PART ONE
Contemporary Issues in Measuring Commodity Costs and Returns

1 Varied Purposes and Implications for Measuring Costs and Returns
B. H. Robinson — 11

2 Measurement Issues Relating to Short-Run Farm Financial Management Decisions
Gayle S. Willett — 19

3 Measurement Issues Relating to Economic Analysis
Glenn A. Helmers and Myles J. Watts — 25

4 Measurement Issues Relating to Policy Analysis
David H. Harrington — 43

PART TWO
Alternatives for Measuring Costs and Theoretical Considerations

5 Theoretical and Practical Considerations in Measuring Costs and Returns
Vernon R. Eidman — 59

6 Comparative Estimation Systems and the Random Coefficient Regression Approach
Robert H. Hornbaker, Steven T. Sonka, and Bruce L. Dixon   67

7 Approaches to Collecting Data on Production Technologies
Carol C. House   81

Discussion
Odell L. Walker   89
Eldon Ball   93
Richard A. Schoney   95

PART THREE
Comparability of Predominant Methods

8 The National Commodity Cost and Return Estimates
Mitchell Morehart, James D. Johnson, and Hosein Shapouri   103

9 The Case of Rice: Similarities and Differences Among Predominant Methods
Leo J. Guedry   129

10 Results of a National Survey on Data and Methods
Karen Klonsky   147

Discussion
DeeVon Bailey   165
John E. Ikerd   169

PART FOUR
Treatment of the Effects of Government Programs

11 Recognizing the Effects of Government Programs in Developing Cost and Returns Statements
B. F. Stanton   177

12 Measuring Cost and Returns: Do We Include Government Program Impacts?
Otto Doering   189

Discussion
Patrick Westhoff   199

PART FIVE
Estimating Costs of Land and Water Services

13 **Market Value Versus Agricultural Use Value of Farmland**
    Lindon J. Robison and Steven R. Koenig   207

14 **Land and Water Costs: A Case of Whose Costs, When**
    Charles H. Barnard and Michael Salassi   229

    Discussion
    Allen M. Featherstone   251
    Michael Duffy   255

PART SIX
Estimating Costs of Non-Human Capital Services

15 **Allocation of Capital Costs in Enterprise Budgets**
    Oscar R. Burt   259

16 **Estimating Costs of Durable and Operating Capital Services**
    Cole R. Gustafson, Peter J. Barry, and Mir B. Ali   273

    Discussion
    Utpal Vasavada   289
    Gregory M. Perry   293

PART SEVEN
Estimating Costs of Human Capital Services

17 **Human Capital Issues in Measuring Costs of Production**
    Daniel A. Sumner   303

18 **Costs and Returns: A Perspective on Estimating Costs of
    Human Capital Services and More**
    Wallace E. Huffman   313

    Discussion
    Carlyle Ross   337
    M. C. Hallberg   341

PART EIGHT
The Future of Commodity Costs and Returns Estimation

19 How Economic Theory Should Guide What We Measure
    Thomas A. Miller   347

20 Similarities and Differences in the Data Needs for Farmer
    Planning, Economic Research, and Policy Analysis
    Arne Hallam   367

21 How the Profession Could Coordinate Itself on This Issue
    R. J. Hildreth   383

APPENDIX
Computing Considerations

    Computing on Personal Computers: Three Case Studies
    Lawrence A. Lippke, Jerry Crews, and James C. Wade   389

Foreword

The papers and discussion recorded in this proceedings volume are the products of yet another conference on costs and returns of producing agricultural commodities. One would think that little new could be added to the numerous past writings, conferences and workshops on this topic, which, incidentally, has been around longer than agricultural economists.

Nevertheless, the intellectual exchange at the conference held in Kansas City in February 1991 was robust and exciting. There were perhaps several reasons for this being a more anticipated--and participated-in--conference than is normally the case. First and foremost, the conference was preceded by some controversy stemming from Congressional interest during the debates leading up to the 1990 Farm Bill in the possible use of cost of production (COP) data to set support prices for some commodities. That interest led to debate among a number of federal and state participants over just what were the costs for producing the commodities in question. The participants in that debate brought to the conference a genuine interest in solving some of the problems at the core of the controversy. Second, this conference attracted a wider range of participants than have traditionally been involved in COP debates. Theorists and researchers on a wide range of topics including land, labor, capital and government policies, along with statisticians, political scientists, extension workers and farmers attended the conference. Thus, the conference brought together practical interests and good science. Finally, the conference was productive because it was unusually well conceived, organized and managed.

Several outcomes of the conference will likely prove of lasting value. The first of these is general recognition and acceptance of the proposition that there are many uses of cost of production estimates, and that the most appropriate concepts and measurements depend in part on the intended use. Uses of COP estimates include, but are not limited to, farm management analysis, supply response modeling, regional and national competitiveness and a variety of public policy uses. In other words, there can be more than one "correct" (or incorrect) estimate of the cost of producing a given commodity or product. To avoid confusion and misinformation, the intended uses and the rationale for the chosen measurement concepts should always be stipulated in published estimates. This proceedings volume provides much food for thought on this topic.

Second, participants in the conference did offer further refinements in cost measurement concepts and methods. This contribution resulted in part from bringing together a more diverse group of economists; i.e., more economists who do not usually work on COP. As a result, the conference cut a wider swath through economic theory than usually occurs in cost of production dialogue and brought fresh new perspectives to some old issues.

Third, a new set of issues enumerated at the conference could bring new relevance and uses for COP data and new conceptual and measurement challenges. For example, the growing concerns about sustainable agriculture and preservation of natural resources could require the internalizing of costs and benefits into private and public decisions once external to those decision frameworks. Measuring total "social" costs of production requires including and measuring costs not traditionally included by agricultural economists.

Finally, a major contribution of the conference was the broad commitment to maintain a continuing effort to improve both the quality of COP data and the appropriateness of use of those data. A specific commitment was made to establish a task force under the auspices of the Economics Statistics Committee of the American Agricultural Economics Association to work on standardization of COP estimating procedures across states and between states and Federal agencies. A concurrent effort has been an initial attempt to interest agricultural economists in other countries to develop comparable COP estimates for cross-country use.

A persistent concern throughout the conference and reflected in the papers in this volume was with the potential for misuse and abuse of COP estimates. For example, two of the most common uses of COP estimates are (1) to determine comparative advantage and trade, and (2) to determine the appropriateness of government commodity price supports. Yet, these are probably the most potentially misleading and abused uses of such data. This potential underscores the responsibility of professional agricultural economists, especially those publicly employed, to explain and interpret COP data in the context of appropriate uses in order to minimize the dangers of inappropriate uses.

All in all, the Kansas City COP conference was an exciting, stimulating and useful venture. Much of that stimulation and usefulness is captured in this timely proceedings volume. It should prove to be a heavily used reference for years to come.

John E. Lee, Jr.

Preface

The chapters in this volume are edited versions of papers presented at a conference entitled "Economic Accounting for Commodity Costs and Returns." The conference was held in Kansas City, Missouri, from February 19 to 21, 1991. The conference was organized through the Economic Statistics Committee of the American Agricultural Economics Association and was sponsored by four agriculturally oriented organizations: The Economic Research Service of the U.S. Department of Agriculture, the Farm Foundation, The Federal Extension Service of the U.S. Department of Agriculture, and the American Agricultural Economics Association. The stated purpose of the conference was "...to explore how different uses of costs and return estimates determine different methods of estimation, and for similar uses, what are preferred estimation methods."

By the conclusion of the conference, several speakers and general conference participants had stated the need for more cooperation among those who develop estimates to increase the consistency of methods. The closing speaker of the conference challenged the group to accept this charge and suggested that the Economic Statistics Committee establish a Taskforce on Standardization of Commodity Costs and Returns. The Economics Statistics Committee did recruit a diverse group of participants to serve on such a taskforce. The first meeting of the Taskforce was held in February 1992. Taskforce leaders plan to have recommendations on standardization in about two years' time. The issues developed in this volume will serve as the basis for the upcoming discussions in the development of standards and the future discussions that, undoubtedly, will follow.

Mary C. Ahearn
Utpal Vasavada

Acknowledgments

The strong support of the Economic Research Service of the U.S. Department of Agriculture was critical to the completion of this volume and the organization of the parent conference.

M.C.A.
U.V.

About the Contributors

Mary C. Ahearn is a section leader, U.S. Department of Agriculture, Economic Research Service, Washington, D.C.

Mir B. Ali is an agricultural economist, U.S. Department of Agriculture, Economic Research Service, Washington, D.C.

DeeVon Bailey is an extension economist, Department of Economics, Utah State University, Logan, Utah.

Eldon Ball is a section leader, U.S. Department of Agriculture, Economic Research Service, Washington, D.C.

Charles H. Barnard is an agricultural economist, U.S. Department of Agriculture, Economic Research Service, Washington, D.C.

Peter J. Barry is a professor, Department of Agricultural Economics, University of Illinois, Urbana, Illinois.

Oscar R. Burt is a professor, Department of Agricultural Economics, University of California, Davis, California.

Jerry Crews is an extension economist, Department of Economics, Auburn University, Auburn, Alabama.

Bruce L. Dixon is a professor, Department of Agricultural Economics and Rural Sociology, University of Arkansas, Fayetteville, Arkansas.

Otto Doering is a professor, Department of Agricultural Economics, Purdue University, W. Lafayette, Indiana.

Michael Duffy is an associate professor, Department of Economics, Iowa State University, Ames, Iowa.

Vernon R. Eidman is a professor, Department of Agricultural and Applied Economics, University of Minnesota, St. Paul, Minnesota.

Allen M. Featherstone is an assistant professor, Department of Agricultural Economics, Kansas State University, Manhattan, Kansas.

Leo J. Guedry is a professor and head, Department of Agricultural Economics and Agribusiness, Louisiana State University, Baton Rouge, Louisiana.

Cole R. Gustafson is an assistant professor, Department of Agricultural Economics, North Dakota State University, Fargo, North Dakota.

Arne Hallam is an associate professor, Department of Economics, Iowa State University, Ames, Iowa.

M. C. Hallberg is a professor, Department of Agricultural Economics, Pennsylvania State University, University Park, Pennsylvania.

David H. Harrington is a deputy director for agriculture, U.S. Department of Agriculture, Economic Research Service, Washington, D.C.

Glenn A. Helmers is a professor, Department of Agricultural Economics, University of Nebraska, Lincoln, Nebraska.

R. J. Hildreth is retired. At the time of this writing, he was Director, The Farm Foundation, Oak Brook, Illinois.

Robert H. Hornbaker is an assistant professor, Department of Agricultural Economics, University of Illinois, Urbana, Illinois.

Carol C. House is a section leader, U.S. Department of Agriculture, National Agricultural Statistics Service, Washington, D.C.

Wallace E. Huffman is a professor, Department of Economics, Iowa State University, Ames, Iowa.

John E. Ikerd is a professor, Department of Agricultural Economics, University of Missouri, Columbia, Missouri.

James D. Johnson is a branch chief, U.S. Department of Agriculture, Economic Research Service, Washington, D.C.

Karen Klonsky is an extension specialist, Department of Agricultural Economics, University of California, Davis, California.

Steven R. Koenig is an agricultural economist, U.S. Department of Agriculture, Economic Research Service, Washington, D.C.

John E. Lee, Jr. is the Administrator, Economic Research Service, U.S. Department of Agriculture, Washington, D.C.

Lawrence A. Lippke is an extension specialist, Texas A&M University, College Station, Texas.

Thomas A. Miller is a professor, Department of Agricultural Economics, Colorado State University, Fort Collins, Colorado.

Mitchell Morehart is a section leader, U.S. Department of Agriculture, Economic Research Service, Washington, D.C.

Gregory M. Perry is an associate professor, Department of Agricultural and Resource Economics, Oregon State University, Corvallis, Oregon.

B.H. Robinson is a division director, U.S. Department of Agriculture, Economic Research Service, Washington, D.C.

Lindon J. Robison is a professor, Department of Agricultural Economics, Michigan State University, East Lansing, Michigan.

Carlyle Ross is a branch head, Alberta Agriculture, Edmonton, Alberta, Canada.

Michael Salassi is an agricultural economist, U.S. Department of Agriculture, Economic Research Service, Washington, D.C.

Richard A. Schoney is an associate professor, Department of Agricultural Economics, University of Saskatchewan, Saskatoon, Saskatchewan, Canada.

Hosein Shapouri is an agricultural economist, U.S. Department of Agriculture, Economic Research Service, Washington, D.C.

Steven T. Sonka is a professor, Department of Agricultural Economics, University of Illinois, Urbana, Illinois.

B. F. Stanton is a professor, Department of Agricultural Economics, Cornell University, Ithaca, New York.

Daniel A. Sumner is Deputy Assistant Secretary for Economics, U.S. Department of Agriculture, Washington, D.C.

Utpal Vasavada is a visiting associate professor, Department of Agricultural Economics, Texas A&M University, College Station, Texas. (At the time of this writing, he was an invited professor, Department of Rural Economics, Laval University, Quebec, Canada.)

James C. Wade is an extension economist, University of Arizona, Tucson, Arizona.

Odell L. Walker is a professor, Department of Agricultural Economics, Oklahoma State University, Stillwater, Oklahoma.

Myles J. Watts is a professor and head, Department of Economics, Montana State University, Bozeman, Montana.

Patrick Westhoff is an adjunct assistant professor, The Center for Agricultural and Rural Development, Iowa State University, Ames, Iowa.

Gayle S. Willett is an extension economist, Washington State University, Pullman, Washington.

Introduction

Mary C. Ahearn

Agriculture is a critical sector in the U.S. economy. Although production agriculture accounts for a small share of our Gross National Product, the whole of the food and fiber system accounts for about one-fifth of our economy and trade in agricultural commodities has a significant and positive influence on our trade balance. Of course, the conditions in agriculture are especially important to the regions of the country where farming activities are concentrated. There are still many communities in the United States where the local economy and identity of its residents are dependent on agriculture. Furthermore, in other countries, especially developing countries, agriculture's share of the economy is more significant than in the United States. The phenomenon of globalization has given agriculture an even greater political importance than its strictly economic significance would merit. This is largely because agriculture tends to be intimately linked to the variety of traditions and cultures of countries, as well as to an obvious concern for food security.

In this age of information, data sets of basic statistics are essential for decision making and analysis at all levels. Estimates of costs and returns for individual agricultural commodities are one of the most basic statistics for agriculture, and the agricultural economics profession is looked to by many to develop these estimates for different types of users. Agricultural economics has continued from its roots to be a very applied social science. Agricultural economists are accustomed to using commodity costs and returns data sets in addressing traditional and contemporary research issues. The most obvious of the emerging issues requiring relevant costs and returns measurement are: regional and international competitiveness; relative subsidization of agriculture across commodities; internalizing social costs of production, e.g., chemical runoff; the relationship between food quality/safety and costs of production; the recognition of the linkages within individual farming systems, e.g., crop rotations; and the implications of the economies of size relationship for farm and community structure.

Many data users would like to assume that an established data set matches the ideal data required for their analyses. A data user who has not had experience with the measurement issues might at first believe that measuring agricultural commodity costs and returns is a straightforward and noncontroversial process. How difficult can it be to measure the cost of producing a bushel of wheat, for example. Simply multiply prices times quantities of inputs, add them up, and divide by the number of bushels produced, right? However, even a short-time or casual user soon discovers that measurement problems are very real and sometimes so major that the data set must be deemed unusable for the particular purpose at hand. For example, go back to the bushel of wheat and consider a few measurement issues that must be addressed. How did the measurement approach deal with the land cost for wheat produced on land that was fallowed in preparation for the wheat production? Or, how does one make regional comparisons of competitiveness based on costs per bushel of wheat when one region suffers an unusual drought or one region produces valuable straw or pasture for livestock? Or, how does one handle the impacts of differing government policies when making international comparisons of costs? Furthermore, if the goal is to evaluate low input sustainable agricultural practices, the whole notion of costs and returns for individual commodities is especially unrealistic. The challenge is even more formidable, perhaps impossible, for the data user who wishes to use estimates from different sources and, most likely, developed with different methods.

In general, the approaches used today for estimating costs and returns were developed at a time when the major use of the statistics was for farmer planning. Given the long history of estimating costs and returns, it is not surprising that the estimation methods are steeped in this tradition. Consistency in methods is important in the production of time series statistics for obvious reasons--if methods are continually changed, the usefulness of the time series is compromised. Therefore, developers of statistics must continually make trade-offs between maintaining consistency over time and improving methods for any one year. The development of this volume is based on the premise that the time is at hand to make the trade-off in favor of evaluating and updating the underlying methods. We need to scrutinize more carefully the usefulness of the statistics produced by the *status quo* methods for addressing the burning contemporary issues. The reader in search of the one "correct" specification of costs and returns will be disappointed by the chapters in this volume. However, this volume seeks to open the black box of standard costs and returns computations and examine its contents.

Two of the most difficult challenges in accurately capturing commodity costs and returns are: (1) allocating costs across farms producing more than one product and for durable inputs, allocating costs across time, as well and (2) determining the most appropriate opportunity costs of owned inputs. How can economic theory guide us in this measurement task? For major crops in the U.S., those costs that require arbitrary allocation decision rules and/or opportunity cost pricing of owned inputs account for upwards of 50 percent of total economic costs. In regards to the allocation issue, some have argued that the first thing that economic theory tells you is that you cannot estimate costs and returns for individual commodities when production is not separable. That leaves some economists to cease further attempts at measurement. In fact, for one strictly concerned with the issue of separability, accurate annual whole farm costs and returns could conceivably be impossible because production may not be separable across years. For example, producers may plant a nitrogen-fixing legume in one year in preparation for the planting of a high nitrogen-consuming crop the next. Critics of those who dwell on the separability issue have argued: (1) the assumptions of economic theory often do not hold and (2) the information has proven useful and will always be demanded, therefore, agricultural economists have a professional responsibility to provide the most accurate estimate possible.

For those not daunted by the separability issue, one must still address what are appropriate opportunity costs of owned inputs. The basic notion of valuing owned inputs at the return that they could earn at the next best alternative is often invoked in cost and returns estimation. A great deal of diversity exists in exactly what rates are viewed as the next best alternative. If there are *n* owned inputs, it is common to value *n-1* of them at "some" opportunity cost and then, based on Euler's theorem, to value the *nth* as a residual. Of course, the conditions of Euler's theorem do not generally hold. These difficult issues and many others are addressed in the chapters.

The book is organized into eight major parts. The first part contains four chapters which introduce the major measurement issues associated with the different end-uses of cost and return estimates. Part 2 contains 3 chapters which look at conceptual issues. This part includes discussant comments as do the next 5 parts of the book. Part 3 includes a review of the predominant measurement methods. A major purpose of including this discussion is to provide a comprehensive review of what is done in the land grant universities and the Economic Research Service (ERS) in a concise fashion before preceding into discussion of several specific cost items. The next 4 parts cover the measurement issues associated with line items in a cost and returns statement that are often, maybe always, controversial. The idea in these parts is to gain a variety of perspectives, so two chapters and one or two discussants address the same issue. Part 8 presents three perspectives on future directions in measuring costs and returns. Finally, the book concludes with a very useful presentation, especially for practicing extension economists, on alternative computing approaches.

Observers of this latest effort to reevaluate costs and returns measurement have often commented with statements such as, "this is not a new issue" and "we have already done that." In Chapter 1 Bob Robinson underlines that view by reviewing the importance of the measurement issues to societies all the way back to ancient times. Robinson also argues that a major source of tension that has plagued costs and returns measurement has been the conflict between the cost accounting approach and an approach based in the economic theory of the firm. In chapter 2, Gayle Willett offers us his view on the issues as they relate to the purpose of farm management and extension activities. Extension economists in state land grant universities are primarily responsible for measuring commodity costs and returns. To extension economists, costs and returns are almost always forecasts since costs and returns are used as aids to farmer planning. This type of cost and return statement is commonly called a budget. The farmer planning purposes of extension economists often dictate the measurement approaches that are used. Willett is not very optimistic that extension economists have sufficient incentives to work towards standardization of their measurement methods, even though their general use of the forecasts is similar.

Glenn Helmers and Myles Watts provide a review of the measurement issues associated with economic research in Chapter 3. Although they do not take us back to ancient times, they do review more than a century of development in this area and express concern that little progress has been made in the last 50 years. Furthermore, they argue that the situation may have even worsened because many data users lack an understanding of the construction of the estimates, in contrast to users of the past. They also alert the reader to the severe limitations of cost of production estimates for economic research purposes, largely because cost of production point estimates are not an acceptable proxy for a commodity's supply function. In Chapter 4, David Harrington discusses the appeal of cost of production for setting policy parameters and details the dangers of doing so, including through the use of a numerical example. Harrington concludes with his ideas on how to break the appeal of cost of production for policy setting.

The introductory chapters are, according to their titles, organized along clear cut lines, however, the content necessarily overlaps somewhat. Readers interested in the historical background should read Helmers and Watts, as well as Robinson. Similarly, readers interested in the measurement issues associated with economic uses of costs and returns should read Harrington, as well as Helmers and Watts.

In Chapter 5, Vernon Eidman provides a conceptual overview of general methods. He starts by establishing the importance of cost and return estimates for a variety of purposes and then comments on the usefulness of the latest effort in ERS to provide greater distributional information on costs and returns. Traditionally, the standard computational procedures have had a geographic area as their unit of observation and yielded only average levels of costs and returns for these geographic areas. For several years now, the ERS has been investing in an approach that will allow for maximum flexibility in the information developed by maintaining the unit of observation at the farm level. Eidman correctly points out, however, that the improvements have not been made comprehensively, so that many of the computational practices of the traditional system are combined with the newer, more flexible approach. An example of this issue is the use of state average input prices to value the farm-level input quantities. The hybrid estimate produced underestimates the variation in costs. A constraint on the comprehensiveness, of course, is the greater data requirements of the farm-level approach to measurement. Eidman also comments on two examples in the literature which use econometric procedures to estimate costs and returns. He concludes that they hold promise as substitutes for the data-hungry traditional approaches, but that not enough information on their viability currently exists to evaluate their potential usefulness.

Rob Hornbaker, Steve Sonka, and Bruce Dixon are well qualified to elaborate on the issue of alternative approaches in Chapter 6 since they authored earlier work on the use of econometric approaches to cost estimation. They summarize that work and discuss the limitations of the approach as a substitute for the traditional approach. They also review alternative estimation approaches more generally. Hornbaker, Sonka, and Dixon have characterized the approaches as falling into three broad categories. This part of their chapter is useful reading for all, but especially for the reader who is relatively inexperienced at commodity-level costs and returns measurement practices. The authors also touch on the issue of trade-offs between statistically representative data and the nonrepresentative panel data available in several states. In Chapter 7, Carol House provides an excellent statistically-oriented overview of the various approaches to data collection and a critique of the statistical and practical advantages and disadvantages of each. She focuses on the issue of data quality and defines three characteristics of quality: accuracy, defensibility, and affordability.

Mitch Morehart, Jim Johnson, and Hosein Shapouri provide an overview of the estimates produced by ERS. These estimates are mandated by law and often requested by policymakers as guides to setting policy instruments. As such, they are the reference point for many, as is evident by the number of authors in this volume who make reference to the ERS methods. This chapter describes the latest methods. For example, the chapter briefly describes the current development of estimates with and without the direct effects of government programs and the long-term project to convert to a farm-level model of computation. However, the reader should also be aware that methods are constantly evolving, and each annual publication should be consulted for a documentation of the methods used for the particular estimates contained in that publication.

In Chapter 9, Leo Guedry details the differences in methods of calculating costs and returns for one commodity in six different institutions. The commodity is rice, and the reason for the focus on rice is because the controversy surrounding the discrepancies of various rice cost and returns estimates was one of the factors that led to the organization of the 1991 conference on costs and returns, as John Lee mentioned in his Foreword to this book. The whole incident is an excellent example of the importance of cost and return estimates and the need for the agricultural economics profession to provide thoughtful, well-documented estimates. It was during the debates for the 1990 farm bill that the Rice Growers Associations in the various states initiated the process to have land grant university economists develop estimates in the ERS format for ease of comparison. Guedry's chapter shows that as many differences exist among the states as between ERS and the states' methods. That point brings us to Karen Klonsky's Chapter 10. Klonsky conducted a survey of the states and ERS for the explicit purpose of reporting to the 1991 conference and in this volume on the variation in approaches to cost and returns measurement. The survey documented what many had expected--approaches varied widely across institutions. But for the first time, her survey results document precisely where the diversity exists. The variation in methods across institutions has been the cause of frustration for many who work in regional research committees and would like to use data across states in a consistent manner. I should also add that this inconsistency is true for whole farm data, as well, which has recently been documented through the activities of the NC-181 Committee on Size and Structure (Casler).

Part 4 includes two chapters on the treatment of the effects of government programs on costs and returns. The chapters in this part are noteworthy in the extent of the agreement reached. Both authors recognize the usefulness of the new ERS approach to report costs and returns both including and excluding the direct effects of government programs. (Planning and data collection for this new approach began in ERS in the spring of 1988, before the controversy arose over the treatment of government payments in rice cost and return estimates.) In addition, in Chapter 11, Bud Stanton establishes the importance of direct payments to the sector's total returns and the effects on the characterization of structure. In Chapter 12, Otto Doering compares the ERS estimates of cash and total economic costs to program provisions, such as target prices, as he discusses the use of cost of production estimates in setting support levels. Doering then moves on to summarize some earlier work which sought to extend the traditional costs and returns statement to include such items as taxes and social costs.

Part 5 addresses the issue of returns to land and water services and the two chapters take quite different approaches. In Chapter 13, Lindy Robison and Steve Koenig begin by describing their concerns with the quality of data on land values and rents. Most of their paper is then directed at attempting to econometrically separate the effects of agricultural use value from the other factors affecting the value of farm land in major farm states over time. In Chapter 14, Charlie Barnard and Mike Salassi take a much more comprehensive view. They begin by classifying the three basic approaches used to value land services and discuss the merits of each. They also review the economics literature on valuing land services and the relevant legislative history on computing costs for land services in ERS' estimates--at one point the exact method was legislatively mandated. They then discuss the pros and cons of the various estimation options.

Part 6 addresses capital costing issues for both operating and durable capital. In Chapter 15, Oscar Burt focuses on durable assets and argues that it is necessary to estimate used asset values implicitly from new prices and a theory of depreciation, since used markets are thin. This issue of used equipment markets comes up again in the discussants' comments. Most of Burt's paper develops the direction the theory should take in this regard in a very mathematical approach. Appropriate measures are defined for the cases of nonstochastic economic life and random economic life of a durable asset. Cole Gustafson, Peter Barry, and Mir Ali discuss the capital costing issue first by reviewing the historical development of accounting in ERS and the economics literature. A major focus of their chapter is on the appropriateness of nominal versus real values. Finally, they present an alternative econometric procedure to allocate returns to capital services.

In Part 7 of the volume, Daniel Sumner and Wallace Huffman address the issue of the cost of labor, or human capital, in a cost and return statement. Like the Stanton and Doering chapters, the general direction of the two chapters on the subject of human capital are quite similar. Unpaid labor is one of the most difficult cost items to measure because the identification of both price and quantity present special problems. Both authors argue that the opportunity cost of unpaid labor should be tied to off-farm work opportunities, rather than the state average hired farm work wage rate as is commonly done. However, in his conclusion, Sumner (Chapter 17) acknowledges the special problem of valuing unpaid labor--that opportunity costs are fundamentally unique to the individual farm. In Chapter 18, Huffman develops the household production model of decisionmaking to show the relevance of the off-farm wage rate as the relevant opportunity cost of unpaid labor. He also provides a brief outline of how the approach could be implemented for costs and returns measurement. Finally, he provides a rather comprehensive summary of his views on trends in labor's share in agriculture and the current situation in agricultural labor markets.

At the time of the development of this volume, standardization of concepts and methods was premature. However, a very genuine interest on the part of many exists to work towards improved and more consistent methods. Part 8 of the volume is an attempt to start down that road by summarizing what has been established in the previous chapters and where disagreement is still evident. In Chapter 19, Tom Miller begins the process by evaluating what economic theory can contribute to the measurement challenge. Miller takes us line item by line item as he evaluates what is customarily done in a cost and returns statement compared to what economic theory would indicate as appropriate. In Chapter 20, Arne Hallam returns to an underlying theme of this volume--that the appropriate method is a function of the end-use of the estimate. He describes the similarities and differences in the data needs by the three major uses of commodity cost and returns estimates. Hallam finds more similarities than differences in the data needs and concludes by arguing for more consistency in methods and the development of a standard of measurement. In Chapter 21, Jim Hildreth observes that little (no?) progress has been made in the past seven decades in addressing issues of economic accounting for commodity costs and returns. He then takes the plunge by recommending that a Taskforce on Standardization be established. As mentioned in the Preface to this volume such a taskforce has now been established by the Economic Statistics Committee of the American Agricultural Economics Association.

Many data developers and users of commodity costs and returns information are extension economists at land grant universities on the "front line" who work one-on-one with farmers as they plan their operation. Providing this interactive service to farmers requires them to provide ready access to estimates and forecasts. The Appendix Chapter describes how computing is handled in three different ways by extension economists in three different states.

Improved measurement of agricultural commodity costs and returns obviously will not independently resolve the important value-laden trade-offs facing society today. However, improved measurement of agricultural commodity costs and returns can help to quantify the cost of the trade-off in terms of the loss of economic efficiency. In order to accomplish this ambitious goal, we need to be aware of how costs and returns are being measured. Are the statistics capturing what we think they are capturing for our purpose at hand? The intensity and diversity of views presented in the chapters of this volume are stimulating and worthy of study.

References

Casler, George L. 1990. "Firm Level Agricultural Data Collected and Managed at the State Level" in Arne Hallam, ed., *Determinants of Farm Size and Structure.* Proceedings of the program sponsored by the NC-181 Committee. January 6-9, 1990, Albuquerque, NM. Ames, IA: Iowa State University.

PART ONE

Contemporary Issues in Measuring Commodity Costs and Returns

1

Varied Purposes and Implications for Measuring Costs and Returns

B. H. Robinson

Introduction

A review of the past often gives us a better perspective of the present. When the subject turns to the contemporary issues of cost of production (COP), history provides a perspective mired mainly in conflict. Indeed, like disputes over national boundaries, human rights, and conflicts between church and state, cost of production also must be considered a major source of conflict throughout history. Disputes over land values and rent, the value of the produce from land, and the subsequent taxes levied have long plagued societies.

A brief annotation of the past illustrates this age-old conflict, where some version of calculating the costs of production has been part of societies since trade developed. When societies shifted from mainly hunting to developing agrarian skills, rules were needed to control access to resources. We might say that rents derived initially from these earliest rules. Passages in the Bible discuss rules established to govern fair rents to landlords. In Rome, Greece, Egypt, and other ancient civilizations, records were kept on rents to landlords--the equivalent of our share rents today.

With the rules and rights came conflicts. In early recorded history, battles were fought for the rights to the scarcest resources--good land and access to water. Peasant revolts in Europe in the Middle Ages were often based on disputes about appropriate or fair rents (taxes) imposed by the Crown, or landed gentry. Similar uprisings in England occurred for the same reason--disagreements over taxes or rents imposed on tenants by landlords who controlled access to scarce resources. The flippant remark of Marie Antoinette to "let them eat cake" is considered to have been based on a disagreement over an excessive tax imposed by the French nobility on tenants for the use of land.

The contemporary issues in cost of production facing us today have their roots in two types of conflict that have a long history. The first is conflict over measurement of cost of production, and the methods employed in that endeavor. The second is a conflict over the use of estimates of cost of production. This historical perspective is important to keep in mind as we contemplate today's issues and the progress made thus far. Today's agricultural inputs are vastly more complex than at any time in the past; economic theory and empirical tests of theory are more sophisticated, and have enabled more precise measurement; and individuals have made considerable progress in articulating the need and purpose for accurate measurement of costs of production. Yet rents to fixed inputs such as land remain a critical and controversial subject.

In short, the job of resolving these contemporary issues rooted in old conflicts is not over. However, two developments have taken us farther along--the development of the agricultural economics discipline and the formal recognition in legislation of the need to develop sound measures of cost of production.

Attempts to Measure COP: The Development of a Discipline

Early economists such as Ricardo and Marx dealt with the issue of costs of production in their treatment of rents. In the late 1800s, however, a new discipline was beginning to emerge with its roots not in economics, but agronomy. The first stage in this metamorphosis was the emergence of farm management specialists. The major item on their agenda was to measure the cost of production at the farm level. Their goal was quite simple: "improve farm decisions by providing a means for assessing management strategies so that greater efficiency and ultimately, higher profits would result" (Taylor and Taylor).

Such were the origins of agricultural economics in the Department of Agriculture (USDA), when William Spillman organized the first Farm Management Branch in the Bureau of Plant Industries in 1902. Later established as the Office of Farm Management in 1905 with Spillman as head, the mission of this group was to provide an understanding of "the economics of individual farms, using cost accounting and farm surveys as methods of finding the most profitable enterprises and systems of farming" (Mighell). Notice that in 1905, at least one of the age-old conflicts appeared to have been resolved--cost of production was to be measured using cost accounting principles and farm survey data.

Land grant universities were also beginning similar work, and cooperative efforts between farm management specialists and educators emerged. Goals shifted from making two blades grow where only one grew before to bringing farmers out of the depression of the 1890s through improved management strategies. Cost of production measurements were considered a key tool in this endeavor.

The issues of use and purpose, though seemingly clear from the perspective of farm specialists in USDA and land grant universities, apparently differed from those of policymakers, so much so that Spillman resigned in 1918 (Mighell). After the United States entered World War I, the focus of farm specialist work inevitably turned to war-related problems such as labor supply and studies of early mechanization and their resulting impacts on profits and supply. Spillman's resignation over a controversy about the use of COP statistics rekindled old conflicts. Political forces were searching for a way to use COP measures as a means to justify an argument that farm prices should be equal to the cost of production plus a reasonable profit (Mighell).

Ironically, the development of agricultural economics actually contributed to the conflict of measurement. As this discipline developed, it took as its foundation the theory of the firm anchored in economics. As agricultural economists turned their attention to the subject of cost of production at the farm-firm level, a conflict thought to be resolved in 1905-- the issue of measurement--was once again ignited. Mixing cost accounting principles with economic principles of production cost decisions and behavior has been like mixing oil and water. And this conflict has only been aggravated over time by the introduction of theories from finance and financial accounting, as well as theories of risk behavior. Many of the dilemmas we face today stem from the disputes presented by mixing these theories.

Disputes Over Use of COP: Political and Economic Issues

The consequence of one dispute between policymakers and specialists in COP measurement has already been mentioned--the resignation of Spillman as head of the Farm Management Office in 1918. From 1918 until the abolishment in 1953 of the Bureau of Agricultural Economics (the predecessor of the Economic Research Service), some form of farm management office dealing with cost of production existed after Spillman resigned. From 1953 until 1961 when the Economic Research Service (ERS) was created, cost of production work was done by the Agricultural Research Service in USDA. Throughout this whole period (1920s through the 1960s), work continued on the development of measures of cost of production as a basis for measuring supply response. A major part of the mission of ERS shortly after its creation focused on regional production adjustment studies in cooperation with land grant universities. These studies were the first of their kind, and led to the expansion of ERS predominantly for the purpose of studying cost of production and production adjustment.

But also during this period, policy changes were taking place and conflicts continued to arise over the purpose and need for accurate cost of production measures. In the early years, Peek's parity pricing concept emerged as farm policy. Policymakers originally sought an income parity measure, but without good measures of cost of production they shifted to price parity since prices were observable, and chose a "good and equitable base period." Over time, policymakers have toyed with using cost of production measures as a basis for farm policy, either directly or indirectly. No one can deny that production cost estimates have often guided policymakers in discussions and, eventually, decisions about what constitutes an appropriate support level. Under provisions of legislation governing farm policy from the 1940s through the 1960s regarding price supports, the Secretary had authority to set and adjust price supports within ranges based on percentages of parity (Lee). The rationalization for raising or lowering support prices within the permissible range was often attributed to cost of production.

Cost of production measures have been used in the administration of several programs within USDA. In the Federal Crop Insurance program, COP estimates have been used to help set rates and determine the types of insurance offered; FmHA has used cost of production measures as benchmarks for loan administration.

Cost of production estimates have often been used as a proxy for the welfare of farmers, and thus to rationalize changes in farm policy and programs. The concept of the "whole farm budget" was developed by agricultural economists to provide estimates of "farmer well-being"--an attempt by the profession to answer the need of policymakers. But the estimates were not deemed statistically reliable or flexible enough to answer pressing policy questions.

Two points emerge from this brief review of past disputes. The first is a lesson that has been learned both by economists and policymakers from decades of debate on farm policy: Measuring the cost of production in agriculture is easier said than done. The second is that if we are to resolve the contemporary issues that face us today, we must make progress on the twin conflicts of methods of measurement and varied purposes for cost of production.

Progress and the Road Ahead

Undoubtedly, many agricultural economists wish the profession could get out of the business of COP. An equal number probably wonder if there is any way to keep cost of production estimates from being inappropriately used in farm policy debates and decisions. But this wishful thinking is comparable to having the agency administrator follow Spillman's example and resign whenever policymakers meddle in ways that economists think inappropriate. The problems will not go away--indeed, they would get decidedly worse if sound economic principles are abandoned and the people that use them turn their efforts elsewhere (not to mention the turnover rate of administrators!).

In any event, progress has been made. We have learned a great deal over the past years and applied this knowledge to expand our theoretical and methodological base to provide better estimates of the cost of production. And in their own way, policymakers have made a contribution, by challenging economists to carefully document our data, methods, and limitations.

Important progress has come from decades of dispute. Until 1973, the United States had no formal program mandate to collect and maintain cost of production data, despite the key role that cost of production has played in influencing the direction of farm and food policy in the nation since the late 1800s. The Agricultural Act of 1973 changed that, directing the Secretary of Agriculture in cooperation with land grant colleges, commodity organizations, and individual farms to conduct a study of cost of production. The study would be directed at wheat, feedgrains, cotton, and dairy and would consider various production practices and establish a national weighted average cost of production for these commodities individually. Most importantly, these studies were to be updated annually, providing continuity and establishing a formal basis for considering and dealing with methodological issues. This formal mandate has been continued, with some revision, in every major farm legislation since 1973 (U.S. Congress).

ERS responded to this mandate and began programs that continue to this day. These programs have gone through considerable revision, experimentation, review by peer and professional groups, and they constitute the basis for our current program to estimate cost of production in cooperation with the National Agricultural Statistics Service (NASS). Improvements have been made in the methodology and data, and ultimately, in the estimates themselves. Many of these improvements are the result of dialogue with many individuals inside and outside ERS, continued refinements in the data available, and guidance received from peers and users.

One particular revision in farm legislation to the original mandate of 1973 that has been a very positive element in our endeavors to resolve the conflicts accompanying cost of production came in 1981. In that omnibus farm legislation, the National Cost of Production Standards Review Board was established. Congress directed that the Board be composed primarily of producers, with representation from varied commodity backgrounds. Congress also mandated that three members of the Board be individuals whose backgrounds demonstrated a unique knowledge of cost of production methodology and application. The Board has met twice annually since 1983, and its mission has been extended in both the 1985 and 1990 farm legislation. The Board has challenged many procedures and methods, and has provided useful guidance in meeting the cost of production mandates.

For those who will continue to address the issue in COP, we must keep in mind that several specific issues still confront us. Many of these issues are not the policymaker's dilemma, but our own, and they stem from the lack of a consistent, underlying theoretical base--what I referred to earlier as the oil and water of economics and cost accounting. Abandoning one theory in favor of another is not always possible, as we have found. Different purposes for COP estimates makes one set of tools preferable over another. The purposes are numerous and varied: planning purposes, micro-level decisionmaking; research; national average estimation versus firm estimation; or for policy purposes.

A host of other issues confronts us that might best be grouped under the heading of calculation problems. These are the difficulties encountered in analyzing the data and providing estimates. Included are questions on imputing and allocating certain prices and costs peculiar to agriculture. A short list of these include:

Data-gathering and decisions:
  • The choice of a survey or engineering approach to data gathering
  • What yields to use for different purposes
  • How to handle or include government program benefits and costs
  • Whether marketing and storage costs should be included

Pricing issues:
  • How to price nonpurchased inputs or farm-produced inputs
  • How to price capital or the service of durable inputs

Cost-allocation issues:
  • Allocating joint costs in a multiproduct firm
  • Allocating land costs
  • How to account for cost variance and distribution

The list is not exhaustive, and the resolution of many of these issues, once again depends on the purpose for which the resulting COP measure will be used.

Implications for Measurement

A major purpose of the chapters in this volume is to recognize that policymakers, farmers, agribusiness, and many other groups look to this profession for sound and credible cost of production information. But a complementary purpose comes from the understanding that differences exist in our approaches to measuring the costs of production, and that different uses for the final estimates dictate different methods of measurement. With that understanding, this volume represents an opportunity to explore how different uses require different techniques, and whatever the use, identifying the best or most appropriate measurement methods.

This is a laudable goal to strive for, and one that we should strive for. Our profession is looked to by many for providing consistent, accepted methods for developing cost of production estimates useful not only to policymakers, but to producers and agribusiness. We should remember the origins of our work with the early farm management specialists, who sought to improve the decisionmaking of producers so that greater efficiency and profitability could be achieved. To argue that cost of production estimates are dangerous because of their potential misuse in farm policy decisions sidesteps the challenge before us to make progress on the perennial conflicts of use and measurement, conflicts that will not go away, even if we do.

Let us hope that the forum provided in this volume gives us the opportunity to educate ourselves, so that we can better educate others who use the information we provide. The editors of this volume should be commended for identifying many of the contemporary issues they have included for discussion. We have both a significant challenge as well as opportunity for making headway on the estimation and use of cost of production estimates.

References

Lee, John E. 1976. "Calculating and Using Costs of Production for Policy Decisions—The Case of the United States." CED Seminar Paper presented at the Instituto de Economia Agricola, San Paulo, Brazil.

Mighell, Ronald. 1973. "Farm Production Economics in ERS and Before." Unpublished CED Working Paper.

Taylor, Henry C. and Anne Dewees Taylor. 1952. *The Story of Agricultural Economics in the United States, 1840-1932.* Ames: Iowa State College Press.

U.S. Congress. 1975. "Costs of Producing Selected Crops in the United States, 1974." Committee Print No. 63-092, Senate Committee on Agriculture and Forestry. Washington, D.C.: U.S. Government Printing Office.

2

Measurement Issues Relating to Short-Run Farm Financial Management Decisions

Gayle S. Willett

Introduction

The preparation and distribution of crop and livestock enterprise budgets for use in farm management decision-making by Cooperative Extension is a long-standing and widespread practice. For example, a recent survey of extension farm management specialists found that all but four states were publishing enterprise budgets (Eck). As one who has participated in these efforts at three land grant institutions over the past 20 years, I can comfortably state that these enterprise budgets are widely used for many purposes by farmers, loan officers, consultants, appraisers, researchers, educators and many others. I also know that budget development procedures differ between most states and that extension specialists and their clientele tend to be pretty satisfied with their particular way of doing things. In fact, suggesting that there is a problem with the way in which an experienced analyst has developed an enterprise budget tends to be about as welcome as a trip to the dentist. Thus, it is with a good deal of apprehension that I attempt to identify some of the measurement issues associated with the preparation of enterprise budgets for use in farm level decision-making. Fortunately, my assignment does not include offering an opinion on how these issues should be resolved!

I would like to begin by noting the typical uses of farm management oriented enterprise budgets. At the same time, I will indicate the particular revenue, cost and return measures that are relevant, given the use at hand. Finally, the principal measurement issues relating to budget preparation will be reviewed.

Enterprise Budget: Farm Management Uses
and Relevant Concepts

An enterprise budget for farm management uses is a projection of revenues, costs, and net returns for an individual enterprise over some specified time period, typically a year. The projections often are made on a per unit basis, e.g., per acre or per head. The procedure used in budget development will, in part, be determined by the intended use of the budget.

The primary uses of an enterprise budget in farm management decision-making include enterprise selection, cash flow planning, marketing, negotiating lease terms, and investment decisions.

Enterprise Selection

Farm managers can make sound economic decisions by selecting enterprises on the basis of greatest returns over variable cost, providing the variability of these returns is compatible with the farm's financial position and the manager's risk preferences. Further, the enterprise choice must, of course, be consistent with the physical environment (e.g., weather and soil), supportable by existing fixed resources and compatible with various institutional constraints (e.g., government commodity programs).

The range of resource costs included in the variable cost category depends on the length of the time period relevant to the enterprise selection decision. The longer the time period, the greater the range of resources appearing as a variable cost in the budget. For example, the variable costs for a spring wheat enterprise to be seeded as a replacement for a frozen-out winter wheat crop may include only the added costs of reseeding, pest control and harvest incurred over a six month period. In contrast, variable costs for a new apple orchard would include the costs of establishing the orchard over a three to five year period, as well as the annual operating costs for a production year. Thus, variable costs will include costs of capital assets (e.g., seedlings, irrigation equipment, frost control devices, etc.), as well as the usual operating costs.

Even though the enterprise selection decision can be based on returns over variable costs, farmers will periodically want to determine the full economic cost of an enterprise and the associated profitability. This follows from the need to periodically replace depreciating capital assets and an interest in evaluating returns to owned resources relative to alternatives. Thus, most enterprise budgets contain both variable costs and the fixed costs associated with already owned resources, thereby permitting both a short-run enterprise selection analysis and a determination of longer-run economic viability.

Cash Flow Planning

Once a decision has been made on enterprise, an annual projection of whole-farm cash flows on a monthly or quarterly basis is generally desired. The cash flow projection can be used to manage cash deficits and surpluses, obtain necessary financing and control the business through a periodic comparison of actual with projected cash flow. The enterprise budget is often an important source of cash flow information. Thus, a complete enterprise budget will also contain an estimate of cash receipts and variable and fixed cash costs.

Marketing

Marketing efforts are more focused and effective when producers have a full understanding of their production costs. More specifically, farmers and their lenders want to use an enterprise budget to estimate variable costs, cash costs, cash costs plus depreciation and full economic costs. Knowledge of these costs and yields gives direction to marketing efforts targeted to achieve returns to cover various levels of costs, including those associated with already owned resources.

Negotiating Lease Terms

An enterprise budget is often used as the basis for negotiating equitable and economically efficient lease terms on capital assets. For example, developing a crop-share agreement based on the relative contributions of the lessee and the land owner requires an estimate of full economic costs (variable and fixed). Further, a lessee attempting to determine the maximum cash rent that can be paid on rented land may want to use an enterprise budget to determine the costs associated with all resources, except cash rent, on the ground in question.

Investment Decisions on Capital Assets

An enterprise budget often serves as a primary source of information for individuals evaluating investments in land, machinery, structures and breeding animals. The items in the budget relevant to the investment decision will depend on the nature of the investment. For example, an investor attempting to determine the returns to land and an associated land value, will want an estimate of returns to land per rotation acre. Thus, it would be necessary to know the revenues and all nonland costs for each crop in the rotation. In contrast, a farmer analyzing the economics of continuing to own or custom hire hay harvesting machinery would be interested in only the operating and ownership costs associated with these items of machinery.

Measurement Issues

With the above farm management uses of enterprise budgets in mind, what are the principal measurement issues that confront the enterprise budget analyst? Although most extension economists with enterprise budgeting responsibilities have resolved these issues, for many of us, they have been resolved differently depending on the clientele uses of budgets, economic environment and analyst experience. The major measurement issues fared by extension economists include: base farm selection, type and quantity of operating inputs, and cost of depreciation, land, and management.

Base Farm Selection

The starting point in developing an enterprise budget is the selection of base parameters for the farm on which the enterprises will be produced. This includes such items as size of farm, mix of enterprises and productivity of resources, including management. More specifically, is the budget entity going to be a large farm, small farm, or average sized farm? Will the farm be specialized or diversified? Will management be average or above average in terms of practices and yields?

Type and Quantity of Operating Inputs.

Recognizing that no two farms have the same enterprise costs, most budgets provide substantial detail about assumed operations, types and quantities of operating inputs and per unit input prices. By providing this detail, users can readily change the assumptions and adjust the costs to reflect their particular situation. The difficulty that arises is getting a good estimate of what is typical or representative for inclusion in the base budget. Attempting to get agreement for example, on pounds of nitrogen, acre-inches of water, or hours of labor to cultivate an acre can be quite difficult. Many analysts rely upon small groups or representative farmers to make these judgements. However, it is not unusual for a group of 10 farmers to have 10 different opinions about what is typical. Machinery variable costs (fuel, oil, repairs, labor) are particularly difficult to estimate.

There may also be some concern over the ability to estimate the variance in enterprise costs and returns. However, it is the responsibility of individual users to adjust published budgets to reflect their unique resource situation. Thus, it is probably fair to conclude that most extension budget analysts are pretty comfortable with providing only point estimates of commodity costs and returns.

Cost of Depreciable Assets

This issue surfaces primarily in the estimation and allocation of the ownership costs associated with machinery, equipment, and facilities. The specific costs in question are depreciation, interest on investment, property taxes and insurance. Given that we continue to experience price increases in depreciable assets due to inflation and technological improvements, there is an issue of whether these ownership costs should be computed using original acquisition cost or current replacement cost.

Moreover, in computing full economic costs for an enterprise it is necessary to allocate ownership costs on assets used in multiple enterprises to individual enterprises. Often this is done on the basis of annual use. For example, if a power unit is used 500 hours per year in all enterprises, this translates to a particular fixed cost per hour. Multiplication of fixed cost per hour times the hours per acre required for the tractor to perform specific operations for a given enterprise provides the basis for allocating tractor fixed costs to that particular enterprise. While conceptually straightforward, this procedure requires an estimate of total farm machine hours, which in turn is a function of farm size and enterprise mix. Given the extensive data needs and the circularity of this process, getting good estimates of annual hours of total farm use and an associated economic life for machinery is a challenge for many analysts.

Land Costs

The procedure for handling costs on owned land differs somewhat among extension analysts. The basic choice is (1) to treat land as a residual claimant to returns after all other inputs (except perhaps, management) have been compensated, or (2) to base the cost on market considerations. Most farmers using extension prepared budgets prefer to see a market-based cost assigned to land. There are two market-based approaches. The first involves multiplying a current interest rate on farm real estate loans times either the land's acquisition price or current agricultural market price. Alternatively, a net rent (crop-share or cash rent) concept may be used. I would speculate that the use of current interest rates and current market prices during the inflationary late 1970s and early 1980s resulted in some pretty unrealistic land costs, and consequently, most analysts are now using a net rent approach.

Unpaid Management Costs

The issue posed by unpaid management is similar to that noted for owned land. Again, the option is to treat management as a residual claimant after a cost has been assigned to all other resources, or estimate a market-based cost for management services. My impression is that most farmers prefer to see a cost for management in the budget. Options for deriving market costs include professional management fees based on a percent (e.g., 5-10%) of either gross revenue or costs, or estimating the value of management if diverted to the next best, similar risk alternative.

Summary

A substantial part of the resources available for use in U.S. extension farm management educational programs is devoted to the preparation and distribution of enterprise budgets. This allocation is consistent with the high demand for enterprise budgets by farmers and other extension clientele. These budgets have been developed to accommodate a wide range of uses and different analysts have adopted different formats and estimation procedures.

While the benefits from a standardized approach to enterprise budget development would be considerable, a lack of agreement on how to handle many of the issues raised above is likely to prevent a high degree of uniformity. After all, enterprise budgeting efforts and professional debate have occurred for close to 100 years. I'm not optimistic that we will ever totally resolve these issues. However, a desirable and attainable goal is that extension analysts will opt to prepare budgets with sufficient detail and explanation that users can make informed adjustments in the budgets to help them effectively address their particular need.

References

Eck, Douglas W. 1990. "A National Survey of Enterprise Budget Development and Use By the Extension Service." Unpublished M.A. thesis, Department of Economics, Utah State University.

3

Measurement Issues Relating to Economic Analysis

Glenn A. Helmers and Myles J. Watts

Introduction

The basic elements of cost are wages on labor, interest on capital used, rent on land and the earnings of management. In placing valuations on these factors, on the ordinary farm, it has not been possible to solve some difficulties. Much of the labor is usually done by the farmer himself or by members of his family. Part of all of the capital may be owned by him. Much of it may represent previous investments which would not be repeated in their present form and which cannot be liquidated. No certain valuations can be placed on these factors of production contributed by the farmer. Rent on land is determined by prices of farm produce, not the value of produce by the rate of rent. Therefore, rent cannot be used as an element in the determination of necessary price. The fourth element, returns on management, varies from farmer to farmer with ability in planning and operating the business.

Even if valuations on the factors of production were agreed upon, there would still be serious difficulties in the way of apportioning them to individual products - unless the farm produced a single marketable crop. The farm is operated and should be treated as a unit. Many expenses apply in indivisible proportions to two or more products, as wool and mutton in the production of sheep, or grain and straw, or cotton and cottonseed.

So-called cost figures obtained under these conditions are without significance as far as necessary price is concerned. The results depend too much on arbitrary methods and change with the method used. "Cost of production" figures are discredited as a basis for price fixing and for tariff determination. They have been used in various public hearings but figures which are more easily understandable and not open to challenge would ordinarily be preferable as well as more pertinent.

When constructed with some care, figures of this same nature are, however, usable as indicators of efficiency in particular enterprises as between farms. Such use should be limited to groups of farms operating under essentially similar conditions and in the same year. In fact, they are seldom used in this manner because of the expense involved in their collection (Hopkins and Taylor).

The above assessment of the status of cost of production in agriculture is 56 years old, yet much of it is as relevant today as then. Since this was written, emphases in agricultural production economics have generally shifted away from firm or micro-based management issues using detailed data secured by survey or records to more empirically oriented studies related to production economics theory and operations research analysis. This includes the use of "representative" decision making situations, studies of resource productivity and technological change, and extensions of these frameworks to policy issues and aggregate industry behavior. Yet the great majority of these studies rely heavily on cost of production "data."

Given the above perspective of 56 years ago, the question remains if theoretical and methodological developments during this 1935-1991 period have reduced the above described problems. Our general perspective is that cost and return budgets continue to face severe problems of conception, estimation, and application. With the increasing proportion of attention by agricultural economists toward analyses constructed upon cost budgets, there is less concern about the conceptual weaknesses underlying enterprise cost budgets than formerly when analysts had greater understanding of their construction. Cost of production budgets are increasingly mechanized. In addition, the emergence of periodically high rates of inflation has led to widespread confusion over budgeting methodology. Hence, it is an important objective to discuss alternative uses of cost budgets, weaknesses and strengths of cost budgets, and improved methodology in budgeting enterprise costs.

In this chapter the perspective of the use of cost and return budgets for economic analysis is emphasized. Our general position is that cost of production concepts are not very useful to the understanding of the economics of agriculture, just as the cost of production of pizza is not very useful to understanding the pizza industry. A more appropriate comparison, given the nature of joint production in agriculture, is the relation of cost of production of pizza to the structural understanding of the restaurant industry. Too great a reliance on cost of production is a danger because of the inherent weaknesses of analyses that follow, the resources devoted to cost of production which would be better used elsewhere, and the limited focus of issues which can result from its emphasis. Cost of production seems, on the surface, to be a useful and basic element to economic analysis. Further, noneconomists relate well to the concept of cost of production, while supply functions, input demand functions, length-of-run and other important issues are less obvious concepts. As a result, cost of production often becomes considered as an end rather than a tool with limited analytic capability.

We do not emphasize management issues of the firm (whether short run or long run) in this chapter.[1] Our emphasis is on the potential for cost of production estimates to aid in the economic understanding of agriculture.

This might be immediately challenged in the sense that some may say that cost of production was never intended to be directed toward economic issues of structure nor is it used today for that purpose and thus, this is a "straw-man" issue. However, price setting justification is one way it is used and some studies have attempted to examine comparative advantage, "supply functions" (costs of different firms), and other economic concepts using cost of production directly. Further, a host of other economic analyses use cost of production budgets (from records, surveys or synthetic estimates) indirectly as data.

Substantial resources have been invested in such studies by the USDA through the ERS and other federal agencies; by states through the land grant universities and state departments of agriculture; and in the private sector. The focus of our concern is with studies funded by government. Given the level of resources committed to such studies and the substantial attention these studies receive in many arenas, including policy, litigation, credit, and management, it is appropriate that they be carefully examined. The examination should include methodology, data base, and applicability. Of course, a thorough examination is beyond the scope of one chapter and our knowledge.[2] Much of our comments will primarily pertain to cost of production and farm income. Others are better able to address alternative uses of the data. We offer observations on the generality of the use of such studies, policy uses, and problems with the current published statistics.

Historical Setting: Implications for Economic Analysis

The historical purposes of cost of production analysis are not easy to assess. Cost of production analysis was a major activity of scientists in land grant colleges and the U.S. Department of Agriculture in the formative years of Agricultural Economics. Indeed, study of cost of production appears to parallel the growth of farm management and production economics--the early emphases of agricultural economics.

A somewhat conventional view of the role of cost of production in the development of agricultural economics is the synoptic accounting-management-theoretical evolution framework.[3] Cost of production is an important issue in this development because early cost of production received much direct emphasis. An impression reported by some literature of early cost of production analysis is that such work was almost completely an outgrowth of natural scientists who developed purposeless accounting ratios, but direct economic and political influences were almost negligible. In such a disciplinary perspective, the early (1890-1920) farm management analysts such as Boss, Hays, and Warren extended a natural science background to enterprise accounting and farm decision making. Also, Taylor (one "exception") from an economics orientation participated in cost accounting research but with stronger interest in the economic relationships involved. The 1920s to 1940s are characterized according to this historical view as a mixture of (1) continued emphasis but declining relevance of cost of production as secured from surveys and farm records and (2) theoretical contributions of those such as Black, Spillman, Tolley, and Ezekial. These emphases included production function analysis and greater emphasis upon the comprehensive and integrated economic relationships of farm units and groups of farm units. The third and remaining period is post 1940s under the leadership of Heady and others in extending production theory and normative farm decision making. Also this period is characterized as one of greater emphasis on risk and uncertainty, aggregate efficiency of agriculture, extensions of area adjustment studies initiated in the 1920-40s period, and general supply response to price. This third period is also characterized by the enormous application of quantitative methodology to applied agricultural economics issues. Marginal productivity theory allowed ready application of statistical validation and operations research studies built on the firm budgeting emphasis of the 20's to 40's.

This characterization of the formative years of agricultural economics when cost of production was such an important focus suggests that cost of production was only an important first step toward management but economic and political uses of cost budgets were largely absent or only periodically emerged. For example, as reported by Pond (p. 14) as well as by Taylor and Taylor (pp. 397-398), Boss indicates that the purpose of cost of production was the improvement of farm management, not price making:

> . . . the effort was not made to find costs for costs' sake or with the expectation of determining an exact cost to be used in price making. Rather the objective was to secure basic data at first hand that could be used in determining which crops under certain conditions gave the greatest net profits when grown for market and which crops could best be worked into farm crop rotations that, over a period of years, would yield the best returns to the farmers. . . . The whole enterprise was aimed at better farm organization, improved farm operation and the development of information that would be useful in teaching school and college classes in farm management and to build up literature on farm management.

Swanson has suggested that only periodic emphases in cost of production work as a basis for agricultural prices or production goals emerged even though the primary emphasis of cost of production was efficiency. This included World War I, the Depression, and World War II.

A somewhat different perspective on the development of cost of production analysis is presented by Hopkins and Taylor in 1935. They suggest that during the depression of the early 1890's cost estimates were made to demonstrate that farm prices were too low. Also, they indicate that prior to the Spanish-American War, the Department of Agriculture developed costs of sugar production to convince farmers that sugar could be grown profitably to make the U.S. independent of sugar exporting countries. They also suggested that cost of production became important during World War I because of concerns of consumers over high prices. However, for wheat, higher administered prices were set to stimulate production. Hopkins and Taylor describe the growth of the post World War I cost of production studies as concern by producers over low prices as well as increased funds available for research purposes. Black indicates that the use of cost of production in setting price "has a close relation to the medieval concept of fair price as a 'just' price" (p. 264), clearly recognizing the presence of the political use of cost of production. Pond recognized that "undoubtedly the large increase in farm cost accounting that took place about 1920 was motivated, at least partly, by the idea that accurate cost-of-production data might have some use in getting favorable price consideration for farm products" (p. 14). However, he then goes on to point out that the agenda of farm management research workers in the North Central region in 1923 included no reference to the use of farm cost of production data in "price-fixing activities" suggesting, perhaps, that this demonstrated that political interest in cost of production was only superficial.

The importance of policy concerns to cost of production over the past century is difficult to fully assess. This is because such concerns can easily be hidden in ostensible concerns over management efficiency and economic analysis. This issue is important because it relates to what the driving forces were in the formation and development of agricultural economics. Contrary to the often-described accounting- management-theoretical view, a more balanced perspective is that broad concerns and policy issues of both farmers and consumers was an important early force in cost of production work and early agricultural economics.

We have just suggested that policy concerns were probably a more important issue in the development of agricultural economics and cost of production than is conventionally thought. In addition the issue of management vs. economics deserves some discussion. Contrary to the often-described progression of management to economics (now ignoring more direct policy concerns), economic issues were a major and early concern of early farm management studies. Further, it is suggested that the inadequacies of cost of production analyses for economic analysis became readily apparent. It is generally thought that early farm management researchers were natural scientists and, thus, lacking in economic logic. However, economic thinking in the early 1900s was strongly influenced by German economists who generally were more directed to institutional frameworks and heavily influenced by forestry, resource economics, and public policy (see Salter, Chapter II). A better perspective of the early period of cost of production analysis was that the economics of the period was not absent but considerably different from the economics of the 1920s to 1940s. The economics of the early period was not the economics of Pareto, Marshall, and Hicks, nevertheless it was economics.

The question thus arises how much concern over economic relationships existed in the early cost of production work? Early farm management studies attempted to understand (1) what economic relationships were involved in the organization of farms to maximize returns and (2) what broader economic relationships were occurring in the agricultural economy. Again, these issues relate directly to cost of production history because cost of production was a primary mechanism of early work. The first issue includes principle discovery, or the isolation of causes of economic success of farms, and optimization principles. The often-described preoccupation of early analysts with estimating costs for costs sake without a larger purpose is not immediately clear when studying their published material.

Warren's textbook of 1913 (preface) defines farm management as "the study of the business principles of farming." Black (p. 11) notes Case, Wilcox and Berg writing in Illinois Bulletin 329 (p. 262) of "certain definite principles that need to be observed in the organization and operation of a farm if it is to be financially successful." Taylor in 1911 carefully describes how crop requirements differ and compete for a farmer's resources and then calculates profit per unit of resource for alternative crops. These are early manifestations of the importance of principle discovery and optimization theory. Later these ideas were extended to production economic theory and normatively oriented optimization (earlier termed the synthetic method and budgeting). Still today, cross sectional economies of size studies could be considered as empirically-derived attempts at principle discovery. Similarly, other analyses attempt to determine principles of financial success from farm records.

The optimization emphasis has had a more clouded history. A criticism frequently made of the analyses of the 1890-1920 period is (1) that analysts were unable to isolate cause-effect relationships because of a lack of empiricism and (2) the applications of principle to individual farms was hindered by a lack of methodological specificity. Cost of production budgets were found important to whole farm optimization whether by budgeting (North Central Farm Management Research Committee) and later in linear programming. Unfortunately, some economic problems such as those described above by Taylor (how is profit per unit described?) were not adequately addressed in the resulting avalanche of programming studies. Such programming studies while correctly having a whole farm framework attempted to extend representative farm conclusions to whole areas including the assessment of resource valuation. The problems were (1) the specific resource constraint driven nature of results and (2) confusion of how much fixity should the analyses contain and consistency of that fixity across enterprises. Of course, today, firm optimization models, whether using simple LP, a risk-income framework, or dynamic optimization use cost of production budgets in their construction. The potential of optimization modeling with specificity to individual farms, however, is still largely only a potential. Little applied work in nongeneral optimization occurs where close attention is paid to the particular cost of production budgets for a specific farm and the particular resource constraints of those farms. An argument can be made that the lack of attention to the imprecision of current optimization models to specific farms is a worse situation than the inadequacies of "rules of thumb" of early farm analysts. Even though allowance is sometimes made for farmers to adjust coefficients (budgets) and right hand sides, this does not solve the problem of what to adjust them to. Readers would be well advised to consider the perspective of those who question the resulting precision and specificity which resulted. Representative of this are the concerns of T.W. Schultz as evidenced in his 1964 article on relevance, particularly micro weaknesses.[4]   Earlier (1957) Schultz stated that "...the journals runneth over with 'results' from linear programming, a new apparatus that is turning out thus far, an undigested mixture of a few insights and many 'numbers' that do not make sense" (p. 335).

The second issue, that of broader economic principles of comparative advantage, supply and demand relationships, and other economic relationships resulting from aggregate behavior was not a second stage of inquiry but also important at the outset. This is obvious in Warren's emphasis upon comparative advantage as well as Black's (p. 95) discussion of supply and demand by region and equilibrium prices, even though a concern was expressed over impracticality. Carver includes in a 1916 book a Department of Labor study comparing the costs of production for a wide range of agricultural crops for the 1829-1830 period as well as the 1895-1896 period (p. 48). The discussion is not devoid of economic content.

In summary, while it is very difficult to sort out the settings in which early farm management was conducted, political considerations were probably a more important factor in the development of cost of production studies than is conventionally thought. Also, the early period of cost of production analysis was not as devoid of purpose and economics as some have later assessed. Rather the economics of that period was different than the later marginal productivity emphasis. Hildreth summarizes the formative years of agricultural economics from the perspective of Taylor and Taylor that the agricultural conditions of the 1880's and 1890's led to the growing economic investigations. Hildreth states ". . . there was a need and a means."

Contemporary Uses of Cost of Production Data
in Economic Analysis

The issue here is the relevance of cost of production used in a direct or indirect manner in yielding structural understanding of the agricultural economy. We include here understanding of the structural impacts of policy changes. A wide range of economic issues are involved here including questions of area change and competitiveness, technical change, resource valuation, leasing changes, input substitution phenomenon, welfare impacts, and structural change of agriculture. Cost of production data are frequently used in such modeling efforts. Some of the more common uses are briefly described below.

Representative Farm Models

Such models are often used to evaluate optimum production adjustments in order to understand how production adjusts to change in types-of-farming areas by aggregating representative farms.

Representative Farms and Policy Alternatives

Here short-run impacts of alternative policy alternatives are examined to try to gain understanding of production, input, and income changes resulting from commodity policy changes and broader policy changes. Usually it is well understood that only first-round impacts can be observed here and that more complex second-round impacts are ignored. These policy alternatives can include commodity programs or other policies such as changes in income tax law or environmental regulation. Changes in technology are often also examined in this framework where only the limited first round impacts are identified.

Quantifying Characteristics of Low Cost Farms

This is similar to early farm management studies where cross-sectional data of cost budgets are examined to yield principles or factors associated with better managed farms, including more recently a greater emphasis on financial components. One difficulty, among others, with this approach is that if short run disequilibria is assumed, a minimum cost of production is not necessarily synonymous with profit maximization.

Macro Studies

Macro studies which use cost of production estimates are generally policy-oriented studies which construct supply functions and resource use changes by the use of area-wide programming models based on short-run adjustments. There are other macro policy models which use cost of production indirectly, particularly for implications of farm income changes.

Economic Problems

In this section we discuss some of the basic problems when attempting to use cost budgets for economic analysis. These problems are not new, and in many ways our discussion is a restatement of Hopkins and Taylor quoted at the beginning of this chapter. Several basic economic problems are examined below.

Cost of Production Point Estimate vs. a Function

Ex post, a cost of production calculation summarizes all input levels multiplied by their prices divided by output. As such, cost of production clearly exists and is a summary, in an accounting sense, of input and output relationships over a specific period of time.

Analyses based on such data, however, do not have the richer set of input/output relationships which involve output supply functions and input demand functions. Further, input supply and product demand functions are essential to our framework of understanding what inputs and what outputs are produced and what affects changes in these across time. Increased estimation of supply functions and analysis using them still hold much potential for greater understanding of agriculture, such as comparative advantage issues. Utilizing only one point along a supply function reduces the analytic capability of a supply function framework.

Structural change in terms of overall output and input, ratios of input use, and input use by area relate to the important economic issues of supply, capital/labor use (and thereby farm size and structure), technological change, and comparative advantage. The limitations of a cost of production calculated in this manner is similar to a setting where a level of expenditure for a product for two points in time demonstrates nothing analytical about the causes of such changes in terms of product supply or demand shifts.

In Hopkins and Taylor broader issues of structural change were not explicitly addressed. However, they still ask, when ordinary persons discuss the relationship of cost of production and price ". . . what is meant by cost, what elements is it composed of, whose cost he had in mind, and under what conditions of production it would generally become evident that he had not given a great deal of thought to these questions. Or perhaps he had considered these to be simple questions which did not require much thought" (p. 416). Marginality and issues of long vs. short run are then discussed.

These issues do require much thought. Our assessment is that a collection of cost of production data points as a basis for analysis of cause-effect relationships and the understanding of structural relationships is totally unreliable. For example, an understanding of the impacts of a policy or a change in technology on different producing regions cannot be understood by examining the resultant costs of production of that crop or a mix of crops by producing region. Some useful insights may be gained, but that is all. Further, it can be argued that examining product prices rather than estimating cost of production is easier and more accurate. This is because only one item is of focus rather than several.

A recent report by Barkema and Drabenstott is illustrative of the deficiencies of cost of production in explaining important structural change issues in the beef industry. Cost of production changes across time are presented, but alternative explanations of demand and supply characteristics for beef and other meats which have been empirically studied allow for a comprehensive understanding of structural change that cost of production budgets can hardly hint of. It should be noted that the authors of the report do not suggest cost of production can uncover those structural issues.

Cost Equals Price

Economic equilibrium provides for the equation of resource cost to output price. If, for example, output demand rises or falls, input demands also rise or fall. Thus, the changed level of output multiplied by its changed price equates to the changed input levels multiplied by the changed input prices. This equilibrium process also occurs when one or more of the input supply functions change or when the production process is changed. Too often, product price is treated as unrelated to cost of production. As will be discussed later in response to permanent economic changes, the long-run time path of adjustment for resource prices is not instantaneous. Further, the supply elasticity of individual inputs strongly impacts relative input changes and their prices. Those resources with least mobility, such as land, absorb most of the long-run impact of changing output prices. However, in the short and medium run all input levels and prices will be influenced by shifts in output demand, the production function, or input supply, not just land.

Of course, any concept of cost of production not relying on opportunity cost of resources will result in major differences in cost vs. price of an output. Hopkins and Taylor are somewhat extreme in their assessment of the response of land costs to output price changes: "Rent on land is determined by prices of farm produce, not the value of produce by the rate of rent. Therefore, rent cannot be used as an element in the determination of necessary price." Still, their underscoring of the effect of changing price of output on input prices demonstrates a linkage of output and input prices often disregarded in discussions of cost of production.

Farm income and cost of production information has been influential in the development of farm programs. In particular, the setting of target prices has been, at least to some degree, based upon cost of production estimates. It is interesting that much of this work has been done by agricultural economists who learned in their first introductory economics courses that in equilibrium, cost is equal to price. Therefore, a rational approach would be to estimate a supply curve. The supply curve would provide the cost of production for various levels of output at the aggregate level. Perhaps such an obvious approach has been discarded because it did not produce the desired results.

Fundamentally, it seems that the input useful to the farm policy debate is determined by the objectives of farm policy. We recognize that those charged with providing input into the debate have no choice but to do so. It does seem that the development of useful context and impact upon alternative objectives is appropriate.

Farm policy programs have been justified on the basis of the following broad categories: morality, farm income subsidy or welfare, and food necessity.

The morality arguments follow several veins but generally point towards the objective of maintaining a substantial portion of the population in agricultural production and that farms should be relatively small, owner-operator firms. Reasonable input into such policy objectives would appear to be the analysis of the policy alternatives on the relative cost of capital and labor and economies of size. It does not appear that the cost of production studies of current farm income measurement information is the most efficient way to evaluate the policy alternatives within these objectives.

The second objective of farm programs is the subsidization of farm income or welfare. If the objective of farm programs is to raise the standard of living through income subsidization of farmers, then the side effects from subsidizing incomes tied to production resulting from allocation inefficiencies would be more undesirable than direct payments to farmers. It is difficult to understand what role cost of production could play in this debate even if the subsidies were going to be tied to production, since it is simply a matter of providing sufficient subsidies in order to raise farm income to the desired level. Of course, in the longer term, linking subsidies to production is self-defeating since the subsidies become incorporated to the value of assets. The next generation of farmers who have to pay for the artificially inflated assets are certainly no better off and possibly worse off. As standards of living of farmers continue to approach standards of living in society in general, we would expect support based on this objective to deteriorate and become less of an issue.

The third objective has to do with the perception that food is a necessity; the desirability of a stable, low cost, high volume of quality food; and security consideration. Again, a supply curve providing a quantity-cost-price relationship is a more appropriate approach. The security argument does not seem very compelling in light of U.S. food exports.

Resource Valuation

The problems of assigning costs to resources not purchased by a producer have long been recognized. These resource costs include machinery, land, operator labor, and management. Current replacement costs of machinery should be used as well as current land values. However, with operator labor and management, serious problems may arise. Also, if risk is included the problem becomes still more complex. Perhaps the labor and management costing problems are best understood in a setting where costs are estimated cross sectionally for various sized farms. In particular, among small-sized farms, a great difference in cost is almost always observed. Some analysts are uncomfortable with these wide differences. In addition to other issues, two are of particular concern here: (1) Has overhead labor been estimated correctly? (2) Has management efficiency been estimated correctly? The assignment of a prescribed level of overhead labor and management can lead to significant overestimates of cost differences among farms. Another way of stating this is to ask why some farms with similar owned resources as others earn more income. Rather than focusing on an answer which suggests they have lower costs, the question needs to be asked, why do they have lower costs? Perhaps "low cost" farms have higher levels of overhead labor and higher costs of management (a higher management value) than is routinely assigned to these farms. Similarly, "high cost" farms may well be the result of overvaluing these labor and management resources. Our suspicions that differences among farms in the use of overhead labor and management productivity are wide, suggests a lack of confidence in our arbitrarily assigning such costs in cost of production work. These are empirical issues that deserve attention.

Joint Output

Little additional comment needs to be made regarding this point because it is generally well recognized that joint output issues cause difficulty in cost assignment. Literature is scant regarding joint output impacts. Some might suggest that this is only a relatively minor issue and not likely to cause major problems. However, we do not know how important this issue is and secondly, even minor errors can have major impacts upon analyses.

Inflation Adjustments

As mentioned earlier, inflation has caused considerable confusion in cost of production budgeting methodology. Errors in the application of inflation adjustments (not simple deflation) to cost of production estimates, as well as resultant farm income statistics, can result in major errors in interpretation of investment worthiness, firm performance, farm credit policy, "fair" prices, and a host of other assessments based either on research or judgment.

The liquidity trap problem which occurs under inflationary conditions for firms with significant levels of credit has been noted elsewhere (Helmers, et. al., Robison and Brake). This includes the issue of proper adjustments of nominal financial performance ratios to project true performance and consistency of the relation of earnings to overall financial performance. As mentioned earlier, where cost of production is used as a basis of price intervention, prices based on real based cost of production can still result in a liquidity trap for some producers when there are significant levels of inflation. Our work related to inflation and budgeting includes Watts and Helmers (1979 and 1981). There are several issues that should be underscored in relation to inflation:

1. The budget must suit the purpose. Nominally based budgets (nominal capital costs and an end-of-year basis) are useful in firm financial and cash flow planning. Real budgets (real capital costs using current land and machinery values) are useful in the analysis of investment worthiness and economic analysis.

2. Real budgets are preferred to nominal budgets for economic analysis and policy. Nominally based budgets are conceptually acceptable, but the linkage to return flows requires considerable effort in adjusting inflating return flows to constant nominal flows. Cost of production budgets (unless explicitly stated) are implicitly understood to be real because the commensurate return flow has not been adjusted. Of course there are some situations where the return flow may be expressed in a constant nominal expression and where a nominally based budget is the appropriate tool for both investment worthiness and financial and cash flow purposes.

3. Budgets, under inflation, are best understood as having a beginning-of-year or end-of-year time point. This is important because the costing processes for some capital items implies an end-of-year basis while other

costs can be conceived of as either beginning or end-of-year. Synthetic budgets have a clear advantage over record-based budgets in this respect.

4. In computing opportunity cost and depreciation, the need to use real interest rates for all capital (long term and short term) and current market values of assets continues to be the most important deficiency in budgeting among universities and USDA.[5] This deficiency cannot be remedied by the simple deflation of cost or return time series statistics which are so important and basic to research and intertemporal understanding of changes in agriculture. Some progress in the use of real capital rates over the past decade seems to have been made. The use of leasing or rental rates for land and machinery is representative of that.

Concerns About Data Quality and Use

To have much confidence in cost of production data and the resulting analysis, consistency is of the utmost importance. Cost of production studies when appropriately aggregated must be consistent with numbers published by USDA in other places, in particular, financial summaries in the Economic Indicators of the Farm Sector. Discrepancies between the agricultural statistics, census data, and other federal government publications, including the Economic Indicators of the Farm Sector, should be justified and explained to the extent reasonably possible. Furthermore, statements and statistics in the Economic Indicators of the Farm Sector must be reconcilable. For example, income should equal the change in net worth, plus withdrawals, less additional investment.

It is our experience that farm income and cost of production studies must be carefully designed to answer desired questions and a study generated for a particular purpose must often be substantially revised to answer other questions. That is, the questions to be answered must first be well defined and then the study designed accordingly. For example, an analysis of credit worthiness is substantially different than an investigation into relative rates of return.

The USDA generates much of the data used in their cost production studies and farm income information from surveys. While distributional statistics are not published, the variance on individual items has to be quite high. Therefore, the applicability of these results to any particular operation must be questioned. Furthermore, the use of these studies as a reasonable norm applicable to an individual operation for credit, marketing, or firm management is of concern. This is not to say that the USDA should be responsible for policing the use of the studies. On the other hand, the potential for misuse seems quite high.

Summary

It appears that problems surrounding the use of cost of production estimates are largely unchanged from earlier decades in this century. Further, it can be argued that problems caused by inflation as well as the general lack of interest in studying those important and empirical problems related to cost budgets, has weakened the accuracy of cost of production budgets. Yet interest remains high in the application of cost budgets in management, price setting, and various research efforts, some of which can be described as efforts to understand structural change of agriculture. Also, there appears to be some growing interest in the use of cost of production estimates for direct analysis of structural or economic change.

Cost of production must be viewed as a means, not an end. For management of specific farms cost of production is a necessary ingredient to enterprise and investment selection as well as financial planning. Because resources are always scarce, the question of how much priority should be placed on the use of public funds for individual farm consulting is a different issue but does touch the issue of the use of cost budgets.

Considerable empirical efforts could improve cost of production budgets. Valuations on owned resources (particularly labor, management, and risk) joint output, and inflation issues are three of several important issues which need more attention.

For policy and economic application, cost of production budgets have far less potential. As a basis of price setting, cost of production budgets have the potential to develop misallocation of resources particularly under conditions where higher product prices result in higher land costs (returns).

For economic analysis, cost of production budgets are used frequently as input to a host of studies of firm adjustments, resource valuation, implications of policy, etc. The deficiencies of cost of production are not acknowledged enough in these studies, particularly problems of inflation. Finally, there is increasing interest in the use of cost of production budgets as direct evidences of economic change.

The important issues of structural change in agriculture require more complex formulations that the comparing of cost of production either across time or among producing areas in response to the many simultaneous changes occurring in the economy. While there are some useful hypotheses generated by study of cost of production budgets, understanding of the complex economic changes occurring in agriculture and their causes requires a larger framework of study.

Notes

The authors acknowledge Alan Baquet for his helpful comments.

1. Economics as used in this chapter refers to forces affecting firms over which firms have no control. Different uses of the terms economics and management are within the area of firm decisionmaking where economics is defined as long-run decisionmaking (full costing) and management is short-run decisionmaking (fixity).

2. Readers should be aware of other literature directed to economic analysis and cost of production. This includes the 1979 USDA workshop proceedings, the 1980 Great Plains symposium (Helmers), and Libbin and Torell.

3. Readers should see a range of authors for a perspective on the emphases of early farm management analysts and the changing emphasis of the agricultural economics discipline across time. These include Heady, Johnson, Salter, Swanson, and Taylor and Taylor.

4. This discussion should not be construed to suggest that the authors believe that empiricism *ipso facto* has resulted in Agricultural Economics "losing its relevance." This view of empiricism is often associated with a perspective that too much emphasis on economic theory and mathematics diminishes the capability of agricultural economists to solve real problems. We believe that a low emphasis on economic theory weakens the science of Agricultural Economics and that empiricism when properly combined with strong theory is helpful to solve the enormously complex societal problems related to food and agriculture. Similarly our discussion should not be construed to suggest that the growth of empiricism based upon marginal productivity was not a generally valuable development for Agricultural Economics.

5. Some find it nonintuitive to use a real opportunity cost on short-term capital. Yet the principle of separating nominal interest rates into an inflationary component and a real interest component is the same whether the time frame for capital use is within-year or multiyear. Perhaps the isolating of the true financing costs for short-term capital is best understood under a setting of borrowing at a nominal rate until year's end. When the loan is retired it is retired in lower valued dollars, thus the true financing cost rate is the real rate. Part of the confusion over this issue probably occurs because care is not always exercised in defining a budget for a time point such as January 1 or December 31 compared to the frequent practice of defining budgets for an entire time span such as a year (Helmers and Watts, 1985). Under inflation, the timing of when an expenditure is incurred within the year is important. Thus, a cost not expected or observed to be incurred at that time point should be inflated or deflated to the defined time point of the budget and a real interest used in association with those costs in calculating interest on short-term capital. It happens that if an end-of-year time point is defined for the budget, the sum of operating costs plus interest on those costs is the same whether the costs are adjusted to an end-of-year basis and a real interest cost applied to these costs or costs are adjusted to a beginning-of-year basis and a nominal interest cost applied. While the sum of the two end-of-year setting cost items is the same for the two methods, the costs of each of the two items is not. Thus, the principle which remains is to calculate short-term capital opportunity interest cost by using a real rate on those short term costs properly inflated or deflated to the time point of the budget.

References

Barkema, A. and M. Drabenstott. 1990. "A Crossroads for the Cattle Industry." *Economic Review*. Fed. Res. Bank of Kansas City, Nov.-Dec.: 47-66.

Black, J.D., ed. 1932. *Research in Farm Management - Scope and Method*. New York: Social Science Research Council.

Boss, A. 1945. *Journal of Farm Economics* 27: 1-17.

Carver, T.N. 1916. *Selected Readings in Rural Economics*. Boston: Ginn and Co.

Case, H. C. M., R. H. Wilcox, and H. A. Berg. 1929. "Organizing the Corn-Belt Farm for Profitable Production." Illinois Agric. Exp. Stat. Bull. 329.

Heady, E.O. 1966. "Changes in the Use of Production Economics in Research," in Production Economics in Agricultural Research, Proc., Univ. of Ill., AE-4108.

Helmers, G.A., ed. 1980. "Developing and Using Farm and Ranch Cost of Production and Return Data." Dept. Agr. Econ. Rep. 104, Univ. of Nebr. (GPAC Pub. No. 95).

Helmers, G.A. and M.J. Watts. 1985. "Developing Inflation-Free Cost and Return Estimates for Risk and Other Economic Analyses." Paper presented at GPC-10 Meetings, Fargo, N.D., May 29-31.

Hildreth, R.J. 1964. "Role of Agricultural Economics as a Discipline." J.S. McLean Memorial Lecture, Ontario Agricultural College, O.A.C. Ref. No. AE 64-5/18.

Hopkins, J.A. and P.A. Taylor. 1935. "Cost of Production in Agriculture." Iowa Agr. Exp. Sta. Res. Bul. 184.

Libbin, J.D. and L.A. Torell. 1990. "A Comparison of State and USDA Cost and Return Estimates." *West. J. Agr. Econ.* 15: 300-310.

North Central Farm Management Research Committee. 1954. "Budgeting in Farm Management Research." Mimeo.

Johnson. G.L. 1963. "Stress on Production Economics." *Austr. J. Agr. Econ.* 7, No. 1: 12-26.

Pond, G.A. 1956. "Farm Accounts as a Source of Data for Farm Management Research." Univ. Minn. Agr. Exp. Sta. Tech. Bul. 219 (N.C. Reg. Pub. No. 64).

Robison, L.J. and J.R. Brake. 1980. "Inflation, Cash Flows, and Growth: Some Implications for the Farm Firm." *So. J. Agr. Econ.* 12, No. 2: 131-137.

Salter, L.A. Jr. 1967. *A Critical Review of Research in Land Economics*. Madison: Univ. Wis. Press.

Schultz, T.W. 1957. "Reflections on Agricultural Production, Output, and Supply" in Baum, Heady, Pesek, and Hildreth, eds., *Economic and Technical Analysis of Fertilizer Innovations and Resource Use*. Ames: Iowa State College Press.

Schultz, T.W. 1964. "Changing Relevance of Agricultural Economics." *J. Farm Econ.* 46: 1004-1014.

Swanson, E.R. 1966. "Economics of Agricultural Production" in *Production Economics in Agricultural Research, Proc.*, Univ. of Ill., AE-4108.

Taylor, H.C. 1911. "The Place of Economics in Agricultural Education and Research." Univ. Wis. Agr. Exp. Sta. Res. Bul. No. 16.

Taylor, H.C. and A.D. Taylor. 1952. *The Story of Agricultural Economics in the United States, 1840-1932*. Ames: Iowa State College Press.

U.S. Dept. of Agriculture. 1979. "Estimating Agricultural Costs of Production: Workshop Proceedings." Economic, Statistics, and Cooperative Service. Report No. ESCS-56, June. Wash., D.C.: Government Printing Office.

Warren, G.F. 1920. *Farm Management.* New York: Macmillan.

Watts, M.J. and G.A. Helmers. 1979. "Inflation and Machinery Cost Budgeting." *So. J. Agr. Econ.* 11, No. 2: 83-88.

Watts, M.J. and G.A. Helmers. 1981. "Machinery Costs and Inflation." *West. J. Agr. Econ.* 6: 129-145.

4

Measurement Issues Relating to Policy Analysis

David H. Harrington

Introduction

The use of costs and returns information in policy analysis focuses on the use of calculated costs of production in setting the parameters for market intervention or income support policies. In the context of U.S. policies, the focus has traditionally been on how to set policies to alter the outcomes the market would normally produce. The keys are setting triggers or targets for intervention in the markets. These keys are necessary, and their effects are of largely the same magnitudes, whether the intervention is in prices (target prices or loan rates), quantities (quotas or market allocations), or incomes (income transfers other than target prices). Similarly, their effects are of largely the same magnitudes whether the intervention is on the product market side or the factor market side. The directions of the side effects may change if the interventions are in the product markets or the factor markets or if the intervention is in the price dimension or the quantity dimension.

Aggregate, Market-Level Focus

Cost of production figures appropriate for policy analysis differ from those appropriate for individual decision-making. At the individual firm level, the factor prices, product prices, and appropriate enterprise definitions are known and invariant to the decisions taken by the firm. The relevant opportunity costs faced by the decision-maker can be specified and will remain invariant to the decisions taken. At the aggregate level, both factor prices and product prices change with the collective action of all firms, and are functionally related to the decisions taken. Further, opportunity costs are no longer known, single values, but are distributions, which are also functionally related to the decisions taken. To treat the market-level and the individual-level relationships as identical in setting policy parameters introduces a fallacy of composition into the analysis--what is possible or advantageous for each individual is either clearly impossible or no longer advantageous for the aggregate.

The Appeal of Cost of Production

In spite of, or perhaps to those unaware of, the dangers of the fallacy of composition, cost of production calculations have a strong appeal for setting commodity policies. To participants in the policy process--Congressmen, commodity interest groups, and policy analysts in Government and universities--cost of production looks like the answer to their prayers because:

- They are commensurate with policy goals. What better way to provide support when it is needed and in the amounts needed than to tie the support levels to costs?
- They adjust automatically to changing cost conditions. If prices paid by producers change, or technology of production changes, so do the support policies.
- They look like objective, value-free calculations--not influenced by the lobbying strength of a commodity interest group. They seem to take the politics out of setting commodity policy.

To producers and asset owners their appeal is based on economic advantage:

- They socialize some of the risks of the market--price, production, or income.
- They privatize the benefits of reduced risk through allowing greater control over assets and more complete capitalization of income streams into wealth increases.

As an example of the latter, consider two requests: (1) "Lend me $1,000 to buy tickets on a $10 million lottery" and (2) "Lend me $1 million, I have the winning lottery ticket." If the lottery is fair, with no house take, the two requests differ in their size and the amount of risk involved. Most lenders would decline the first, small request as a bad risk but accept the second, large request as a good risk. A policy guaranteeing his/her cost of production puts a producer in the second category.

In the remainder of this chapter, I will examine some issues in the use of cost of production calculations in setting policy parameters and discuss who gains and who loses from cost of production pricing policies. My conclusions are that, far from being an objective, value-free measure, cost of production calculations yield arbitrary and manipulatable numbers. Instead of adjusting to changing economic conditions, cost of production criteria actually change the economic conditions in undesirable ways. Moreover, any benefits to producers and asset owners are short-lived at best, because the cost of production basis for support eventually has to be discarded. I will conclude with a discussion of how to minimize the dangers of cost of production pricing and how to break the appeal of cost of production in the policy process.

Problems in Cost of Production for Policy Analysis

The three major problems with using cost of production estimates for policy analysis are arbitrariness in calculation, circularity or bias in calculation, and the feedback loops and cost spirals.

Arbitrariness of Calculations

The existence of joint costs and joint products in agriculture preclude any objective criteria for: (1) Allocating costs of factors that are either durable or used in more than one production process (for example double-cropped land) or (2) Allocating costs among two or more products that are produced in the same production process. At the individual firm level you can overcome the second problem by judiciously defining the enterprises to include the joint products in the bundle of outputs of the enterprise. At the aggregate level you can't do this as easily because not all producers have the same mix of joint products. Policy decision-makers, as well as analysts, have trouble thinking in terms of the cost of (say) an acre of wheat that produces grain, straw, wheat pasture, and Government transfer payments in varying amounts. As a result, the common practice is to net everything back into cost per bushel; and this introduces a degree of arbitrariness into the aggregate cost of production calculations.

The first problem of joint costs is not tractable at either the individual firm or the aggregate level. There is no objective way of allocating joint costs to the different time periods or products involved. Obsolescence depreciation is a good example of a cost that cannot be uniquely associated with production of any crop in any year. In the absence of an objective criterion for allocation of these costs, any allocation is arbitrary.

Estimates of costs vary by their degree of arbitrariness. More specifically, there are three characteristics of costs that determine the degree of arbitrariness of the estimate. (1) Whether the quantities of the input are directly known for an enterprise or whether those quantities must be indirectly allocated among years or enterprises. (2) Whether costs are cash or noncash. Prices of inputs are known for cash costs but they are not known for noncash costs. (3) Whether the commodity of interest is produced jointly with another product. The only costs known with certainty and uniquely associated with a single commodity are those whose input quantities are directly known, are cash costs, and are for a commodity where production of another commodity does not occur jointly. For all other costs there are increasing degrees of uncertainty and arbitrariness. At most 50 percent of crop production costs and considerably less than 50 percent of livestock production costs fall into the known, single category. All other costs are, to some extent, manipulable by the assumptions and conventions used to assign the costs.

Specialized Resources and Circularity or Bias in the Calculations

Agricultural production has many specialized resources that can only be used effectively in agriculture, and thus derive their value from conditions in agriculture. As most recently pointed out by Pasour, the values of these specialized resources are not independent of demand and price conditions for the products they produce. Land, breeding stock, and the inventory of durable equipment are three notable examples of items which derive their value from conditions in agriculture. The independence of supply from demand is one of the basic conditions for the existence of a unique supply function for a commodity. Without this independence, a supply-demand cross implies neither equilibrium nor uniqueness. It can wander off in any direction, but usually it wanders towards higher prices as the supply function shifts upward through capitalization of increased returns into the opportunity costs of specialized factors. A supply function that is "contaminated" with specialized factors should more properly be termed a supply response function because the factor markets adjust to feedback from the demand function, thus destroying the *ceteris paribus* assumption of a true supply curve.

To overcome the problems of unknown cost components, the most notorious of which are for the specialized resources, cost of production calculations frequently resort to rewarding factors by their distributive shares valued at their opportunity costs. The assumption is that in equilibrium the firm will maximize its profits by adding factors until their marginal value products are equal to their opportunity costs. The distributive shares of the factors are assumed to exactly exhaust the total value product.

But, by Euler's Theorem the distributive factor shares will exactly exhaust the TVP if the total value function is homogeneous of degree $n = 1.0$. If $n > 1.0$ the TVP is more than exhausted, and if $n < 1.0$, the TVP is less than exhausted (Heady, pp 407-408). Homogeneity of degree $n = 1.0$ exists only at the low point of the long run average cost curve or at the boundary between stage I and stage II of the production function. It is unlikely that all firms are operating under either of these conditions.

In general, $n > 1.0$ is a more likely situation in aggregate because of lumpiness of several factors and the resulting existence of economies of scale, at least among small firms (Miller, Rodewald, and McElroy). Thus, in the presence of economies of scale ($n > 1.0$), the common assumption of cost of production calculations allocates more returns to factors than is generated in the TVP. Therefore, the markets cannot be rewarding factors this way. Contrariwise, if $n = 1.0$ the total cost of production will exactly equal the revenue received for the product; and the unit cost of production will exactly equal the price received for the product. Therefore, cost of production calculations are either circular arguments if $n = 1.0$, or upwardly biased cost estimates if $n > 1.0$. Thus, at the market level they are either not terribly useful, or else they are highly suspect.

Feedback Loops and Price-Cost Spirals

A feedback loop is built into the price and cost determining system if calculated costs of production are used to set prices or policy parameters. If such a feedback loop does not exactly mimic the market pricing behavior for the specialized resources then a price-cost spiral ensues with the supply response function constantly shifting as formula-mandated changes in returns are capitalized into the opportunity costs of specialized resources. Furthermore, even if the feedback loop does exactly mimic the market pricing behavior for the specialized resources, introducing the feedback loop effectively off-loads all demand based considerations in the determination of the product price. Producer prices and quantities will be determined by the intersection of the supply response function and the administered price, irrespective of what the market will absorb at that price. Inclusion of direct Government payments in cost of production calculations can introduce another cost-payment spiral; but that is the subject of another chapter so we only mention it in passing here.

How do cost-price spirals occur?   Cost-price spirals occur in one of two ways:

1. Formula-led spirals, if the cost of production calculations do not exactly mimic the market-based pricing behavior of specialized resources, creating ever-higher supply response functions, or
2. Policy-led spirals, if the cost of production calculations do exactly mimic the pricing of specialized resources, moving prices to a higher level on the same supply response function, but requiring policymakers to intervene before the next adjustment takes place.

Formula-led spiral. Table 4.1 illustrates how a formula-led spiral can occur (Harrington). If one uses the same assumptions to value farm production assets as used to calculate cost of production then because of the circularity discussed earlier the cost of production will always come out exactly equal to the price received for the product, as in Tables 4.1 and 4.2.

However, if farmers or investors in farm assets value their time, management, or assets differently than assumed in the calculations; or if the factor markets exhibit a different implicit rate of return than assumed in the calculations, the calculated cost of production will differ from the market price. If the calculated cost of production is then used to set the price a formula-led spiral results. For example, suppose that farmers or farmland owners, possibly because of expected capital gains on their assets, were willing to accept a 5 percent return on their assets instead of the 10 percent assumed in the cost of production formula. The $47,379 return to assets implied in the $3.82 per bushel support price would be capitalized into a value of assets of $947,580. Bidding for ownership of the land, either to operate or to rent out, would rapidly push land prices to this level. If the increased land value is used in the cost of production calculations for the subsequent year, the production costs as calculated by the formula will rise to $5.56 per bushel. If support prices are set at this level, the residual cash flow attributable to assets will increase to $94,758, which would then be capitalized by the behavior of the market into an asset value of $1,895,160. The process will go on without end until the feedback loop is broken by changing the policy or the calculations.

This example may seem to be somewhat dramatic because the assumed rate of return was twice that observed in the factor markets, causing the spiral to double asset values every year. However, comparing nominal borrowing rates with observed rates of return to assets in the farm sector can give even more dramatic differences between implicit market rates of return and assumed opportunity costs. If the error in mimicking the actual behavior of the asset markets were 20 percent then the value of assets would spiral upwards at 20 percent per year. Such an error could occur from using a theoretically justified real rate of interest in place of the observed rate of

TABLE 4.1 A Formula-Led Spiral: Hypothetical Data for a 640 Acre Wheat Farm

Panel I	Panel II
Calculated Cost of Production	Market Valuation of Assets, Period I

Panel I		Panel II	
Cash production expenses	$35,175	Market price =	$3.82/bu.
+Capital cost allowances	10,266		
+Business taxes paid	1,119	Total revenue	$103,713
+Opportunity cost of labor	3,885	-Capital cost allowances	-10,266
+Opportunity cost of mgt.	5,889	-Business taxes paid	-1,119
+Opportunity cost of	$473,790	-Cash prod. expenses	-35,175
capital @ 10%	47,379	-Opportunity cost of labor	-3,885
		-Opportunity costs of mgt.	-5,889
= Total costs	$103,713		
		= Net annual return to assets 47,379	

Full economic costs of production:
$103,713/27,150 bushels = $3.82/bu.

Return capitalized @ 5% = $947,580

Panel III	Panel IV
Calculated Cost of Production	Market Valuation of Assets, Period II

Panel III		Panel IV	
Cash production expenses	$35,175	Market price	$5.56/bu.
+Capital cost allowances	10,266		
+Business taxes paid	1,119	Total revenue	$151,092
+Opportunity cost of labor	3,885	-Capital cost allowances	-10,266
+Opportunity cost of mgt.	5,889	-Business taxes paid	-1,119
+Opportunity cost of	$947,580	-Cash prod. expenses	-35,175
capital @ 10%	94,758	-Opportunity cost of labor	-3,885
		-Opportunity costs of mgt.	-5,889
= Total costs	$151,092		
		= Net annual return to assets 94,758	

Full economic costs of production:
$151,092/27,150 bushels = $5.56/bu.

Return capitalized @ 5%=$1,895,160

Source: Adapted from Harrington.

return from current income to agricultural assets. If the behavior of factor markets is not exactly commensurate with theory then a spiral can occur.

Policy-led spiral. A policy-led spiral is illustrated in Table 4.2. Using the same data as in Table 4.1 except substituting a market-observed implicit rate of return to assets of 5 percent for the hypothetical 10 percent opportunity cost of capital gives a total cost of $80,023 and a cost per bushel of $2.95. If policy-makers deem the price ought to be higher and set it at $3.82 as in

TABLE 4.2 A Policy-Led Spiral: Hypothetical Data for a 640 Acre
        Wheat Farm

Panel I	Panel II
Market Valuation of Assets, Period I	Caluculated Cost of Production

Panel I		Panel II	
Market price =	$2.95/bu.	Cash production expenses	$35,175
		+Capital cost allowances	10,266
Total revenue	$ 80,023	+Business taxes paid	1,119
-Capital cost allowances	-10,266	+Opportunity cost of labor	3,885
-Business taxes paid	-1,119	+Opportunity cost of mgt.	5,889
-Cash prod. expenses	-35,175	+Opportunity cost of $473,790	
-Opportunity cost of labor	-3,885	capital @ 5%	23,689
-Opportunity costs of mgt.	-5,889		
		= Total costs	$80,023
= Net annual return to assets 23,689			
		Full economic costs of production:	
Return capitalized @ 5% = $473,790		$80,023/27,150 bushels = $2.95/bu.	

Panel III	Panel IV
Market Valuation of Assets, Period II	Calculated Cost of Production

Panel III		Panel IV	
Market price =	$3.82/bu.	Cash production expenses	$35,175
		+Capital cost allowances	10,266
Total revenue	$103,713	+Business taxes paid	1,119
-Capital cost allowances	-10,266	+Opportunity cost of labor	3,885
-Business taxes paid	-1,119	+Opportunity cost of mgt.	5,889
-Cash prod. expenses	-35,175	+Opportunity cost of $947,580	
-Opportunity cost of labor	-3,885	capital @ 5%	47,379
-Opportunity costs of mgt.	-5,889		
		= Total costs	$103,713
= Net annual return to assets 47,379			
		Full economic costs of production:	
Return capitalized @ 5% = $947,580		$103,713/27,150 bushels = $3.82/bu.	

the previous example, then the residual cash flow attributable to assets will
rise to $47,379, the value of assets capitalized at 5 percent will rise to
$947,580, and cost of production will again exactly equal the price received
for the product as the market bids this residual return into the value of the
assets. Notice that the spiral stops here--as soon as the value of assets
adjusts to the new residual return to assets. The new value of assets reflects
a point on the previous supply response function, but it is no more unique
nor necessary than the asset values resulting from the formula-led spiral.

The process is just more self-limiting, the process stops unless the policy-makers intervene and raise prices again.

Who Benefits from Cost of Production Pricing?

The primary incidence of cost-price spirals is on the wealth of land owners. Land captures most of the increased cash flow and bidding for ownership or annual control (rental) capitalizes the increases in cash flow into changes in wealth that can greatly overshadow the annual returns to land from current income.

More generally, the gainers from cost of production pricing are the owners of the assets when the process starts and the factor markets are learning how to adjust to the new policy. Owners of assets benefit more if their assets are (1) specialized to the production of the supported commodity, (2) have few substitutes in production of that commodity, and (3) are factors having inelastic or fixed supplies.

Land meets all of these characteristics and hence captures most of the benefits for its owners. At the other extreme are operator labor and management--they are not specialized, have nearly perfect substitutes, and very elastic supplies. Consequently, the land ownership function of a farm reaps essentially all the benefits, while the operation function gains little if any of the benefits.

Who Loses from Cost of Production Pricing?

The losers from cost of production pricing include consumers and users of the product, if market prices are increased to reach the price target and consumers are unable to switch to substitute products. If direct payments are used to achieve the price target the losses are borne by the taxpayer. In a more general equilibrium sense, we should note that anyone who has a competitive role in the factor markets is harmed by the cost of production pricing policies, and anyone who has a complementary role in the factor markets is benefitted. The opposite is true in the product markets-- competitors are benefitted and complements are harmed. If the cost of production figures are used to set quota or allotment policies, the effects in each market will be just the opposite of the effects of pricing policies.

Nevertheless, the large losers in cost of production pricing will be the owners of assets when the merry-go-round stops--as it inevitably must. The resource owners and their lenders will suffer capital losses that may even endanger their survival as the values of assets revert to expectations in a post cost of production world.

How To Break the Circle?

We'll consider five ways that have been proposed to break the circle that gives rise to cost-price spirals. All have been used in various countries in the past, and many have been used in combinations. Not all of them actually work to prevent price-cost spirals.

Mimic the Factor Markets

If cost of production calculations are made with consistent series on the values of assets and the rates of return to those assets then the calculations should track the historic operation of the factor markets. This method of calculation precludes the introduction of formula-led cost-price spirals. ERS currently does this by using an observed composite rental rate to value land services, and using observed current market values and observed rates of return in agriculture to value nonland fixed capital in its series. Another method which ERS has used in the past is to use current market values of assets and a real rate of interest. This method is consistent with capital budgeting theory, but it does not exactly mimic the behavior of the factor markets. It could introduce a minor cost-price spiral as a result of the small errors in tracking the behavior of the factor markets. Any method which uses nominal interest rates, or current acquisition costs of assets in place of the consistent, market mimicking values will introduce a formula-led cost-price spiral of the magnitude cited.

Leave Land Out of the Formula

Another way to prevent spirals is to leave out a factor or factors. The argument is that land, management, and probably operator labor are residual claimants and are thus product price determined rather than product price determining. Not all agricultural economists agreed with this as a solution. Sandy Warley and Phil Raup discussed it at length in a verbal exchange reported in a previous cost of production symposium (U.S. Dept. of Agriculture, pp. 30-31). ERS employed this argument until it was directed by the Congress to include returns for these factors in its calculations. We didn't convince very many people for very long. Only the 1977 Farm Bill excluded land returns in adjusting support prices for changes in costs. Groenewegen and Clayton used a similar argument to partition costs into short run variable costs and all other costs. They proposed supporting 100 percent of short run variable costs and 50 percent of other costs above short run variable costs. This, they proposed, would prevent any cost-price spirals.

However, if returns to specialized durable factors make up a large proportion of costs and the return to these factors is calculated with current market values and nominal interest rates, as done by some analysts, the errors in methods could easily make the calculated allocation to these factors more than double the actual return that is determining the behavior of the markets. In this situation the Groenewegen-Clayton adjustment would not prevent a spiral.

Cap the Support Prices

Capping the support price can break the circle and stop a cost-price spiral at any level at which it is exercised. An absolute cap simply replaces the cost of production calculation and breaks the circle directly when the cap is reached. Moving averages of market prices act like moving caps; they slow down a spiral but do not stop it. To my knowledge, no combinations of cost of production pricing with a price cap or price damper have been used in major U.S. commodities. However, some U.S. marketing orders have used parity calculations and price limits as parameters in their market diversion policies. Canadian marketing boards may have used similar policies with calculated costs of production.

Cap The Supported Quantities

By capping the quantities for which supports are paid you leave the market to determine the price at the margin and the quantity to be produced, giving the appearance of decoupled payments. This has been used by the Canadian Dairy Commission since the 1960s and 1970s in their Subsidy Eligibility Quotas and later in their Market Sharing Quotas. It is also the basis for the Production Entitlement Guarantee (PEG) proposal for GATT by Blandford, De Gorter, Gardner, and Harvey. Aspects of decoupling and capping supported quantities are evident in the frozen program yields and the 0-92 provisions in U.S. commodity policy. The decoupled nature of the payments is only true in the short run. In the long run, the existence of the subsidy payments within the quota attract or retain more resources in farming than would otherwise be the case, and affect entry, exit and scale of farms. Also, if the cost of production formula is used to set the target price within the quota, spirals are still possible. Excess resources in agriculture will tend to depress the over-quota market price; production will tend to contract toward the quantity covered by the quota; and quotas, if transferable, will capture the increases in cash flow instead of land. The potential for a spiral still exists if the quota is treated as a factor in the cost of production formula. If it is left out, this scheme is simply a repeat of the proposal to leave out a factor.

Fully Decouple the Payments

Fully decoupled payments are made on the basis of criteria other than production or resource control in both the short run and the long run. They can only do this by breaking all connections between the policy and the market. If they break all connections with the market they can no longer alter the outcomes that the market produces, which was the reason for the policies in the first place. Fully decoupled payments are not sustainable in the long run--precisely because they are decoupled. However, they do break the circle of a cost-price spiral and may provide a way to back down gracefully from policies that have distorted market relationships.

How to Break the Appeal of Cost of Production in Policy

There appears to be a "Gresham's Law" of cost of production estimates. Bad estimates (for policy setting purposes) chase out good. This is likely due to three things: (1) Proper and justifiable cost of production calculations for individual business decision-making are based on the use of opportunity costs that are usually higher than the market-observed implicit rates of return. (2) The subtle distinction between the proper methods for calculating cost of production figures for aggregate policy decision-making versus individual business decision-making has not been well explained by us or anyone else yet. Most participants in the policy process think that what is right for the firm is right for the aggregate. (3) The people who stand to gain in the short run from cost of production pricing are organized, localized in relatively few electoral districts, and have considerable influence on the policy process. Those who stand to lose are none of these things.

To counteract these tendencies, agricultural economists can review the historical experiences of experiments in cost of production pricing. Each of these experiences, and others throughout the world, deserve study to determine what the effects of the policies were on all the affected groups. They have been unalloyed disasters for consumers and taxpayers in almost every case.

- In the United States, the performance of the peanut subsector bears study. The price of in-quota peanuts has been adjusted by cost of production figures for several years. The prices of in-quota peanuts and additional peanuts both appear to be widely distorted.

- In British Columbia a system of dairy quotas and indexing of milk prices by consensus cost of production estimates from a panel of farmers was instituted in the early 1970s. Within a year the value of the quotas exceeded the value of the dairy herds in British Columbia and costs had risen to accommodate this increase.
- The Canadian Egg Marketing Agency instituted quota systems and cost of production indexing of egg prices at about the same time. Egg prices in Canada have hovered significantly above prices in the United States ever since.
- Finally, the Canadian Dairy Commission instituted the systems of Subsidy Eligibility Quotas and Market Sharing Quotas with prices and direct payments indexed by cost of production calculations. Significant surplus disposal problems ensued and government payments rose rapidly for several years.

Secondly, departments of agricultural economics should at least mention and describe the problems of such policies in their production, marketing and policy courses. With the exception of the work of ERS and a few works by university agricultural economists, such as Marshall Martin, and E.C. Pasour the literature and curricula seem to have forgotten this area of policy formulation. This is strange because the analytic tradition and some of the major conclusions go all the way back to David Ricardo. Moreover, cost of production pricing proposals seem to surface at least once each generation.

Third, we as a profession, should continue to stand firm with logic, analysis and persuasion. The proponents of cost of production pricing arguments have a deceptively simple idea that they are backing. The arguments against it are subtle and not easily understood, except from an economist's unique training and point of view.

The alternative is to let it happen. If the forces of the policy process rally behind the deceptively attractive policy of cost of production pricing, and the economics profession offers weak, inconsistent, and/or contradictory evidence in the policy debates, then such policies will be enacted. Three major effects will become evident over time: (1) Policy-makers will be faced with unacceptable budgetary costs or consumer income transfers to the wealthy. The farm operation function will capture little of the transfers; purchasers of assets after the policies go into effect will gain little; but owners of assets when the policy is instituted and while the market is adjusting to them will gain the most. (2) Resource owners when the policies are ended stand to lose the most. If the policies are lavish enough to be unsustainable, whoever owns the assets when the expectations change about continuation of the policies will take considerable capital losses. Their continued survival as firms may be jeopardized and they may jeopardize their lenders as well. (3) Moreover, if such policies are allowed to "just happen" the economy and the policy process will be on a completely different trajectory. The theory of the second best, the rent seeking behavior of the policy participants, and the transaction costs may preclude us from ever getting back to the more desirable policy trajectory.

References

Blandford, David, Harry DeGorter, Bruce Gardner, and David Harvey. 1989. "There is a Way to Support Farm Income with Minimal Trade Distortions", *Choices*, First Quarter: 20-25.

Groenewegen, John R., and Kenneth Clayton. 1982. "Agricultural Price Supports and Cost of Production", *American Journal of Agricultural Economics* 64: 271-275.

Harrington, David H. 1983. "Costs and Returns: Economic and Accounting Concepts", *Journal of Agricultural Economics Research.* 35, No.4: 1-8.

Heady, Earl O. 1952. *Economics of Agricultural Production and Resource Use.* Englewood Cliffs, NJ: Prentice-Hall.

Martin, Marshall. 1977. "Cost of Production: The Concept and Some Implications for Its Use in the Determination of Target Prices and Loan Rates." Department of Agricultural Economics, Purdue University. Station Bulletin No. 162.

Miller, Thomas A., Gordon E. Rodewald, and Robert G. McElroy. 1981. "Economies of Size in U.S. Field Crop Farming." Economic Research Service, U.S. Department of Agriculture. Ag. Econ. Rpt. No. 472.

Pasour, Jr., E. C. 1980. "Cost of Production: A Defensible Basis for Agricultural Price Supports?" *American Journal of Agricultural Economics* 62: 244-248.

U.S. Dept. of Agriculture. 1979. "Estimating Cost of Production: Workshop Proceedings." Economics Statistics and Cooperatives Service, ESCS-56, June.

PART TWO

Alternatives for Measuring Costs and Theoretical Considerations

5

Theoretical and Practical Considerations in Measuring Costs and Returns

Vernon R. Eidman

Introduction

Agricultural economists have a long history of both forecasting costs and returns of producing agricultural commodities and of recording the income and expenses that occurred during some historic period. Farm management texts commonly discuss these topics as they relate to the individual farm business. The data sources and methods for preparing cost of production estimates for use in analyzing aggregate production and policy issues have been discussed regularly in the literature. (For example, see USDA, 1979). In general these discussions focus on methods to estimate the private costs and returns of the enterprise in current dollars for a producer or group of producers. This chapter continues in that tradition, avoiding the knotty issues of estimating social production costs. It also bypasses four areas needing further discussion: removing the indirect effects of government programs from cost estimates, estimating costs of land and water services, estimating the costs of capital services and estimating the costs of human capital inputs. Each of these four areas are being treated in later chapters.

The task here is to review the current status of costs and returns (CAR) estimates for individual commodities in terms of the issues and opportunities that face us today. My comments focus on the data and methods used by the Economic Research Service to estimate the costs incurred and the returns generated in producing a commodity over geographic areas, such as states, regions and the U.S. The initial comments relate to potential uses of such estimates and the format in which the CAR are being presented. This is followed by a discussion of several concerns I have with the data sources and computational procedures currently being used to prepare CAR.

The final section comments on some of the newer allocation procedures that promise to replace current methods and reduce the amount of survey data required.

Potential Uses

A recent publication notes that the USDA production costs and returns estimates have been designed to inform policy makers of what costs and returns would be in the absence of the direct effects of government programs (Salassi, *et al.*). The discussion notes a second purpose is to provide information on the profitability of producing the major agricultural commodities. In addition, many economic studies of aggregate supply response, the impact of changes in technology, relative competitiveness across states, regions and countries, and the impacts of various types of governmental policy upon agricultural producers are frequently based in part on CAR estimates. Given the heavy reliance on this data, we should consider approaches to prepare CAR estimates as accurately and completely as feasible.

Average annual costs and returns for large areas are of limited use to individual producers. However, costs and returns estimates with appropriate information on the distribution of costs may be of value to farm operators and their advisors as they evaluate their competitive position. Knowing how their costs compare to other producers in the state, region and country may be of use in planning a long-run strategy for the business. Thus, moving to a format that presents the distribution of costs has the potential to increase the number of users. In addition, the individual observations from the Farm Cost and Return Survey (FCRS) may be of use in developing enterprise budgets for smaller geographic areas. Finding methods for agricultural economists in the experiment stations and extension services of the various states to access the relevant survey data would help bridge the gap between enterprise budgets prepared by the individual states and the USDA estimates.

Presentation of Costs and Returns Estimates

One of the major problems in estimating costs and returns of production is that both the costs and the returns vary from farm to farm and from one part of the country to another. The data required to estimate variability in the commodity prices received by producers has not generally been collected by FCRS. The survey does collect data to estimate production costs and yields at the farm level. In addition to summarizing the average CAR at the state, regional and national levels, procedures currently being implemented summarize the individual observations as a cumulative distribution of economic production costs per unit of output (CDPC). This is a very useful method to present a more comprehensive picture of the economic costs of producing a commodity (Ahearn, *et al.*). Two relationships are graphed for each commodity. One CDPC shows the cumulative percentage of production by economic cost level, while a second displays the percentage of farms by cumulative cost level. With both CDPC's plotted on the same graph one can relate the cumulative percentage of producers to the cumulative percentage of the commodity produced during the historic period. In addition, characteristics of the sample farms making up segments of the CDPC can also be summarized. Thus the location, crop production characteristics, farm size and other characteristics can be reported for low-cost, mid-cost and high-cost producers.

The CDPC's themselves may overwhelm some policy makers, and other noneconomists. These groups could be served by summarizing average costs of production by segment of the distribution and presenting the related data on the location and characteristics of producers. The complete CDPC's should be of particular interest to economists using the data for research. It also will provide an opportunity for many economists at the state level to analyze the competitive position of specific producer groups with other producers in the region and the nation.

Users of CAR could be better served by augmenting the presentation of CDPC for economic costs in 3 major ways. The first is to present both variable cash costs and economic costs using the CDPC format. The CDPC for variable cash costs is of interest for many research applications. In addition, data on producer characteristics by segment of the CDPC for both cost levels would indicate to what extent producers with low variable costs also have low economic costs of producing the commodity. Variable cash costs are likely to be impacted less than overhead costs by the indirect effects of government programs. Thus comparing the characteristics of low, mid and high cost production for the two relationships may provide some insight into how economic costs will change with adjustment in commodity, environmental and other programs. The second suggestion is to provide producers' characteristics for more intervals on the CDPC. In general, doing so for each of four to five intervals that include 20 to 25 percent of the producers would provide a clearer indication of how producer, farm and production system characteristics change as one moves from low cost to high cost production. The third suggestion is to provide more data on some of the environmental characteristics of production systems. The FCRS currently collects the brand name of herbicides, insecticides and fungicides applied. Knowing the brand names, the pesticides could be categorized based on appropriate criteria such as toxicity, leaching and surface run-off. Collecting the quantities applied or assuming the recommended rate is used would permit summarizing pesticide characteristics along with other characteristics of producers by segment of the CDPC.

Data Sources and Computational
Methods for Cost Estimates

Presenting the CAR estimates in the form of CDPC places greater demands on the data sources and the estimation procedures. Presenting the data in this format focuses attention on the entire distribution (and particularly on the tails of the distribution) rather than on the mean value. This immediately raises questions about how accurately the estimates represent the cost levels at each point on the distribution, from the lowest cost to the highest cost producers. My review of the procedures and data sources currently being used suggest several areas that should be evaluated carefully to determine how current procedures may bias the resulting shape and level of the CDPC. In particular, I am concerned about the collection of input prices, the estimation of machinery, equipment and drying costs and the allocation of overhead costs and the allocation of fixed labor supplies.

Input Prices

The FCRS collects data on the quantities and/or price of variable inputs and services used in producing crop and livestock enterprises. The total expenditure is collected for many inputs and the average cost per acre or unit of livestock is calculated directly without recording the price and quantity. This approach should capture the combined variation of both the quantity and the price paid for inputs. However, for fertilizer and perhaps a few other inputs, FCRS records the quantities applied and the cost per acre is obtained by applying average prices paid as reported in *Agricultural Prices*. This procedure will miss the pecuniary economies farmers realize in purchasing during the off-season or purchasing larger quantities. Fuel prices are another important variable input where reliance on average prices paid to calculate the cost per farm may bias the shape of the CDPC. Using these average prices can be expected to reduce the variation in costs recorded by the CDPC.

Operating Costs of Machinery and Irrigation Equipment

FCRS collects data on the characteristics and annual use of the machinery and irrigation equipment on the farm. The machines used for each field operation, the number of irrigations and the water application per irrigation are also recorded. Standard engineering procedures are applied to estimate fuel, lubrication and repair costs for the crop enterprise on the farm. While the procedures are used widely, it seems appropriate to evaluate how accurately they represent the quantities of fuel, the cost of lubrication and the cost of repairs farmers are buying. Given the magnitude of fuel and repair costs, significant errors in these estimates could bias the variable cash costs and the resulting shape of the CDPC.

Drying costs for corn, sorghum, sunflowers and rice are estimated based on commercial rates and specification of fuel types and the initial moisture content. Many farmers producing corn and perhaps other crops dry the crop on the farm. Costs of removing a given amount of moisture vary significantly by type of drying system. Like the use of average input prices, estimating drying costs with commercial rates is likely to understate the variation in production costs.

Allocation of Fixed Cash Expenses and Capital Replacement

The CAR estimates have used several methods to allocate general cash overhead expenses and the interest paid on both operating and real estate loans. It is my understanding that a common system has been established to allocate these costs across all commodities for future estimates. When the transition to the system is completed, these costs will be allocated in proportion to the relative value of farm production. Thus, an enterprise that generates 40 percent of a sample farm's value of production will be allocated 40 percent of the general overhead and cash interest expense on the farm. Real estate taxes and insurance are allocated on a per acre basis. Property tax on machinery and equipment is calculated per hour of annual machine use at the farm level and allocated to each enterprise in proportion to the hours of machine use. Insurance costs on machinery, equipment and buildings are allocated in an analogous manner.

Capital replacement costs for machinery, equipment, and vehicles is calculated on a per-hour rate. It is allocated to each enterprise based on the hours of use per acre or per unit of the livestock enterprise.

Any allocation procedure for joint costs is somewhat arbitrary. The system evolving has the virtue of allocating these costs in a consistent manner across alternative crop and livestock enterprises, which is of some comfort in comparing CDPC across regions and commodities.

My major concern with these allocation procedures for fixed cash expenses and capital replacement is that certain categories of costs may be omitted from the calculations and that these omissions may be greater for some types and sizes of farms than others. Fuel, lubrication, repairs, insurance and capital replacement costs associated with overhead vehicles are likely to be omitted. The costs of maintaining the farmstead, and the costs of maintaining set aside and conserving acres also are unlikely to be included in the cost allocation process. Perhaps the costs on set aside and conserving acres should be omitted since the government payment income is omitted. However, it appears the definition of economic production costs intends to include the expenses of operating overhead vehicles and the costs of maintaining the farmstead. Their omission is particularly bothersome because they are likely to be a larger proportion of joint expenses on farms producing both crops and livestock. Furthermore, the proportion of these expenses is likely to increase as the size of the farm decreases. Their omission will result in an underestimate of the costs on crop/livestock and smaller farms. The issue to be examined is the magnitude of any omissions and whether the omissions are likely to bias the shape as well as the level of the CDPC.

Labor

Current FCRS procedures ask the respondent for the total unpaid operator and family labor used by month throughout the year. Then the survey asks for the percentage of this labor that is devoted to the crop or livestock enterprise for which detailed data are being collected. However, no effort is made to have the operator allocate labor among all of the enterprises on the farm. Thus, it is unlikely this procedure will produce good quality estimates of the amount of unpaid labor devoted to the enterprise.

Recent Methodological Developments

The recent application of regression procedures to estimate enterprise variable cash costs from data on total quantities used on the farm promise to reduce data collection requirements for CAR estimates (Hornbaker *et al.* and Just *et al.*). Just, *et al.* contrasted two methods to estimate variable input costs for individual enterprises, a profit maximization approach and a "behavioral" or positive approach. Their profit maximizing approach requires annual data on the total amount of each variable input used on the farm, the revenue distribution among enterprises and the input prices. In contrast, the positive approach estimates the amount of input use per enterprise using annual farm data on total use of the variable input and the amount of land devoted to each crop. These methods were applied to multiyear data on a sample of farms in one area of Israel. Two variable inputs (irrigation water and other) were allocated among five crops. The Hornbaker *et al.* method predicts the input costs for the crop enterprises based on the total cost of the input for the farm, the amount of land devoted to each enterprise, and some variables indicating size and other characteristics of the farm. This approach was applied to data over a two-year period for 161 Illinois farms to allocate five categories of variable inputs among three crops: (a) fertilizer and pesticides, (2) seed, storage and drying, (c) power and equipment, (d) hired labor and (e) other expenses.

These two studies report reasonable success in allocating inputs among the crop enterprises suggesting that further evaluation of the positive approach is justified. In particular, the procedures could be evaluated with a wider range of enterprises and a more diverse set of farms. Some of the variable input categories, such as fuel and repairs, should be allocated across livestock as well as crop commodities if these procedures are to be applied to whole-farm data. It would be useful to evaluate the accuracy of the estimated allocation by contrasting them with both the allocations farmers recorded, and with the procedures currently used to make the allocation for the CAR estimates. Some initial tests of this type could be conducted using records that include the farmers' recorded allocations. Additional tests with FCRS data could be conducted where the survey has collected the allocations to individual enterprises. These tests should help clarify where use of the regression approaches will reduce data collection costs while maintaining or improving the accuracy of variable input estimates at the enterprise level.

The problems reported in applying these methods suggest the regression procedures may replace current allocation methods gradually. They may also represent an appropriate way of responding to significant pressure to reduce survey costs.

Summary

The CAR estimates are a major source of data that are used widely in economic research and policy formulation. The recent efforts to increase their accuracy and the informational content should make the series more valuable to current users as well as expand the number of users.

Efforts to present information on the way production costs vary and the characteristics of production units by cost level should be expanded within the limits of the sample survey to provide statistically valid results. Using the CDPC format to present regional and national economic cost of production estimates increases the informational content. The informational content could be further enhanced by presenting both variable cash costs and economic costs for each of these regions in the CDPC format. Providing information on the environmental characteristics for low, medium and high cost producers would also enhance the value of the data presented.

Some of the data used appear to be inadequate to support development of accurate estimates within the CDPC format. The use of state average price levels for some inputs, the methods used to estimate machinery, equipment, irrigation and drying costs and the possible omission of some categories of overhead costs are of particular concern. The impact of each of these on the accuracy of the CAR estimates should be evaluated and appropriate adjustments made if the evaluation suggests doing so is important to the accuracy of the CAR estimates.

The accuracy and cost savings associated with using regression methods to allocate variable cash costs should be evaluated. The recent literature suggests that the use of positive approaches, rather than basing the allocation on a normative framework, significantly improves the accuracy of these results. While the published results are very encouraging, clearly more testing of the procedures than have been published are warranted before these procedures are used to replace current methods of estimating variable input expenditures for individual enterprises at the farm level.

References

Ahearn, Mary, Mir Ali, Robert Dismukes, Hisham El-Osta, Dargan Glaze, Ken Mathews, Bill McBride, Robert Pelly and Mike Salassi. 1990. "How Costs of Production Vary." U.S. Dept. of Agriculture, Economic Research Service. Agriculture Information Bulletin No. 599. Washington, D.C.: Government Printing Office.

Hornbaker, Robert H., Bruce L. Dixon and Steven T. Sonka. 1989. "Estimating Production Activity Costs for Multioutput Firms with a Random Coefficient Regression Model." *Amer. J. of Agric. Econ.* 71: 167-177.

Just, Richard E., David Zilberman, Eithan Hochman and Ziv Bar-Shira. 1990. "Input Allocation in Multicrop Systems." *Amer. J. of Agric. Econ.* 72: 200-209.

Michael Salassi, Mary Ahearn, Mir Ali and Robert Dismukes. 1990. "Effects of Government Programs on Rice Production Costs and Returns, 1988." U.S. Dept. of Agriculture, Economic Research Service. Agriculture Information Bulletin No. 597. Washington, D.C.: Government Printing Office.

U.S. Dept. of Agriculture. 1990. "Economic Indicators of the Farm Sector: Costs of Production-Livestock and Dairy, 1989." Economic Research Service. ECIFS 9-1.

_____. 1990. "Economic Indicators of the Farm Sector: Costs of Production-Major Field Crops, 1988." Economic Research Service. ECIFS 8-4.

_____. 1979. "Estimating Agricultural Costs of Production-Workshop Proceedings." Economics, Statistics and Cooperative Service. Report No. ESCS-56.

6

Comparative Estimation Systems and the Random Coefficient Regression Approach

Robert H. Hornbaker, Steven T. Sonka, and Bruce L. Dixon

Introduction

Understanding the behavior of costs is a key element in the strategic assessment of business alternatives (Porter). Cost is a dynamic concept and its behavior is of interest when the size and organization of firms differ. Also of interest is how costs change over time as economic conditions vary. The value of knowing cost behavior is just as high in farm as in non-farm settings. Within the agricultural world, "cost of production" is critically important to the producer considering both short and long run alternatives. Non-farm agribusiness decision makers strive to estimate production costs for farm commodities so as to predict future demand for purchased inputs and future supply of agricultural output. Increasingly, government policy makers require cost of production information for international negotiations as well as for calculation of domestic farm income support levels.

As has been noted in numerous review papers in the literature of agricultural economics, this diversity of interest guarantees that no one estimate of cost of production will be satisfactory to all who have legitimate interests in that estimate. By extension then, no single estimation system is likely to be most preferred for all possible situations. Therefore our goal should not be to determine a single best approach but rather to increase our understanding of the strengths and weaknesses of the alternatives.

This chapter is organized into two major components. The first compares alternative estimation systems relative to both historical work in agricultural economics and to the changing agricultural industry. In the second section, we review an approach that addresses a major problem in the calculation of commodity production costs. That approach, use of random coefficient regression (RCR), allows the analyst to use actual farm data in the estimation of commodity production costs, even when the farm accounting system or a survey instrument does not directly collect data by commodity.

Alternative Estimation Approaches

Not surprisingly, a review of the agricultural economics literature reveals a long tradition of concern relative to the estimation of production costs. Interestingly, however, there does seem to be a marked increase in the level of interest in production cost estimation at the policy level. For policy purposes, agricultural economists historically preferred to concentrate on farm size and structure issues. Production cost was important as a variable that was needed for empirical assessments of size and structure, not so much as a policy variable in its own right. Throughout the 1980s, the intertwined effect of the farm financial crisis, associated use of target prices as a policy variable, and issues of international protectionism and competitiveness have greatly increased the direct policy interest in production cost estimates for major agricultural commodities.

Despite this change in emphasis, there is much that can be learned from prior literature relating to farm costs, size and structure. One useful collection of review papers was presented at workshop associated with the 1983 Meetings of the American Agricultural Economics Association at Purdue University (Center for Agricultural and Rural Development). Two types of estimation problems are identified in such reviews. The first relates to the sources and underlying meaning of the data used in the estimation technique. The second estimation issue refers to choice of technique.

Although not the primary assignment of this chapter, underlying data issues are important. The need to impute values to key production factors such as labor and land exists regardless of the estimation technique. The appropriate means to collect and then use pecuniary sources of scale advantages is likely to be increasingly important over time. Further, the changing nature of the agricultural industry will raise new, intriguing and challenging questions. The expanding importance of quality differentials, for leaner animals or specific types of crops, complicates the analyst's task in estimating production costs. To what extent, for example, should food grade or white corn be treated as a different output than #2 yellow corn? These issues, and several others, exist across the range of estimation techniques available.

Traditionally, we divided estimation approaches to size and structure research into positive and normative categories. The positive approach involved analysis of actual farm data, either directly through econometric techniques or more indirectly through observing which sizes and types of firms survived over time. Normative approaches used tools such as budgeting, economic engineering, or mathematical programming to indicate how relative costs should behave with changes in size or structural variables.

Although appropriate for size and structural analyses, the positive-normative categorization seems less relevant for attempts to estimate costs of commodity production. Often production costs are estimated using a combination of data from positive and normative sources. Therefore the following three categories seem more descriptive: direct reliance upon expert judgement, analysis based upon special survey results, and manipulation of accounting based farm firm data. For this discussion, the survey approach will include both routine data collection efforts of the U.S. Department of Agriculture and targeted surveys conducted as special studies primarily in university settings. The following paragraphs will identify some strengths and weaknesses of those approaches.

Often the value and importance of an activity is not truly appreciated until that activity is gone or becomes less common. Between the end of World War II and the start of the 1980s, Land Grant institutions made very effective use of expert judgement to produce high quality estimates of production costs. That expert judgement was packaged (usually) within the farm management extension specialist(s). That individual (or those individuals in some states) used many sources of knowledge and information to derive commodity specific costs. Although presented as rather mundane budgets, these estimates were based on actual farm experience and data as well as expertise of other extension specialists in related disciplines. The most effective farm management specialists at least informally tested their estimates against a sample of actual farm operations.

As we are all aware, disciplinary elitism on the one hand and budget restrictions on the other have combined to greatly reduce our real capacity in the farm management extension area. And that capacity is now spread over a wider agenda (low-input sustainable agriculture, environmental concerns, etc.) further reducing our capability to generate expertise-based costs of production. The rise of computerized budget generators and other computer aided means to estimate production costs is an innovative means to combat this problem. Indeed, we are likely to see an "expert system" to estimate costs (if one doesn't exist already) before the end of the 1990s. That expert system could incorporate sophisticated normative modeling as well as heuristic rules of an individual or group of experts.

Although greater use of information technology is a positive development, we need to realistically assess those tools in the context of our declining "real" knowledge base. One danger of computer-aided production cost estimators is that they provide answers that tend to be unqualified as to their actual appropriateness. (In theory expert systems can be developed which would indicate their lack of knowledge but such systems are not likely to be the norm in the near future.) Instead, we tend to get attractive printouts of cost estimates from computer aided sources, rather than the qualified judgement and knowledge that came from direct counsel with the farm management extension specialist. Although those qualifications often frustrated the analyst user, they also warned the analyst about inappropriate use of those costs.

The other two methods of estimating costs, analysis based upon special survey results and manipulation of accounting based farm firm data, are probably more conceptually similar than they are different. Both rely upon observations of actual farm actions where possible and then supplement the data derived from those observations with theory and information from other sources where necessary. Data that describe cash operating expenses usually are obtained directly. In some cases (i.e., fertilizer) quantities are obtained from the producer and then combined with average input costs. Producers may have greater success recalling amounts rather than costs, particularly when costs are shared with landowners. However, the presence or absence of pecuniary economies is arbitrarily eliminated by this approach.

With either survey or accounting based data, valuation issues loom large for inputs that are not in the cash operating expense category. Attributing appropriate annual expenses to machinery and equipment is a subject of seemingly endless debate in the economic literature. When all is said and done, however, the analyst will have to make a valuation choice that is not universally satisfactory. Similarly, the problems of assigning costs to the use of operator labor and land, traditional residual claimants in agriculture, are conceptually intriguing but the actual choice made "to get a number" will not be totally satisfying. Indeed, whether survey or accounting based data are used, we need to recognize that our resulting cost estimates actually are determined by cash operating expenses and analyst assumptions regarding the remaining inputs.

The survey and accounting based approaches are not totally equivalent, however. Both have strengths and weaknesses. The capability to derive a representative sample is a major differentiating factor for these two data sources. Even in relatively large farm record systems, very small and very large operations tend to be underrepresented. Therefore, the survey based approach has a far better potential to achieve statistical representativeness.

Although having data that are statistically representative is a key concern, accuracy and reliability of the data also are significant issues. In the survey approach, the respondent is expected to use an existing data source or to recall values to complete the questionnaire. The presence or absence of a trained enumerator is, of course, a major factor in obtaining valid data. Important roles of the enumerator are to assist the respondent and to assess the quality and usability of the respondent's answers. Even with the trained enumerator, reliability and validity of responses are concerns. This is particularly true relative to the year-to-year variability of production agriculture. Matching expenses to output, a problem when production activities are not matched to the accounting year, would seem to be better done within the survey framework. However, if the respondent relies upon financial records to respond to the survey, the matching problem may again occur.

Accounting system data would seem to have the value of greater consistency and reliability. Because the respondent usually is compiling the data for their own business purposes, the motivation exists to accurately record observations. Consistency is enhanced when using systems where a professional is employed to assist the producer with data collection, compilation and interpretation. Differences in accounting and economic interpretations of cost often can be overcome if the system is consistent in its procedures. Finally, year-to-year variability in performance on farms is a major issue in many types of production agriculture (Sonka, Hornbaker, and Hudson; Garcia, Offutt, and Sonka). Access to several years of data from a panel of producers provides an important means to remove year-to-year variability and also to measure the effect of that variability on cost of production.

Another difficulty in estimating commodity production costs is that most farm operations produce more than one commodity. The farm management extension specialist could use expert judgement, input from other disciplinary specialists, and observation of actual operations to derive single commodity cost estimates. As noted previously, we have lost much of our expertise in the Land Grant system to perform the difficult and time-consuming actions needed to derive such estimates. A procedure that has the potential to estimate single commodity costs using data that can be obtained from survey and/or accounting based sources is described in the following section.

The Random Coefficient Regression Approach

The appropriate specificity level of the cost estimates varies with the intended use of the information. For many policy users, aggregated state, region or national estimates may be sufficient. In contrast, producers may need to differentiate costs by individual fields to implement the most profitable crop and environmentally stable selection, rotation, and production plans. While methods exist for estimating both types of information, often a particular method is appropriate only for specific problems or applications.

In this chapter, an alternative means of estimating production activity costs and technical coefficients for mathematical models is presented. The method presented here is random coefficient regression (RCR). This section of the chapter incorporates a summary of the estimation methods and results also presented in Hornbaker, Dixon and Sonka, and Dixon and Hornbaker.

Commodity activity costs are directly available only when the farm operators record cost or quantity information by commodity. In such situations, calculation of technical coefficients and per acre costs, therefore, is straight forward. However, in most cases where farm records or survey information are available, data are recorded for the total farm and not by individual commodities.

RCR is an appropriate estimation method in these cases because the activity costs and the technical coefficients for the sample of farms are stochastic identities. The identities are stochastic because the per unit costs and the technical coefficients vary from farm-to-farm within the sample. The technical coefficients are not uniform among the cross-section of farms, therefore, the technical relationships and the costs have a distribution over the sample population. Estimation of the means and variances of these distributions is examined below. The method applies to both estimating activity costs or technical coefficients.

Assume a firm produces p different products and performs q activities. At least one, and perhaps all, of these activities are required to produce any of the p products. The cost to the firm of ith activity, $i=1,...,q$, can be written as an identity:

(1) $$TVC_{ij} = \sum_k c_{ijk} Q_{jk}$$

where

   TVC_{ij} is the total variable cost to firm j of activity i,

   c_{ijk} is the variable cost of activity i for firm j per unit of product k,

and

   Q_{jk} is the level at which firm j produces product k.

   In the case of technology coefficients, a linear technology for a firm may be written as:

(2) $$\sum_k a_{ik} x_k = b_i$$

where

   a_{ik} is the technical coefficient for resource i on output k,
   x_k is the level of output k, and
   b_i is the available resource i.

Since (2) characterizes a firm's technology, the linear constraints are identities. For a given resource b, each firm uses an allotted amount in its production processes and this amount will exactly equal the summation of the technical coefficient for the resource times the amount of the products produced with that resource. For example, the identity for a farm that uses 18,000 pounds of nitrogen at the rate of 1.2 pounds per bushel of corn and 1.4 pounds per bushel of wheat is written as: $1.2 \, x_1 + 1.4 \, x_2 = 18,000$, where x_1 and x_2 are the bushels of corn and wheat, respectively. The farm identity likewise exists if the estimated technical coefficients are per acre of output, rather than per bushel of output. Because the linear technology equations for individual farms are identities, the variation or distribution of the a_{ik} is from farm-to-farm.

Production Activity Costs (Hornbaker, Dixon and Sonka)

Various factors can cause the cost coefficients, c_{ijk}, to vary from firm to firm. If such values are known, they can be modeled as exogenous variables. Exogenous variables are relevant in many empirical applications and their use can improve the explanatory power of the estimation procedure. Assuming the relevance of exogenous variables, the activity costs in (1) may be rewritten as functions of exogenous variables in the following form:

$$(3) \qquad c_{ijk} = B_{1ik} + B_{2ik}z_{2jk} + \ldots + B_{mik}z_{mijk} + v_{ijk},$$

where
      B's are parameters to be estimated,
      z's are explanatory variables, and
      v_{ijk} are random error terms with mean zero and constant variance.

In the application presented here, the exogenous variables were allowed to vary across products. To estimate the activity costs (3) is substituted into (1) to provide

$$(4) \qquad TVC_{ij} = \sum_k (B_{1ik} + B_{2ik}z_{2ijk} + \ldots + B_{mik}z_{mijk} + v_{ijk})Q_{jk}$$

An estimate of c_{ijk} for activity i and output k for firm j can be computed from the estimated B's. Using the method of Swamy and Tinsley, predictions of the v_{ijk} for each observation can be obtained. The predicted c_{ijk} is computed for each firm j as follows:

$$(5) \qquad \hat{c}_{ijk} = \hat{c}_{ijk} + (\sum_h a_{hk}Q_{jh})[\sum_h (\sum_k Q_{jk}a_{hk})Q_{jh}]^{-1}(TVC_{ij} - \sum_k Q_{jk}\hat{c}_{ijk})$$

where

a_{kb} are the elements of the covariance matrix A which is the covariance matrix of the v_{ijk} describing the covariation of v_{ijk} among the p products, and

$\tilde{c}_{ijk}$ are the best linear unbiased predictors (BLUPs)

$\hat{c}_{ijk}$ are the estimates of the mean c_{ijk} based on the generalized least squares estimates of the B's

The BLUP essentially allocates that component of TVC_{ij} not explained by the estimated activity costs and Q_{jk} over the p products. Therefore, (5) provides a predictor of separate enterprise costs for each firm based on its actual total variable cost. This approach provides both the average activity costs useful to policy makers and agribusiness and estimates for individual producers which they may use for farm planning.

Empirical Application. In this empirical application, the cost equations were estimated for total operating expense, five specific production activities and three different crops: corn, soybeans, and wheat. Only the total variable cost and the estimate of fertilizer and pesticide costs are presented here.

The sample data span a ten-year period (1976-85) of annual observations from 161 Farm Business Farm Management Association records in Illinois. To maintain a relatively homogenous data sample, farms were included in the sample only if they had 95% or more of their land base in corn, soybeans, wheat, or set-aside and less than 5% of the total cash income originated from livestock-related enterprises in any given year.

A priori we hypothesize that the total operating expense per acre is a function of farm size, enterprise yields, soil quality, location, enterprise mix, and the level of management. The estimated models presented here are for total operating expense and fertilizer and pesticide expense.

$$(6) \qquad TOE = [B_{11} + B_{21}(1/TA) + B_{31}(CYLD)]Q_1 + [B_{12} + B_{22}(SA/TA)]Q_2 + [B_{13} + B_{23}(1/TA)]Q_3$$

$$(7) \qquad FP = [B_{11} + B_{21}(CYLD)]Q_1 + B_{12}Q_2 + B_{13}Q_3$$

where

TOE is total cash operating expense,
FP is fertilizer and pesticide expense,
TA is tillable acres,

CYLD is corn yield, and

Q_1, Q_2, & Q_3 are acres in corn, soybeans and wheat, respectively.

Models for additional expense categories, not presented here, were for seed, storage and drying (SD), power and equipment (PE), hired labor (HL) and other expenses (OT). The sum of the five individual categories is equal to the total cash operating expense (TOE).

Nine sets of models were estimated with each set using two-year moving-average data. The average estimates for five of the nine sets of TOE and FP models are reported in Table 6.1. Percent root mean square errors (Pindyck and Rubinfeld) are between 1.2% and 2.1% for TOE and FP. The RCR method provided reliable estimates for the costs reported here and for the other expense categorizes for corn and soybeans. However, the other expense category estimates for wheat were less reliable. On average, only 34 of the 161 farms produced wheat. There appears to be a problem with the reliability of the estimates if only a few farms produce an output. However, the same problem may exist with OLS methods. It seems, from our experience, that the reliability of the estimates is improved with increased homogeneity of the sample farms.

Moreover, we note that individual estimates are less reliable if a large number of the sample farms report no expense for a particular category. In this analysis the percent root mean square errors for hired labor (HL)

TABLE 6.1 Per Acre Cost Estimates from the RCR Model for Total Operating Expenses and Fertilizer and Pesticide Expenses [a]

Year	Corn	Soybeans	Wheat	
		Total operating expenses (1977 dollars)		
1976-1977	146.5	45.5	54.5	(1.3)
1978-1979	130.7	63.8	53.0	(1.3)
1980-1981	133.5	50.5	90.6	(1.2)
1982-1983	120.9	50.4	56.5	(1.2)
1984-1985	107.2	66.9	32.0	(1.4)
		Fertilizer and pesticide expenses (1977 dollars)		
1976-1977	50.3	1.7	30.7	(2.0)
1978-1979	43.7	8.0	39.5	(2.1)
1980-1981	58.1	16.1	40.6	(1.5)
1982-1983	53.9	8.1	35.5	(2.0)
1984-1985	48.8	12.0	5.2	(2.0)

[a] Numbers in parentheses are the percent root mean square error.

Source: Adapted from Hornbaker, Dixon and Sonka.

TABLE 6.2 Mean Best Linear Unbiased Predictors (BLUPs) by Acres [a]

	Acres			
	0 - 300	300 - 600	600 - 900	900 +
Average number of farms per year	18	88	36	19
Total operating expenses (1977 dollars)				
Corn	144.9	129.3	124.3	122.3
Soybeans	55.8	52.4	54.6	56.9
Wheat	-51.9	55.3	99.5	116.0
	(1.6)	(0.6)	(0.8)	(1.2)
Fertilizer and pesticides expenses (1977 dollars)				
Corn	50.6	51.7	51.3	53.1
Soybeans	9.6	8.2	9.8	8.7
Wheat	27.9	29.9	30.9	29.3
	(2.3)	(0.9)	(1.4)	(1.6)

[a] Numbers in parentheses are the percent root mean square error.
Source: Adapted from Hornbaker, Dixon and Sonka.

ranged from 2.7% to 14.5%. Many of the farms in the sample reported no hired labor expense.

Estimates of the mean BLUPs are reported in Tables 6.2 and 6.3 by size categories. These values are the individual farm predictions averaged by size categories. As seen here, estimates for the medium size farms are more accurate than for the largest and smallest. Also, there again is a problem with negative and unreliable estimates for the commodity (wheat) grown on a small percentage of the farms.

Conclusions from Hornbaker, Dixon and Sonka. The random coefficient regression model provides a reasonable approach for estimating costs of production over a population of multiproduct firms because it allows average variable costs to vary from farm-to-farm. The RCR model also utilizes data as they are empirically collected by most farmers. Thus, an important advantage of the RCR method over other approaches is that enterprise specific estimates can be provided for firms which do not keep input data for each output. Moreover, the approach can be used to provide estimates for farms of similar size and type which keep no production cost estimates.

TABLE 6.3 Mean Best Linear Unbiased Predictors (BLUPs) by Sales [a]

	Gross farms returns ($1,000)			
	0 - 75	75 - 150	150 - 225	225 +
Average number of farms per year	21	80	40	20
	Total operating expenses (1977 dollars)			
Corn	138.8	129.4	125.1	125.4
Soybeans	54.7	52.8	53.0	59.0
Wheat	-51.9	55.3	99.5	116.0
	(1.6)	(0.6)	(0.7)	(1.2)
	Fertilizer and pesticides expenses (1977 dollars)			
Corn	49.7	51.8	51.3	53.5
Soybeans	9.0	8.4	9.6	8.6
Wheat	27.5	30.0	30.5	30.3
	(2.1)	(0.9)	(1.3)	(1.6)

[a] Numbers in parentheses are the percent root mean square error.
Source: Adapted from Hornbaker, Dixon and Sonka.

A shortcoming of this approach, and of many others, occurs when few farms produce one or more of the commodities for which estimates are desired or when a large number of zero observations exist for individual expenses. These conditions can result in negative estimates for individual farms. The problems of negative estimates and small sample size were later addressed in the paper by Dixon and Hornbaker.

Technology Coefficients (Dixon and Hornbaker)

In this study we first argue that random coefficients regression, is for most applications, the correct way to estimate the coefficients of deterministic, mathematical programming models from sample data. Second, technology coefficients in most mathematical models are usually non-negative. If the coefficients are viewed as varying across firms, predictions of each coefficient on a firm-by-firm basis should be non-negative; not just the estimated means of the coefficients. We have noted, in the prior study, that this was a problem when the actual observations for individual firms are zero or very close to zero.

The issues addressed above were tested with 12 Monte Carlo experiments. The observations for the experiments were generated as stochastic identities as shown in equation (2). Moreover, the results were generated for 2 samples sizes, 25 and 50, therefore, a total of 24 experiments were conducted. Two of the hypotheses tested in the Monte Carlo experiments were : (1) inability to estimate the covariance matrix precisely does not seriously affect the estimates and (2) constrained RCR estimators are more efficient estimators than OLS constrained to have non-negative estimates.

Conclusions from Dixon and Hornbaker. The coefficients of linear constraints that typically appear in mathematical programming problems are more realistically conceptualized as stochastic identities. As such, estimation by least squares (or other constant coefficient methods) given a sample of observations of only resource levels and activity vectors is theoretically inappropriate. For small samples the results of the Monte Carlo experiments indicate RCR, with coefficient variances and BLUPs restricted to be non-negative, to be superior to restricted ordinary least squares. Moreover, the inability to estimate the coefficient covariance matrix precisely is not sufficient reason to reject RCR estimation methods.

A further benefit of conceptualizing technology constraints as random coefficient models is that the non-negativity constraints implied by properties of the programming model can be used to generate more *a priori* constraints on the estimates than can a constant coefficient framework. Since each element of the constraint matrix must be non-negative, the predicted realization of the coefficient for each observation can be restricted to be non-negative instead of simply requiring the mean coefficient to be non-negative.

Summary

This chapter has two purposes. The first is to compare alternative estimation systems relative to historical work in agricultural economics and the changing agricultural industry. Second, an approach that addresses a major problem in the calculation of commodity production costs is presented and critiqued. That approach, use of random coefficients regression (RCR), allows the analyst to use actual farm data in the estimation of commodity production costs, even when the farm accounting system does not directly collect data by commodity.

It is noteworthy that policy driven interest in estimation of production costs, as a goal in itself, seems to have increased over the last decade. Traditionally, agricultural economists have categorized methodologies for study of farm size and structure (which used production costs as an important information item) as being either positive or normative in nature. Estimation of production costs does not tend to neatly fit those categories. Often a mixture of positive and normative elements is employed. Therefore a three-fold categorization was employed for this chapter. Those categories are direct reliance upon expert judgement, analysis based upon special survey results, and manipulation of accounting based farm firm data. The marked decline in the experience base needed to derive expert based estimates and the strengths and weaknesses of the three approaches are noted in the first section of the text.

The random coefficient regression (RCR) approach, reviewed here, provides a framework for estimating production costs and technical coefficients for mathematical models. This approach has been shown to be superior to ordinary least squares when costs or technical relationships vary from farm-to-farm. Moreover, RCR provides mean estimates and estimates for individual farms in the sample. Random coefficient regression is, therefore, a positive approach which can provide, from farm accounting records, the estimates needed by today's farmers and policy makers.

References

Center for Agricultural and Rural Development. 1984. *Economies of Size Studies.* Ames, Iowa: Iowa State University.

Dixon, B.L. and R.H. Hornbaker. 1989. "Estimating the Technology Coefficients in Linear Programming Models." Abstract published in *American Journal of Agricultural Economics* 71: 1343.

Garcia, P., S.E. Offutt, and S.T. Sonka. 1987. "Size Distribution and Growth in a Sample of Illinois Cash Grain Farms." *American Journal of Agricultural Economics* 69: 471-6.

Hornbaker, R.H., B.L. Dixon and S.T. Sonka. 1989. "Estimating Production Activity Costs for Multi-Output Firms with a Random Coefficient Regression Model." *American Journal of Agricultural Economics* 71: 167-177.

Pindyck, R.S., and D.L. Rubinfeld. 1976. *Econometric Models and Economic Forecasts.* New York: McGraw-Hill Book Co., p. 316.

Porter, M.E. 1980. *Competitive Strategy.* New York: The Free Press.

Sonka, S.T., R.H. Hornbaker, and M.A. Hudson. 1989. "Managerial Performance and Income Variability for a Sample of Illinois Cash Grain Producers." *North Central Journal of Agricultural Economics* 11: 39-47.

Swamy, P.A.V.B., and P. Tinsley. 1980. "Linear Prediction and Estimation Methods for Regression Models with Stationary Stochastic Coefficients." *Journal of Econometrics* 12: 103-42.

7

Approaches to Collecting Data on Production Technologies

Carol C. House

Introduction

A major purpose of this book is to explore the concept of "quality" in the economic accounting for commodity costs and returns. A basic first step in that exploration is to examine the sources of data that are being used in cost of production estimates, and see whether they provide a suitable foundation for quality analysis.

We begin by asking the question "what does the economic analyst need from a cost and returns data set?" In my experience as a consulting statistician, I have found that the analyst generally has a good idea of the quantity and detail needed in data, but less understanding of what is needed to assure quality. Therefore let me answer my own question from a quality perspective. What does the analyst need from a cost and returns data set? He or she needs data which can produce accurate and defensible estimates of commodity costs and returns for specifically defined enterprises, while keeping within cost constraints. Quality data are therefore accurate, defensible, affordable; and they target the desired population. These are the key characteristics we will look for as we explore data collection alternatives.

It is essential that quality be built into the data collection process from the beginning. Although there are many factors in that process that are important for assuring quality data, this chapter will focus on two of the most critical: statistical inference and non-sampling errors. Statistical inference determines whether, and to what extent, the results from the study can be generalized to a broader set of farming operations. An examination of non-sampling errors will show where things are likely to go wrong in a data collection process and how those errors can affect the resulting quality. To illustrate these concepts with concrete examples we will explore four common approaches to collecting cost and returns data and compare their relative potential to produce quality data. These four approaches are: (1) probability surveys of producers, (2) non-probability surveys of producers, (3) producer panels, and (4) records keeping services. The chapter will conclude by examining the tradeoffs that are often needed between affordability and the other aspects of quality.

Statistical Inference

The first two steps in data collection are to identify the group, or population, which is to be studied, and then select a sample that can provide information about that group. Both steps affect the statistical inference that can be made from the resulting data. We will discuss each step in order.

Targeting the Population of Interest

The *target population* for a data collection activity is the group about which the analyst wishes to draw conclusions or make inference. Common examples of target populations for cost of production studies could be all farms, all farms producing corn in a certain geographic area, or all large progressive farms growing corn in that same area. Any such group could be legitimately targeted depending on the analytical needs and budget of a particular study. The problem arises, however, when an analyst targets one population which is needed for his or her analysis but chooses a data collection method that actually focuses on some other group.

Consider the following example. A political pollster wishes to predict the results of an upcoming election, so he conducts a survey to find out how people will vote. The target population is the group of all registered voters who will vote in the election. The pollster uses a telephone survey contacting a random sample of names from the registered voter roles. By using this particular approach to data collection the pollster focuses not on the target population, but on the group of registered voters with telephones, regardless of their likelihood to vote in the election. This second group is called the *sampled population*, and is the one about which the pollster will actually be able to make inferences. The demographic characteristics of the two populations (the one about which the pollster wants to make inference and the one about which the pollster will be able to make inference) are not the same. Studies have shown that the likelihood of voting in an election or of having a telephone each differ by age, sex and race. If those same demographic characteristics are important factors in choosing between the two candidates, then this survey will create a biased forecast of election results. For accuracy, defensibility, and to target the appropriate population, there must be a match between the target population and the sampled population.

To further illustrate this point, consider the problem of estimating costs and returns using data from one of a variety of records keeping services. Approximately eleven percent of farming operations across the U.S. currently use such a service. The percentage differs by type and size of farm. Dairy farms report the most use (20 percent), and beef, hog and sheep producers the least use (eight percent). Farms with over $100,000 gross value of sales are four times more likely to use a records keeping service than smaller farms. Even for those large farms, less than thirty percent use such a service (Willimack). Because such a small part of the farm sector use this type of bookkeeping, only very limited inferences can be drawn from these types of data sets. Although these data sets have a number of very good qualities, they do not allow the analyst to make statistically defensible inferences about the entire farm sector or even a sizable portion of it.

Selecting a Representative Sample

The second step in the data collection process is to select a representative sample from the targeted population in such a way that valid inferences can be maintained. Two general types of samples are possible: a statistical sample or a judgment sample. A *statistical (or probability) sample* is selected when each unit in the population has a positive and knowable chance of being included in that sample. The probabilities of inclusion in the sample are used to create sample weights, which in turn convert the sample estimates into population estimates. Statistical theory then helps the analyst describe certain characteristics of these estimates, such as variability, precision (one component of accuracy), and efficiency. This theory provides the base for defensibility.

A *judgment sample* is selected from the population through some other method, usually the subjective decision of a one or more individuals. This means that at least some units in the population had no chance of selection and/or it is not possible to determine what the selection probabilities (and thus the sample weights) are.

Of our four common approaches to data collection, only the probability surveys of producers use statistical samples. Non-probability surveys of producers and the producer panels both depend on judgement samples. Use of data from a records keeping service usually does not specifically involve sampling. A statistical sample is generally considered to be superior to a judgmental sample. Hansen, Hurwitz and Madow, in their classic sampling text, raise the question of whether a judgement sample will be representative of the population: "Another survey method which may also involve serious biases, ... is the selection of typical persons, cities, or areas to represent the whole population. ... The fact that such a purposive selection can be 'validated' by comparing its results with those of other surveys with respect to certain characteristics of the persons involved does not insure that it will be adequate for these or some other characteristics at a different date."

Yates illustrated problems with the accuracy of judgmental selection in his classic experiment back in the 1930s. He displayed 1200 stones on a table and asked 12 people to pick three samples of 20 stones each that would represent the size distribution of the entire collection. Thirty of the thirty six estimates were heavier than the true average. In addition, the judgment samples contained a larger percentage of stones whose weights were close to the average and a much smaller percent of stones near the extremes in size than did the population in general. Thus his study showed not only a consistent tendency to overestimate the average, but also to underestimate the variance of the distribution.

In his book, *A Sampler on Sampling*, Williams further discusses both the accuracy and defensibility of estimates from a judgment sample. "The accuracy of judgment samples cannot usually be determined. They are not necessarily inaccurate, but if they are accurate the accuracy is usually unknown and depends upon the expertise of a specific individual (selecting the sample)." Defensibility requires an empirical assessment of that accuracy.

Non-Sampling Errors

A second factor related to quality of data collection is the potential to make errors during the data collection process. Statisticians call these errors "non-sampling errors" to differentiate them from sampling errors. Sampling error occurs because only a sample rather than the entire population has been contacted. It is measurable through the variance and can be reduced by increasing the size of the sample. Non-sampling errors are generally unplanned errors that are hard to measure or control, and can lead to serious biases. There are several types of non-sampling errors. I will expound on two of the most important, non-response errors and response errors.

Non-Response Errors

Non-response occurs when one or more sampled farmers can not or will not provide some or all of the survey data. Non-response reduces the effective sample size and thus the quantity of information in the subsequent data set. The problem of reduced data is solved easily and routinely by selecting a slightly larger sample than actually will be needed. A far more serious problem occurs if the characteristics of the non-respondents differ considerably from those of the respondents. When this happens the representativeness of the sample (and hence the accuracy and defensibility of the estimates from that sample) come into question, especially if the non-response is sizeable.

The designers of a probability survey of producers must have a procedure in place to adequately adjust for non-response. A common non-response adjustment is to increase the sampling weights attached to those respondents which are identified as "like" the non-respondents, where "like" is defined by some objective criteria. If the population is stratified so that "like" operations are grouped together before sampling, this process is fairly straightforward. Another common procedure is to impute data for the non-respondents, again using information from "like" respondents. The effectiveness of these procedures depends on the accuracy of the assumptions made about the non-respondents, and misclassificatiom can lead to serious biases. Neither adjustment procedure is a good substitute for obtaining accurate data from the entire sample.

The effect of non-response on the other approaches to collecting cost and returns data is harder to characterize. For producer panels, the actual effect may be greater than is evident initially. On the surface there may be little non-response. However, a farmer who is asked to be on a panel probably has been identified by one or more experts as someone who, among other characteristics, is likely to participate. If he does in fact refuse, he is replaced by someone with similar characteristics who is willing to participate. Thus a producer panel is composed of producers who are willing and able to provide information and excludes those producers who are not. As with surveys, if the characteristics of the non-participants differ substantially from the participants, there can be serious biases.

Some non-probability surveys of producers ignore non-response, others attempt to adjust for it using statistical procedures like those described earlier, still others target only likely respondents. Non-response is generally not an issue in using data from a records keeping system.

Response Errors

Response error occurs when a respondent attempts to provide accurate information but fails to do so. The questionnaire designer is more often to blame for this type of error than the respondent. Analysts want to know things about farming operations that farmers do not know themselves, or that they keep track of in a different way. Example: how much fuel was consumed by a particular tractor on the soybean crop? The farmer knows the total fuel used on the operation, and that he used three different tractors on his soybean crop. He may do some calculations in his head and come up with an educated guess that is in fact fairly correct, he may do those calculations and come up with a value that is not correct (a response error), or he may refuse to guess (non-response). In any given survey a poorly conceived question will produce all three with few clues to distinguish the response errors from the accurate information. Another common cause of response error is the ambiguously worded question which each respondent interprets differently. The resulting data set becomes a mix of inconsistent responses.

There are two ways to minimize this type of response error. The first and best way is to ask meaningful questions that farmers can answer correctly. The records keeping approach to data collection ranks high in this category. These systems set up orderly processes of information retrieval, using concepts familiar to farmers and collecting information those farmers need to monitor their operations and pay taxes. These systems include training, instructions manuals, and repetition--all of which leads to quality responses. Their disadvantage is that they may provide only a portion of the detailed information needed by the analyst.   A second way to minimize this type of response error is to consistently interpret questions for the farmer when he appears unable to respond. The interviewer or panel moderator plays the role of the interpreter. A well-trained interviewer or moderator knows when responses are inadequate and can effectively probe for more information when needed.   A poorly trained interviewer can cause irreparable harm to the consistent and accurate interpretation of questions.   Producer panels present a unique opportunity for proper interpretation of questions. Panel members interact with one another as well as with the moderator to make sure that they understand the concept in question. This interaction makes the producer panel one of the best way to get responses to fairly detailed questions involving complex economic concepts.   They are also excellent ways to pretest questionnaires to be used later in probability surveys where the inference can be broader. The unknown element in a panel is the group interaction affect.   The dynamics of small group interaction can be drastically different depending on the personalities involved.   Group interactions can affect responses and hence data quality.

A second major type of response error results from the respondent's inability to accurately recall information. The best solution to this problem is to retrieve data directly from the farmer's own records. Clearly, this is the concept behind using records keeping services for data collection. Farmers responding to surveys or participating in panels should be asked to refer to their own records when answering questions on cost and returns. Johnson, Banker, and Morehart report that approximately seventy percent of Farm Cost and Return Survey respondents used records to assist with their responses. However, records keeping among farmers is inconsistent. Twenty three percent reported that they did not keep formal records of any kind. Diary type surveys are sometimes used to combat recall errors. Under that approach, the farmer is contacted early and asked to keep the specific types of records needed. This approach can be very effective if the farmer has incentive to keep the records up to date.

Summary

What is the best approach to collect data on production technologies? From the point of view of statistical inference the answer is unambiguous. The analyst needs to be able to make statistical inferences from the data to the target population. To assure that inference will be valid, one needs probability sampling from a sampled population that closely matches that target population. Of the four approaches to data collection we are exploring, only the probability survey of producers provides this for a broad array of target populations in the farm sector. When we consider non-sampling errors we get more mixed results. The potential for creating biases by not including farmers reluctant to participate in surveys, panels or records keeping services is a universal problem. There are, however, statistical techniques that can moderate non-response problems, and their use is critical for quality data collection. Response errors can cause many problems. Records keeping services have the potential to deliver some of the most accurate data for the items they collect. Surveys conducted with a trained interviewer and producer panels with a moderator both have the ability to provide good quality responses when farmers use their own records. Both have potential response error problems.

Now we need to consider cost constraints. Will the data collection method that provides the best "overall" quality for cost and returns information also provide the best "quality per dollar?" Cost is the great compromiser, the thing that always pulls us back from achieving the goal of the best quality data. The Farm Costs and Returns Survey costs approximately $5.5 million dollars. That translates to over $300 per completed interview. The other methods we have discussed are significantly less expensive. Even if we were to conclude that a large scale probability survey provides the overall best quality data, is it possible to get more quality *per dollar* using some other method? Hansen, Hurwitz and Madow specifically address this dilemma and conclude that the answer depends on what the data will be used for and how important it is to have reliable results:

> *"How, then, is one to know which type of method to use? A suggested answer to this question is the following: If it is important that reliable results be obtained, and if a fairly heavy loss may be involved if the wrong action or decision is taken ..., then a method for which the risk of error can be controlled should be used if possible. On the other hand, if conditions are such that only fairly*

rough estimates are required from the sample, and important decisions do not hinge on the result, ... it may be that the biases of a low cost nonrandom method will be considerably less important ...(and this method) may be expected to produce results of sufficient reliability more economically ..."

Therefore, identifying the best approach to collecting data on production technologies is dependent on the answer to my first question: what does the analyst need from a cost and returns data set? If reliable results are very important, a probability based approach with sufficient effort extended to control non-sampling errors is essential, even if the cost is high. In other situations, sufficient reliability may be achieved through more economical approaches.

References

Hansen, Morris H., William N. Hurwitz, and William G. Madow. 1953. *Sample Survey Methods and Theory.* New York: John Wiley & Sons.

Johnson, James, David Banker, and Mitchell Morehart. 1990. "Financial Recordkeeping by U. S. Farmers and Ranchers." *Agricultural Income and Finance Situation and Outlook.* U.S. Dept. of Agriculture, Economic Research Service. AFO-37, May.

Williams, Bill. 1978. *A Sampler on Sampling.* New York: John Wiley & Sons.

Willimack, Diane K. 1989. "The Financial Record-Keeping Practices of U. S. Farm Operators and Their Relationship to Selected Operator Characteristics." Paper presented at the Annual Meetings of the American Agricultural Economics Association, Baton Rogue, LA.

Yates, F. 1937. "Applications of the Sampling Technique to Crop Estimation and Forecasting." Transactions of the Manchester Statistical Society Session 1936-1937.

Alternatives for Estimating Costs and Theoretical Considerations: Discussion

Odell L. Walker

Vernon Eidman reviewed the current status of cost and returns (CAR) estimates, particularly those prepared by the Economic Research Service (ERS) of the U.S. Department of Agriculture (USDA). Hornbaker, Sonka, and Dixon provided a brief historical overview of cost estimation efforts and described and evaluated a random coefficients regression (RCR) model for allocating whole farm cost data to individual commodities. House emphasized quality issues to guide data collection for estimating commodity cost and returns. The authors are to be complemented for focusing on important issues and making key contributions to our understanding.

Eidman introduced practical issues and suggestions for improving CAR estimates. He chose not to remind us of theoretical problems in seeking to estimate the cost and return for a commodity. However, he recognized the problem implicitly by emphasizing the desirability of having information about the distribution of production costs for a commodity.

The 1978 Conference on Estimating Agricultural Costs of Production emphasized theoretical problems early in the session (Stovall and Hoover, Miller). Perhaps we have become more pragmatic about the job since then. We know that it is not possible to specify a really meaningful, single value for commodity cost or return for the U.S., a state area or even a farm. Part of our job is to educate users of the empirical estimates to appreciate limitations and practical uses of the CAR estimates.

Clearly, individual firms have different cost functions and time and fixity make a difference. Simple estimates of commodity costs and returns probably overlook interrelated supply and demand for numerous products and inputs. For example, the cost of producing wheat in the Southern Plains in any year depends, among other things, on the price of stocker cattle (e.g., higher stocker prices lead to more fertilizer use and pull some land from cash wheat and other uses), input prices (e. g., nitrogen and operating capital) and government programs (e.g., 0-92 and set-aside and flex requirements). Let's be sure we have a little guilt (i.e., lingering doubt) about ignoring those problems as we proceed with practical matters leading to CAR estimates.

I concur with Eidman's endorsement of the cumulative distributions of commodity production costs (CPDC) produced by Ahearn *et al.* and like his suggestions for improvements. However, the problem I see with CPDC is its failure to identify the components of variability in the distributions. Casual users would assume that differences in costs of production among farmers are systematic. That is, they might be thought due to resource endowments, business size, managerial skills and location. However, the random component (e.g., from weather, pests, and yield in a given year) may be the most important source of differences in production costs. Long ago, Miller suggested that the cost distributions should be based on normal yields rather than actual yields and I suspect that he wanted to remove that random component of cost variability.

We also must question how the CPDC are to be interpreted. A high cost of production for a commodity may reflect reality of economic forces working in a given production area or on a particular farm. We can not jump to the conclusion that the high cost producers are inefficient, will move away from producing the commodity or are unworthy of governmental payments. A better guess is that the commodity is produced, even at high cost, for some rational reasons which we probably need to discover. Or, back to my original point, maybe the high cost is a random event.

Eidman makes useful observations and suggestions for data and computational methods used for USDA cost of production estimates. Part of this emphasis is on validation of critical cost components, such as fuel and repair costs estimated from engineering equations. It is true that years and then decades go by in which we use about the same data and methods without validation and updating. Machinery, which comprises a third or more of most crop expenses, is a good example. Also, Eidman is concerned that some overhead cost items may be overlooked.

Methods of charging for certain classes of inputs will be the subject of later chapters. However, I will comment on one to make a point about the need for cooperation among all interested parties. The ERS, USDA methodology charges for machinery, equipment and breeding livestock capital replacement using current acquisition prices (McElroy). In contrast, the Farm Financial Standards Task Force, American Bankers Association makes a strong recommendation for using original cost for capital recovery. They argue that the latter is consistent with usual accounting methods across industries and avoids effects of wide swings in current market prices for capital items.

Without deciding now about the relative merits of the alternative capital costing methods, we can at least observe that consistent industry practices are desirable. Users build our commodity cost estimates and other data into firm income, balance sheet and cash flow financial statements. To avoid confusion of clientele groups and to assure an appropriate joint effort with accountant, farm finance and other groups, a very broad consensus is needed on methods. The Financial Standards Task Force also proposes an agricultural data base. Clearly, members of this Conference need to talk to other groups about cost and returns issues.

One alternative commodity cost estimation method mentioned in Eidman's chapter is the topic of Hornbaker, Sonka, and Dixon. The method allocates known total farm operating costs for a sample of farms to component (activity) and total costs for individual enterprises by means of a stochastic identity. Total production of the commodity is known for each farm but enterprise component costs are not known. Allocations are obtained for individual farms and the whole sample (e.g., a region).

Using my earlier terminology, systematic variation appears to be "explained" in the RCR methods by means of exogenous variable such as farm size, yields, soil quality, location, enterprise mix and management level. Can we presume that random variation not explained by the exogenous variables is due in part to production conditions which vary from year to year and from farm to farm? (Hornbaker, Sonka, and Dixon say that the reliability of the estimates is improved with increased homogeneity of the sample farms). We should know more about the interpretation of the small mean square errors they report. Would the method help us do a better job of identifying components of variability in commodity cost estimates?

Hornbaker, Sonka, and Dixon had ten years of data from 161 farmers. That is a great data luxury which most States and USDA do not have. The availability of data and the messy approach required to estimate the covariance matrix (Hornbaker and Dixon) appear to be potential problems. If the method proves to be an efficient alternative, we must determine how to generate such data across the U.S. Perhaps the Farm Cost and Returns Survey data can be designed for and used with the method.

House addresses the data quality issue by comparing the statistical inference and nonsampling error characteristics for four ways of obtaining data--probability sampling, non-probability sampling, panels and records. She recognized that tradeoffs between quality and cost/time considerations may be necessary. The key quality loss in failing to use a probability sample is in the ability to make defensible statements about the target population. She suggests that the non-sampling errors may actually be lower and/or easier to control with panels and records sources.

The nature of the CAR estimation job requires that we obtain data from several sources. For example, certain engineering/scientific relationships are best obtained from scientists. However, those relationships have parameters which change with individual farm/area situations. Practices, inputs, and yields are also farm/area specific. Farm and area prices and actual cost calculations probably require some normalization to purposes and procedures prescribed for the CAR analysis. How do we judge the statistical inference properties of the composite result?

If we have limited ability to make probability statements about a particular estimate anyway, could we settle for data sources which minimize non-sampling problems and provide objectiveness, conceptual soundness and repeatability? The predominate use of CAR data is as one bit of information in a private or public decision process. "What if" calculations such as breakeven analysis broaden that process. Politicians interpret and add other considerations before choosing specific policy parameters such as target and support prices. That decision level probably uses more emotional anecdotes than statistical inference.

As options are discussed, it is important to identify needs and uses for cost and return estimates, decide which ones agricultural economists can satisfy best, and design educational programs to help assure that the information is used properly and productively. I think agricultural economists are best at forward looking cost and return budgets. Forward looking budgets are also most interesting for decision making. They require our economic analysis skills such as identifying variable as opposed to fixed factors, projecting and interpreting price information and handling technical agriculture questions. To be consistent with other industries in reporting historical costs, a greater role for accountants may be desirable, although our agricultural expertise is certainly needed.

References

Ahearn, Mary, *et al.* 1990. "How Costs of Production Vary." Economic Research Service. U.S. Department of Agriculture Information Bulletin No. 599. Washington, D.C.: U.S. Government Printing Office.

Financial Standards Task Force. 1990. "Recommendations of the Farm Financial Standards Task Force--Exposure Draft, March 1990." American Bankers Association, Washington, D.C.

Hornbaker, Robert H. and Bruce L. Dixon. 1989. "Estimating Production Activity Costs for Multioutput Firms with a Random Coefficient Regression Model." *American Journal Agricultural Economics* 71: 167-177.

McElroy, Robert G. 1987. "Major Statistical Series of the U.S. Department of Agricultural, Volume 12: Costs of Production." Economic Research Service, U.S. Department of Agriculture. Washington, D.C.: U.S. Government Printing Office.

Miller, Thomas A. 1979. "Cost of Production Variability--Issues and Problems." *Estimating Agricultural Costs of Production Workshop Proceedings.* Economic Research Service, U.S. Department of Agriculture. Washington, D.C.: U.S. Government Printing Office.

Stovall, John G. and Dale Hoover. 1979. "An Overview of the Cost of Production-- Uses, Concepts, Problems and Issues." *Estimating Agricultural Cost of Production Workshop Proceedings.* ERS, USDA. Wash., D.C.: U.S. Govt. Printing Office.

Alternatives for Estimating Costs and Theoretical Considerations: Discussion

Eldon Ball

The subject of this book is economic accounting. More specifically, we are concerned with measuring the economic cost of producing an individual commodity. Yet statistics of income by type of farm document the prominence of the multiproduct farm. In 1982, for example, farms primarily engaged in the production of cash grains derived almost 20 percent of total cash receipts from the marketings of other crops and livestock. Marketings of crops accounted for a similar percentage of cash receipts of farms primarily engaged in livestock production.

There are several reasons to believe that the output decisions made for each product on a multiproduct farm are interdependent. A producer's output decisions will be interdependent whenever the output quantities are technically related, i.e., the case of joint production. Sheep raising is a classic example. Two outputs, wool and mutton, are produced in varying proportions by a single production process. Thus, neither output decision is independent of the other. Of course, if two products are produced in fixed proportions, the analysis of a single output can be applied. However, it is unlikely that the fixed proportion case will be met in practice.

If production of two or more outputs is joint then not all of the inputs can be allocated across the outputs in a meaningful way. Consequently, the production of an individual commodity and its resource use cannot be modeled independently of other production activities on the farm. Consider, for example, a producer who sows a grain crop in autumn, grazes it during the winter, and harvests grain from it in early summer. As a practical matter, it would be very difficult to determine what fraction of the fertilizer, pesticides, and labor inputs should be allocated to each of the two production activities.

Even in the absence of joint production, there are other reasons to believe that the producers' output decisions are interdependent. Without joint production, the output of each commodity depends only on the input quantities allocated to that production activity. If, however, the aggregate amount of any input is fixed, then the output decisions will still be interdependent since each production activity will compete for use of that scarce input. Alternatively, suppose that producers face a credit constraint. Then each production activity will compete for the use of the scarce dollars needed to finance those activities and again output decisions will be interdependent.

As a final example, suppose production is non-joint, all inputs are variable, and there are no credit constraints. In addition, suppose the producer uses agricultural chemicals on two or more crops and that quantity discounts are available in buying these chemicals. Then the cost of growing one crop depends on the aggregate amount of chemicals purchased. Output decisions will again be linked, in this case, through the interaction of input costs.

The foregoing discussion has strong implications for the analysis of agricultural production and economic behavior and for the type of data required. Analysis of product-specific activities independent of other farm production cannot capture the critical substitutions between inputs or outputs. In technical terms, a separate production (cost) function should not be specified for each commodity produced since such a production (cost) function is not well defined.

Thus collection of data on all activities on the farm and the use of multiproduct behavior models significantly expands the capability to conduct policy analysis. Furthermore, the collection of aggregate data by farm should not preclude using it for the more descriptive analyses that also are a necessary component of the efforts of the Economic Research Service. The converse, however, is not true. Unless data are collected on all activities on the farm, they cannot be used to evaluate the impacts of policy alternatives because they do not address output substitution possibilities.

Alternatives for Estimating Costs and Theoretical Considerations: Discussion

Richard A. Schoney

Introduction

Cost of production (COP) studies invariably suffer from two important limitations. First, everyone wants costs of production (COP) information, but few people are willing to produce COP estimates. Secondly, almost everyone who does produce COP estimates has their own unique method of calculating them. Thus, it should not be surprising that there is a continual interest in reviewing estimation methods. In reviewing the three chapters, the following framework of three questions is used:

(1) What is the question/problem?
(2) What is the relevant model?
(3) What is the most appropriate data collection technique?

These three simple questions help explain the diversity of models and collection techniques.

What Is the Question/Problem?

Carol House reminds us that the first step in data collection is to determine the appropriate question. In addition to the traditional extension, teaching and policy purposes, there are a wide variety of other end users of COP with widely differing purposes. These questions might include the following: Are high cost producers poor managers? What portion of costs and returns are within an individual operator's control? What about COP and the environment? Are there economies of size? What are the characteristics of the COP distribution? Do they change over time? How has technology impacted on COP?

What Is the Relevant Model?

There seems to be fundamental disagreement among practitioners and theorists on a number of different points. For example, some maintain that COP is impossible to determine when more than one product is produced. Other thorny issues include ex-ante versus ex-post data, carryover effects, costs (and valuation) of risk and fixed cost valuation and allocation. There is particularly widespread disagreement on the latter. However, in practical terms, if the farm firm is separated into two separate problems--operation and investment--some of the problems of allocation and valuation of fixed resources become more tractable. The operation of the firm in the short run can be considered as an optimization of profits subject to fixed resources; this suggests that fixed costs should be allocated in proportion to their shadow price or marginal value product. The valuation of fixed resources, particularly land, can be based on modern capital investment analysis under uncertainty and the capital asset pricing model. It is unfortunate that none of the chapters address this issue, because without an explicit model of the firm with a clearly defined objective function, then COP procedures are reduced to a series of arbitrary calculations and thus, are open to question.

What is the Most Appropriate Data Collection Technique?

In general, enterprise cost data collection techniques seem to fall into one of two approaches: bottom up and top down. The bottom up techniques are generally synthetic or engineering in nature and hence, tend to be more normative in nature. The top down techniques tend be based on whole farm data, usually accounting data, and generally are more positive in nature. The following sections assess some of the intrinsic differences between these two fundamentally different approaches.

Bottom Up Techniques

The bottom up approach typically features budget generators and/or Delphi panels in constructing costs by starting at the very lowest activity level and building up costs based on detailed "recipes" (Kletke, McGrann *et al.*, Lent and Campbell). The reduction of production to a series of fixed and pre-specified coefficients is the heart and genius of the budget generators: farmers and production specialists relate very well to the physical delineation of crop production. Likewise, there are no "surprise" results as most costs can be recreated with a simple hand calculator.

Budget generators are sometimes combined with Delphi panels in order to develop "consensus budgets." In my opinion, these are highly unreliable methods in that many participants compromise by incorporating too many inputs and too many machines. For example, one participant may prefer machine color while another prefers another; the compromise may be to include both. Likewise, many extension budgets are based on detailed production recipes constructed by state production specialists. These budgets represent not only "target" yields but often "safety first" input levels. The result is often unreliable COP estimates that lack credibility with farmers.

The Top Management Model. In general, budget generators and bottom up approaches suffer from three general problems:

(1) They produce unreliable estimates of fuel, electricity, and repairs
(2) They have an open system of allocating fixed costs
(3) They are unable to allocate general overhead costs

With many traditional budget generators, there is no guarantee that individual costs sum to the farm total. These problems were directly addressed in the development of the Top Management Model (Schoney, 1984). Farm costs such as fuel, repairs and fixed costs are viewed as a "pie." The problem, then, is to discover a process where the pieces sum to the "pie". In the Top Management Model, a system of calibration and direct and indirect allocation procedures are developed (Schoney, 1986).

Estimates of fuel, electricity, and repairs based on engineering equations are notoriously unreliable for any given individual. In order to compensate for individual differences in the Top Management system, a calibration procedure is used to scale computer estimates to individual whole farm estimates.

Many budget generators rely on individual use estimates in order to directly allocate fixed costs. However, there is strong evidence that individual use estimates are unreliable (Klemme, Schoney and Finner); there is no guarantee that internally generated total hours or acres match user estimates. Moreover, if a machine is not used, there is no way to directly allocate a fixed charge to a given activity or enterprise. This allocation problem is resolved in the Top Management Model by a combination of direct and indirect fixed cost allocation procedures. Direct allocation procedures are based on direct hour, acre, or bushel usage generated by direct internal accounting procedures. The indirect method is used when no direct use can be traced to an individual recipe. This is the case with machines such as a snow blower used on the farmstead or a pick-up truck which is used to travel to town. In these cases, an indirect method is used based on total estimated indirect use. Total costs including both variable and fixed costs are estimated as overhead costs and allocated to each cropping recipe according to its proportion of total direct cash costs. Thus, all fixed costs are accounted for regardless of whether or not the machine or building is used and the summation of individual fixed costs must match the farm total.

In addition to fixed ownership costs and financial leases, there are a number of costs such as fuel and repairs for the pick-up truck as well as general business expenses such as the accounting, legal, insurance and license fees which are allocated in the Top Management Model using the indirect cost method.

Top Down Models

The top down models are more positive in nature with fewer implicit assumptions. Carol House is correct in that accounting data holds rich promise as a source of data. However, it is misleading and dangerous to assume that accounting data are without error. Most farmers do not maintain accrual accounts because cash-based tax accounting procedures provide them with substantial income tax advantages. Moreover, accounting methods of depreciation and valuation are arbitrary and unreliable for COP work; they are at best a compromise between the federal government and the tax filer. In Canada, we have found not only sizeable errors between undepreciated capital costs and fair market value, but, because of the various options associated with the Canadian system, systematic biases which depend upon the income status of the tax filer.

Accrual whole-farm accounting data are likely to be more reliable and cheaper to collect than bottom up data. However, the problem with top down Data is that enterprise allocation procedures tend to be unreliable. This is where the Hornbaker, Sonka, and Dixon approach of random coefficient regressions (RCR) seems to have much promise, particularly in fixed cost allocation; their model would seem to follow from profit maximization subject to fixed resources. Nevertheless, they still have the problem of fixed cost valuation. Unfortunately, they have not developed the theoretical underpinnings of their model nor have they addressed fixed costs. Furthermore, they have chosen a sample sufficiently uniform as to avoid empirical problems but in doing so they have also failed to prove the real merits of the RCR model over conventional techniques.

Summary and Conclusions

In summary, Carol House makes the case for more scientific data collection. This is a timely caution, easily forgotten. She also reminds us that data collection must be problem- and model-driven. However, because of varying costs and degrees of complexity, COP estimation is also a management problem, with a cost-effective solution likely including a combination of data collection approaches. Potential cost-effective data sources include detailed farm practices surveys, whole-farm accounting data and linking of other data bases including census and tax filer information. Ultimately, however, COP data must be judged not only by data quality but the relevancy and appropriateness of the COP model and the questions being addressed.

References

Kletke, Darrel. 1979. "Operation of the Enterprise Budget Generator." Agricultural Experiment Station Research Report P-790, Oklahoma State University, Stillwater.

Klemme, R.M., R.A. Schoney, and M.F. Finner. 1981. "Estimates of On-Farm Machinery Usage and Costs." American Society of Agricultural Engineers Paper No. 81-1516, Chicago, Illinois.

Lent, G. and B. Campbell. 1984. "A Computer Program for Generating Crop Enterprise Budgets." Economics Information Report no. 84-04. Economics Branch, Ontario Ministry of Agriculture and Food, Toronto, Ontario.

McGrann, J.B., K.D. Olson, T.A. Powell, T.R. Nelson. 1985. "Microcomputer Budget Management System User Manual." Department of Agricultural Economics, Texas A&M University, College Station.

Schoney, R.A. 1986. "New Developments in Farm Planning Models: The Top Management Model." Paper presented at the Meetings of the Canadian Agricultural Economics and Farm Management Society, July 6-10, Saskatoon, Saskatchewan.

Schoney, R.A. 1984. "Using Extension Workshops as a Means of Collecting Farm Level Data." Paper presented at the Meetings of the American Agricultural Economics Association, August, Ithaca, New York.

PART THREE

Comparability of Predominant Methods

8

The National Commodity Cost
and Return Estimates

Mitchell Morehart, James D. Johnson, and Hosein Shapouri

Introduction

The U.S. Department of Agriculture (USDA) has been engaged in the estimation of costs of production of farm commodities since the early 1900s. The Bureau of Agricultural Economics, USDA organized in 1922, was established, in part, to study farmers' costs of production (Mighell). Various cost of production studies have been conducted in individual states throughout much of the twentieth century, as well. Early studies involved complete cost accounting with detailed labor records and regular visits to account-keeping farmers. Relatively long-term projects were maintained at Minnesota, Illinois, and Cornell with USDA support and at other locations as well. Most studies conducted after 1950 were largely short-term projects based on survey data collected from farmers. These studies were designed to examine cost structures for particular commodities with the primary objective of assisting farmers in planning and improving their management performance. Although not intended for policy analysis, these data were often cited in policy debates or hearings on agricultural legislation.

In the early 1970s ERS, in cooperation with several Land Grant Universities, established a systematic approach to the development and maintenance of farm enterprise budgets. The Firm Enterprise Data System (FEDS) consisted of farm and ranch enterprise budgets, whole farm budgets, and the processing and distribution of firm budgets (Krenz). The enterprise budget system utilized a computerized budget generator developed at Oklahoma State University to process input data into completed crop and livestock budgets and to facilitate storage and updating. This standardized set of budgeting procedures enabled cost estimates to be developed which were comparable across commodities and regions.

Recently, ERS has developed models which estimate commodity cost and returns at the farm-level (Glaze). Under the Farm-Level Budget Model (FLBM), enterprise costs and returns are calculated for each farm surveyed and summarized at the state, regional, or national level based on the appropriate weight for each observation. Under FEDS, cost and return estimates are calculated as if all production for a commodity is represented by a single average acre in the State. The FLBM is a much more flexible procedure which allows examination of distributional issues associated with various cost or return structures, for example, see Ahearn, *et al.* and McBride.

USDA's estimates of commodity cost and returns are currently published in the *Economic Indicators of the Farm Sector* series (USDA, ERS).

Characteristics of Cost and Return Accounts

The primary purpose of cost and returns accounts constructed by USDA is to provide a national weighted average of actual costs and returns on an historical basis. Given this purpose and legislative guidelines, there are five characteristics that underlie estimates of costs and returns and affect accounting procedures adopted by USDA. These characteristics relate to the combined operation-landlord, the per-acre or per-output accounting, the treatment of multi-output and joint production, the exclusion of marketing costs and returns, and the treatment of participation in Government programs.

Combined Operation-Landlord Costs and Returns

All resources used in the production of a commodity are considered, regardless of ownership. Therefore, an ownership cost for all land, whether owned, rented, or financed, is imputed in a consistent manner. Interest expense is treated as a cash expense because interest is generally paid to those other than the combined operation-landlord entity. Variable expense items, such as fertilizer, represent the combined costs of both the farm operation and landlord. The integrated measurement of costs also means that estimates of cash expenses do not include cash and share rent paid by the farm operation to the landlord. What is a rental expense to the farm business is offset as income to the landlord.

Per-Acre and Per-Unit-of-Output Accounting

Costs and returns are estimated on a per-planted-acre basis for all commodities except sugar cane and livestock. A planted acre is a unit that is comparable, and in some cases substitutable, among crops. Costs and return accounts for livestock are estimated on a per-head basis and on a per cwt. of milk or meat basis.

Costs for field crops are included only for acreage that was planted with the intention of being harvested for grain. Similar limitations apply to other commodities. Thus, corn and sorghum silage and small grains planted as cover crops or solely for grazing are not considered. Per-acre estimates are preferred for making comparisons across crops because yields differ across commodities and not all commodities are measured in the same units of output, for example, bushels or pounds. Costs are included for production that is abandoned because of crop damage. Land costs for double-cropped acreage are divided between the two crops. Two years of land costs are counted when crops are planted on fallowed land.

For livestock, the unit of accounting is the breeding unit for beef cattle (cow), dairy (cow), and sheep (ewe). For livestock enterprises also producing the final product, costs are accounted in per cwt. such as hogs, milk, and fed cattle. When using costs per breeding unit all associated costs are included such as bull or ram costs, costs during gestation, pre-gestation, and lactation phases. Only when an intensive feeding regime is followed using concentrated feeds for fattening are the costs transferred to a feeding enterprise. Cost for harvested feeds are charged at market rate, or cost of production for grazed forages if not purchased at market rates. Items used to produce feed are not included in similar items in the livestock account. For example, the proportion of a tractor used to produce feed is included in the feed.

Multi-Output and Joint Production

Most agricultural operations consist of more than one enterprise. This multi-output production means that costs associated with inputs, such as machinery, equipment, and buildings are shared by the various enterprises on the farm. Some commodities are always produced jointly, that is, one commodity cannot be produced without the other, for example, oats and oat straw. Joint production means that the costs and returns of the primary and secondary commodity cannot be separated.

Allocation of shared resources and their associated costs among the various products complicates the process of estimating separate average costs in any multiple product firm. There does not exist a theoretically justified procedure for allocation of costs.

For joint production, where inputs cannot be separated among products, value of the secondary commodity is included with the returns of the primary commodity. Assuming that secondary product revenue equals secondary product cost, a consistent accounting of total costs and returns requires that the value of production of secondary crops be included in total returns.

Exclusion of Marketing Costs and Returns

Costs of production for crop commodities are separated from the costs of marketing by measuring production costs to the point of first sale or storage, if the crop is not sold immediately after harvest. Costs of drying and hauling the crop to the elevator or processor are included. All crops, except sugar, are valued at their time of harvest.

Livestock and dairy costs of production include marketing charges from farm gate to the point of first sale. Marketing charges for dairy includes expenses for hauling milk, hauling dairy stock to and from market or between farms (ranches), commissions, coop assessments, advertising, National Dairy Promotion, and import and export charges. Marketing charges for livestock includes hauling charges for moving stock to and from market or between farms and ranches, marketing charges paid for purchase and sales of livestock, such as commissions, yardage, inspection, insurance, feed, and auction fees.

Government Programs

Government agricultural programs affect production costs and returns by supporting crop prices and farm incomes and influencing crop acreage in production. The Government intervenes directly in agricultural markets by limiting the import of products, by making direct purchases, and by establishing marketing quotas. The Government supports prices through the Commodity Credit Corporation's (CCC) nonrecourse loans. The Government also supports farm income with direct payments to farm operators and landlords for some commodities.

Livestock and livestock product prices are not supported by government programs, with the exception of dairy, wool, and mohair. Federal dairy price supports, milk marketing orders, import restrictions, domestic and international food aid, and state milk regulations affect the industry. The government indirectly supports the price of manufacturing milk grade received by farmers by buying butter, cheese, and nonfat dry milk at predetermined prices. Dairy price support levels are directly linked to projected annual government purchases. The "National Wool Act of 1954" enacted by Congress states that wool is an essential and strategic commodity. The act was designed to support wool prices at a level fair to both producers and consumers. The incentive price is tied directly to shorn wool and expressed in cents per pounds. Payments for the program are financed through CCC. Payments for wool production are two kinds: on shorn wool and on the sale of unshorn lamb.

Indirect effects of programs cannot be excluded from costs and return estimates since they influence the markets for production inputs, commodities and producer behavior. For example, participants forgo current income from their acreage that is set aside, which may lead to increased output on the acreage in following years because the land has been fallow or planted to legumes. Both participants and nonparticipants are affected when the supply of a crop is restricted and prices rise.

Cost and return estimates produced by USDA have traditionally excluded the direct effects of income support programs from the cost and return accounts. Estimates which exclude the direct effects of government programs are more useful for policy makers interested in knowing the costs and returns that come from the market place. Including the direct effects of programs in cost of production estimates that were used for policy purposes (such as establishing support levels) would create a built-in, escalating spiral (Ericksen and Harrington). With the 1988 rice questionnaire, USDA began to collect information on costs incurred for participating in the government income support programs with the intent of producing estimates including the direct effects of the program (Salassi, *et al.*).

Data Sources

Prior to 1984, the USDA collected cost of production data on special surveys dedicated to that purpose. The major shortcoming of these surveys was that they were conducted with a non-probability sample design. Since 1984, the primary source of data used by USDA to develop cost of production estimates has been the probability-based Farm Costs and Returns Survey (FCRS). The FCRS is an integrated survey that combines multiple versions of a questionnaire to obtain, simultaneously, data needed to estimate income, balance sheet, and enterprise cost accounts for each farm surveyed in addition to information on the structural attributes of the farm business. It is conducted each winter in the 48 contiguous States. The survey is administered through personal interview by trained enumerators who contact approximately 24,000 farms and ranches.

The cost of production versions of the questionnaires are designed to produce precise cost of production estimates for a specific commodity and representative coverage of other desired information. Each questionnaire or observation represents a number of similar farms, the particular number being the survey expansion factor, which is the inverse of the probability of the observation being selected in the sample. This allows an enterprise cost and return account to be developed from observed input data by size of farm, degree of specialization, and other attributes that may affect enterprise costs and returns.

Commodities included in survey design are dictated by the scope of the cost of production program (legislative mandate and agency goals). The expense of data collection prohibits detailed surveys of every major field crop and livestock enterprise in every year. Enterprise cost data have been collected on a four or five year rotating cycle with priority given to the legislatively mandated commodities, wheat, feed grains, cotton, and dairy (see Table 8.1). As a result, cost estimates have been updated in non-survey years only for changes in prices.

Additional information is required to supplement the basic technical information collected through the Farm Costs and Returns Survey. The majority of secondary data sources are part of USDA's overall survey program administered by the National Agricultural Statistics Service (NASS). Most of the information gathered is for individual states. These secondary data are also used to update cost and returns estimates between survey years.

For crop enterprises these additional data items include crop yields, harvest-month crop prices, micronutrient prices, electricity prices, natural gas prices, the U.S. average annual interest rate on six-month Treasury Bills, the rate of return to production assets in agriculture, and the wage rate for all agricultural workers (Federal Reserve System; USDA, ERS; USDA, NASS).

Estimation of livestock costs and returns also requires additional price and quantity data available from other surveys conducted by ERS and NASS. These additional data include livestock prices, milk prices, milk yield, milk cow and livestock numbers, land value, wage rates, price and quantity data for dairy and livestock feeds, and prices paid for other agricultural inputs (USDA, ERS; USDA, NASS). The sources for the majority of secondary data such as energy prices, short and long term interest rates, land values, and wage rates that are used in estimating costs and returns for livestock and dairy enterprises are the same as crops.

Estimation Methods

The Farm Enterprise Data System (FEDS) has served as the basis for the calculation of cost of production estimates since the early 1970s. The FEDS

TABLE 8.1 Farm Costs and Returns Survey Cycle for Cost of Production

1984:	1985:
Rice	Cow-calf
Burley tobacco	Hogs
Pasture and forage	Fed cattle
Sugar beets	Dairy
Sugar cane	Fall potatoes
1986:	**1987:**
Wheat	Corn
Sorghum	Barley
Soybeans	Cotton
Sunflower	Peanuts
Sheep	Fed Cattle
	Flue-cured tobacco
1988:	**1989:**
Flax	Dairy
Oats	Wheat
Sugar beets	Burley tobacco
Sugar cane	Oranges
Rice	Grapefruit
Hogs	Onions
1990:	**1991:**
Soybeans	Corn
Sorghum	Peanuts
Cow-calf	Flue-cured tobacco

budget generator currently used by ERS has the state or regional level as the lowest level of aggregation. As a result, state or regional average information on production practices, prices, and quantity of inputs has been assembled to derive estimates through FEDS. This approach to the development of enterprise cost estimates has limited the extent to which cost differences can be explained and described.

ERS has recently developed an alternative approach to the estimation of enterprise costs and returns. This process relies on models that directly incorporate farm level survey data. These Farm Level Budget Models (FLBM) estimate an enterprise cost and return account for each observation based on reported input levels and technology. To estimate state, regional, or national cost and return accounts, each farm is weighted using survey expansion factors. As with the FEDS, the FLBM is updated in nonsurvey years using secondary data sources to construct appropriate indices. As FLBMs are constructed for the various crops, estimates generated under the two systems have been scrutinized. Any differences were found to largely be the result of the level of aggregation. Results for two of these comparisons, for rice and wheat, are published in Glaze and Glaze and Ali. Estimates based on the FLBM are being phased in as the models are built for the individual commodities.

Both the FEDS and FLBM are organized in a similar fashion, combining survey data, secondary price and production information, and field efficiency and other engineering relationships through computer processing to derive national cost and return estimates. Most of the accounting formulas are identical between the two approaches. Aside from the level of aggregation there are some differences in the treatment of specific cost items between these approaches. Those differences most worth noting involve machinery use, and capital replacement: The FEDS uses a machinery complement based on State or regional averages of commonly used equipment (up to 100 pieces). The FLBM uses a machinery complement that is unique to each farm surveyed. Unlike the FEDS, the FLBM is not limited in the number of pieces of machinery that can be used. Thus, smaller and older machines, which tend to have lower costs, can be included. The FEDS estimates capital replacement costs based on the average age for the machinery, while the FLBM uses total useful life. The difference is the result of dividing the purchase price less salvage value by the manufacturer's estimate of hours of useful life (FLBM) or by the product of the hours of annual use multiplied by the years of assumed useful life (FEDS).

Structure of the Accounts

USDA provides information on production costs and returns in the form of a commodity account, listing gross value of production, variable cash expenses, fixed cash expenses, capital replacement, allocated returns to owned inputs (or opportunity costs), total costs, and two measures of residual returns (Tables 8.2 and 8.3).

Total cash expenses reflect the short-run variable and fixed costs incurred. Short-run variable costs give the minimum price needed to raise and harvest a crop with a given yield on an average acre of cropland. If price drops below this level then it is optimal not to produce a crop at all. Economic costs reflect the breakeven long-run average price necessary to continue producing a crop. Farmers can continue to produce at costs below this level, but could not replace their capital as it wears out or could not receive their opportunity costs return on their owned factors.

TABLE 8.2 Typical Crop Production Cash Costs and Returns Statement Per Planted Acre

Item	Year One	Year Two	Year Three
	- Dollars -		
Gross value of production (excluding direct Government payments):			
Primary crop			
Secondary crop			
Total, gross value of production			
Cash expenses:			
Seed			
Fertilizer			
Lime and gypsum			
Chemicals			
Custom operations			
Fuel, lube, and electricity			
Repairs			
Hired labor			
Purchased irrigation water			
Technical services			
Drying			
Ginning			
Total, variable cash expenses			
General farm overhead			
Taxes and insurance			
Interest on operating loans			
Interest on real estate			
Total, fixed cash expenses			
Total, cash expenses			
Gross value of production less cash expenses			

Harvest-period price (dollars per bushels)
Yield (unit of production per planted acre)

Gross value of production less cash expenses is the difference between gross value of production and total cash expenses. It is a measure of short-term returns to production and the amount of cash that would be left after all cash expenses, including interest payments, have been made. This measure indicates what would be available to cover longer term costs, such

TABLE 8.3 Typical Crop Production Economic Costs and Returns Per Planted Acre

Item	Year One	Year Two	Year Three
	- Dollars -		
Gross value of production (excluding direct Government payments):			
Primary crop			
Secondary crop			
Total, gross value of production			
Economic (full-ownership) costs:			
Variable cash expenses			
General farm overhead			
Taxes and insurance			
Capital replacement			
Operating capital			
Other nonland capital			
Land			
Unpaid labor			
Total, economic costs			
Residual returns to management and risk			
Harvest-period price (dollars per bushels)			
Yield (unit of production per planted acres)			

as capital replacement, or to retire debt. Rational growers will not produce unless at least variable cash expenses are expected to be covered in the short run.

Residual return to management and risk is the difference between gross value of production and total economic costs. Because of variability in crop prices and yields, returns to risk may be either positive or negative in any given year. However over the long-run they are expected to average near zero. Returns to management and risk, therefore, should be positive. For cost of production estimates that exclude the direct effects of Government programs the returns to risk and management may be lower than if participation was considered in the estimates.

The enterprise cost and return accounts are often referred to as commodity, or enterprise, budgets. However, these statements are not budgets or projections, but historical estimates. Cost and return estimates are based on actual production practices, levels of input use, yields, and prices as measured in surveys of farmers and suppliers.

General Approach

For a year in which a survey is conducted there are four general approaches to estimating costs: direct costing, allocation of whole-farm costs, valuing of input quantities, and indirect costing. The methodology for updating costs in nonsurvey years is dictated by the approach adopted.

Direct costing is achieved by simply summarizing responses to a question about the amount paid for a particular item. This method is best suited for estimating components of some variable costs. In nonsurvey years these costs are adjusted using indices constructed to reflect year-to-year changes in prices.

Allocating whole-farm expenses occurs for inputs that are not associated with production of a particular commodity. Expenses incurred by the whole-farm for a particular input are allocated to an enterprise based on its share of total value of production. These expenses, in nonsurvey years, are updated by the percentage change in total farm overhead and interest reported in the FCRS.

Valuing quantities of inputs requires survey data of the physical quantities of inputs used in production. Costs are estimated by multiplying quantities used by state or regional level prices. Quantities are held constant in nonsurvey years and multiplied by new state-level price information.

Indirect costing involves the combination of survey information and engineering formulas. Detailed information is collected on the survey regarding the machinery complement used in production. In addition, data are obtained on hours of machine use, times-over the acreage, age of machine, size of machine, and type of fuel used. This information is used to support equations of technical relationships which describe fuel consumption, repair requirements, and replacement costs. In nonsurvey years, these items are updated by using prices and other secondary data. Engineering formulas are modified periodically to reflect technological advances.

The detailed definitions presented below are intended to reflect the most current USDA procedures. For some commodities, primarily those which have not been surveyed recently, there may be other methods used for specific cost components.

Gross Value of Production

USDA commodity cost and returns accounts include estimates of value of production for each crop. Value of production excludes income from crop insurance indemnities or disaster assistance payments and includes the value of secondary products. Secondary products are such things as wheat straw, peanut hay, wheat pasture, cottonseed, and sugar beet tops. For crops, the value of production is calculated as the yield per planted acre valued at the average harvest month market price. Yields per planted acre are published NASS yields or determined at the farm level from survey results. Data from *Agricultural Prices* and *Crop Production* (USDA, NASS) are used to update value of production.

Costs and returns for livestock and dairy enterprises include estimates of cash receipts from sale of primary and secondary products for each enterprise. Cash receipts are calculated by multiplying volume of feeder and meat animal and/or milk produced per breeding unit or per unit of output by their relative annual average prices received by farmers. In contrast to crop production, milk and livestock products are produced all year round, and there is no specific season for livestock and dairy production. In addition, livestock products are processed for consumption after they have left the farm. Cash receipts for sheep enterprise also include government payments for wool and unshorn lambs. Primary products for livestock and dairy enterprises are fed beef, feeder beef, slaughter hogs, feeder pigs, slaughter lambs, feeder lambs, and milk. Secondary products are cull cows, cull sows, cull ewes, wool, cull dairy cows, and dairy calves. Data from *Livestock Market News* and *Agricultural Prices* are used to update the value of production (USDA, NASS).

Two additional components of returns must be calculated when the direct effects of government programs are considered when formulating crop enterprise accounts. The first component is the deficiency payment. Deficiency payment per yield unit is defined as the difference between the target price and the higher of the loan rate or market price. Farms participating in government programs must establish a program yield, on which the payment is based for each crop enrolled. Deficiency payments per planted acre of crop enrolled in government programs is calculated by multiplying the program yield per acre by the deficiency payment per yield unit. In addition, if a 0/92 or 50/92 program exists, producers can reduce planted program acreage but still receive deficiency payments on a portion of that acreage. Therefore, the total deficiency payment includes payment on planted program acreage as well as program acreage not planted but eligible for payments. The second component of returns associated with participation in Government programs is marketing loan proceeds. Payments to producers under provisions of a marketing loan are calculated as the difference between the announced loan rate and the marketing loan repayment rate.

Variable Cash Expenses

Items considered variable expenses for most crop enterprises would include seed, fertilizer, lime and gypsum, chemicals, custom operations, hired labor, fuel, lubrication and electricity, repairs, irrigation water, drying, cotton ginning, freight and dirt hauling charges for sugar beets, and technical services such as insect scouting and soil testing. Direct costs associated with participation in government programs such as maintenance of set-aside acreage are included with each relevant component of variable costs for accounts that are constructed to reflect the direct effects of participation.

Seed. Seeding rates per acre and the proportions of seed purchased or homegrown are determined from survey results. Commercial seed prices are obtained from *Agricultural Prices* for most crops (USDA, NASS). In some cases, price information is collected directly on the survey. Homegrown seed is valued at the previous year's market price for the crop plus an allowance for cleaning and treatment. Seed costs per acre are estimated as the amount of seed applied per acre times price plus the cost per acre of using homegrown seed. Seeding rates and the proportion of seed that is homegrown are held constant between surveys. Prices paid for seed are updated to estimate seed costs in nonsurvey years.

Fertilizer. Fertilizer costs are calculated by multiplying prices paid per pound of primary nutrient (nitrogen, available phosphorus, or water soluble potash) by the pounds of nutrient applied per acre. Fertilizer application rates are obtained from the survey (regardless of application method), while prices are obtained from *Agricultural Prices* (USDA, NASS). Nutrient application rates are held constant between survey years. In cases of custom application, the cost of materials is separated from other charges and added to obtain an estimate of total fertilizer costs per acre. Quantities of secondary nutrients and other micronutrients are obtained on the survey. Prices for these items are published in the Chemical Marketing Reporter (Van).

Lime and Gypsum. Rates of application of lime and gypsum are obtained on the survey along with the cost of these ingredients. In nonsurvey years, prices for these inputs are obtained from *Agricultural Prices* (USDA, NASS).

Chemicals. Crop chemical costs are estimated from per acre expense data reported in the survey. In the year of a crop survey, producers are asked to report expenses for chemicals and pesticides, insecticides, herbicides, fungicides, and harvest aids used on the crop. In nonsurvey years per acre chemical expenses are adjusted by the change in the index of prices paid for agricultural chemicals published in Agricultural Prices (USDA, NASS). This procedure does not consider variation in quantities used from year to year as a result of new chemicals, infestation of insects, disease outbreaks, or unusual weed problems.

Custom Operations. When machinery and labor are hired together to perform a production operation, the costs are treated as custom expenses. Fertilizer application, land preparation, seeding, spraying, cultivating, harvesting, and hauling are the types of custom operations for which data are collected. Custom expenses are adjusted between surveys by the index of prices paid for farm services and cash rent published in *Agricultural Prices* (USDA, NASS).

Fuel, Lube, and Electricity. These costs consist of the energy consumed by machinery, vehicles, and irrigation equipment in the production of a specific commodity. Information on machinery use, size, and total acres covered in various field operations are obtained on the FCRS. Engineering performance equations are used to determine the hours of machine use. Fuel consumption is then calculated based on an hourly rate for each machine. Prices for fuels in each state are adjusted for any refundable federal or state excise taxes. Prices for natural gas are obtained from the Department of Energy and electricity rates come from the Edison Electric Institute. Lubrication costs are assumed to be 15 percent of fuel costs. The FCRS also collects detailed information on the irrigation process such as well depth, pumping rates, and type of distribution system. Engineering formulas are used to estimate energy requirements for irrigation systems. Fuel and electricity prices are updated annually.

Repairs. Repair costs are also based on estimates of hours of machine use. A repair rate per hour of use is calculated based on the list price and age of machine used in production along with engineering relationships derived from field tests. Accumulated repair costs are divided by accumulated hours to give a rate per hour. Per acre estimates of repair costs are then calculated by multiplying hours of use in production times the repair rate. List prices for machinery are updated annually from producer price indexes for machinery published by the Department of Labor.

Hired Labor. Information on total expenses for contract labor, cash wages, bonuses and other cash benefits are obtained from the FCRS. The operator is asked to provide the percentage of each item that applies to the particular commodity. Hired labor costs are adjusted between surveys for the change in agricultural wage rates at the state level using data from *Farm Labor* (USDA, NASS). Estimated wages for the operator, other unpaid labor and any noncash benefits for hired labor are not included as out-of-pocket cash expenses, but are included in the estimate of total economic costs.

Purchased Irrigation Water. The cost of water purchased from irrigation districts or for pumping water from a private associations are listed as cash expenses. This information is collected directly on the FCRS. The expense of pumping water from on-farm sources of water are included in fuel, lubrication, and electricity; repairs; capital replacement; and labor costs. In the case where a producer pays for water with a share of the crop the expense for water is reflected in the landlords (waterlords) share of the crop received. As before, the producers cash costs for pumping this water is included as a cash expense.

Technical Services. Expenses incurred for items such as land surveying, pest scouting, and soil testing are treated as technical services. These costs are obtained on the FCRS. In nonsurvey years, the costs are updated by the change in the index of farm services and cash rent published in *Agricultural Prices* (USDA, NASS).

Drying. Information on custom rates for drying are collected on the survey for corn, sorghum, peanuts, and rice. For drying operations conducted on the farm, information on the type of fuel used and initial moisture content are collected on the FCRS. Engineering relationships are used to determine the amount of energy required to remove moisture. The costs associated with drying are reflected in fuel, labor and other components of variable costs.

Ginning. Cotton ginning costs per acre are calculated by multiplying ginning cost per bale by cotton yield per acre. Data on ginning charges come from ERS staff reports. Cotton yields are obtained from *Crop Production* (USDA, NASS).

Freight and Hauling. Freight and dirt-hauling charges are estimated for sugarbeets. These data are obtained from the FCRS. Allowances are made for deductions made by processors to equalize freight costs among growers and to minimize dirt hauled to the sugarbeet processing plant. Hauling allowances, which are paid to some sugar beet and sugarcane growers, are listed separately and subtracted from variable cash expenses.

Miscellaneous Costs. Cost for items such as baling wire; levee gates (rice and sugarbeets); deductions by processors for testing, association dues, disease control, and research activities, and any other expense not listed are considered miscellaneous costs. Data for these items are collected on the FCRS.

Variable Cash Expenses Specific to Livestock and Dairy

Livestock and dairy cash expenses include cost items for feeder beef and pigs, feed and forages (grains, concentrates, hay, other hay, silage, pasture, etc.), milk haling and marketing, veterinary and medicine, fuel, lube, electricity, machinery and building repairs, hired labor, livestock hauling, and technical services such as veterinary services and Dairy Herd Improvement Association (DHIA). Direct costs associated with dairy price supports (dairy assessment) are included in the cash expenses in the costs and returns for milk enterprise.

Feeder Beef and Pigs. Feeder beef and feeder pigs used per unit of production are determined from survey results. Prices for feeder beef and feeder pigs are obtained from *Livestock Market News* and *Agricultural Prices* (USDA, NASS). Feeder costs per unit of production are estimated as the amount of feeder used per unit of production times prices of feeder. Amount of feeder used per unit of production remain unchanged between surveys. Prices of feeder cattle and feeder pigs are updated to estimate feeder costs in nonsurvey years.

Feed and Forages. Volume of feed and forages used per unit of breeding or unit of production are calculated from survey results. For most livestock enterprises, the largest single component of variable expenses is feed. The FCRS furnishes the amounts and types of feed used and how much was grown by the producer for the livestock enterprise. The quantity includes feed fed to both breeding and replacement stock. The full cost of purchased forage and the variable costs of forage production is reported in these line items.

The other costs (such as taxes and insurance, capital replacement charges, and returns to operating capital, nonland capital, and land) for the crop machinery and cropland associated with homegrown livestock forage production are included in specific line items. The homegrown forage costs are estimated annually from separate forage production budgets. The price of purchased forages were reported in the FCRS and are updated between survey years by indexing according to changes in the prices reported in *Agricultural Prices* (USDA, NASS).

Concentrates. The concentrates expense includes all purchased and homegrown grains, nutritional supplements, and premixed feeds in dairy enterprise budgets. Contrary to the treatment of forage costs, all concentrates are valued at market price, as reported by producers for purchased concentrates in the FCRS. The price and quantity of concentrates fed is updated between surveys by indexing according to the change in the quantity fed per hundredweight of milk and the value of grains and concentrates fed to milk cows, respectively, as reported in *Milk Production* (USDA, NASS). The level of feeding also is adjusted to year-to-year changes in output per cow between surveys.

In the other livestock enterprise budgets, the market price for grains, concentrates, and supplements is used. These market prices are adjusted between survey years by indexing according to changes in prices paid for the respective commodity as reported in *Agricultural Prices* (USDA, NASS). The quantities of grain and supplements are obtained from the FCRS and are not adjusted between survey years.

Veterinary and Medicine. Information on total expenses for veterinary services and miscellaneous medical services and supplies include technician fees, branding, castrating, hormone injections, pregnancy testing, artificial insemination, breeding fees, sheep shearing, sprays, dips, dusts, ear tags, chains. These expenses are obtained from the FCRS. In the survey, the operator is asked to provide the percentage of each item that applies to the particular commodity. Veterinary and medicine expenses are adjusted between surveys for the changes in the index of prices paid by farmers for agricultural chemicals and wage rate obtained from *Agricultural Prices* (USDA, NASS). This procedures does not consider variations in quality and quantities used from year-to-year as a result of new disease outbreaks, or unusual problems.

Livestock and Milk Hauling, Marketing, and Bedding. Data on total expenses for livestock and dairy marketing, livestock and milk hauling, and bedding are obtained from the FCRS. In the survey the operator is asked to give his best estimate of the percentage of the expenses (or dollar amount) for each item used in production of a particular livestock enterprise. These expenses are updated between surveys by their relative changes in the index of the prices paid by farmers for these items obtained from *Agricultural Prices* (USDA, NASS).

Shearing and Tagging Expenses. Information for shearing and tagging for sheep enterprises are taken from the FCRS and are updated by changes in the wage rate. Ram death loss accounts for the number of rams that need to be replaced to maintain the flock. The quantity is obtained from the FCRS and is valued at the market price for sheep as reported in *Agricultural Prices* (USDA, NASS).

Dairy Herd Improvement Association (DHIA) and Artificial Insemination (AI). Information on expenses paid for DHIA and AI are collected from the FCRS. In the survey the operator is asked to provide either the dollar amount paid for DHIA and AI or expenses for DHIA and AI as a percentage of total veterinary service and medical services and supplies. Between the surveys DHIA and AI expenses are adjusted by index of prices paid by farmers for fuel and energy, auto and trucks, and wage rates obtained from *Agricultural Prices* (USDA, NASS).

Dairy Assessment. Dairy assessments were authorized by the Omnibus Budget Reconciliation Act of 1982, the Dairy and Tobacco Adjustment Act of 1983, Food security Act of 1985, and the Food, Agriculture, Conservation, and Trade Act of 1990. In general, assessments are collected on per hundredweight of milk marketed.

Fixed Cash Expenses

Fixed expenditures include personal property taxes, real estate taxes, insurance, general overhead, interest on operating loans, interest on long-term debt for which the operator or landlord would be responsible regardless of whether a commodity is produced. Unlike variable expenses, fixed costs associated with a particular commodity enterprise are more difficult to obtain from farmer surveys. Most of these items are purchased for the farm as a whole, paid for or billed to the farm in a lump sum amount, or used for the entire range of farming activities. Since these inputs pertain to the entire farm business their costs must be divided among enterprises based on an arbitrary allocation rule. USDA currently bases its allocation of fixed cash expenses (except taxes and insurance) among enterprises based on each commodity's share of the total value of farm production.

General Farm Overhead. Overhead costs are comprised of expenses for utilities (excluding water and electricity for irrigation), farm shop and office equipment and supplies, accounting and legal fees, blanket insurance policies, fence maintenance and repairs, motor vehicle registration, and any other general expenses attributable to the entire farm business. This information is collected on the FCRS annually then allocated to a particular enterprise. Per acre estimates of general farm overhead are updated between cost of production surveys by calculating an index relating average general farm overhead for all U.S. farms. The general farm overhead expense is included in the fixed cash expenses category for the livestock and dairy costs and returns budgets except for commercial feedlot budgets. However, electricity and liability insurance costs already accounted for in fuel, lube, and electricity for hog, farmer feeder, and cow-calf enterprise accounts are eliminated from general farm overhead.

Taxes and Insurance. Taxes for crop enterprises include personal property taxes on machinery and real estate taxes. Tax costs for machinery are based on the purchase price and average tax rates. Tax costs based on current machinery prices are adjusted by an index reflecting a 4-year lag in prices paid indexes for machinery. The choice of a four year lag is based on the assumption that high-cost items in the machinery complement are owned by the operator for eight years, so, on average, machinery is four years old. Taxes on farm real estate are estimated as a rate paid per acre of farmland owned from FCRS data. State-level data from *State Tax Handbook* are used to update tax rates between surveys (Commerce Clearing House). Insurance costs associated with the machinery complement are estimated using a procedure similar to that for taxes. Machinery prices are updated annually from data in *Agricultural Prices* (USDA, NASS).

For livestock and dairy enterprise accounts, tax expenses are based on the current value of new machinery, equipment, and building lagged back to the base year. Land taxes are based on the value of land and the real estate taxes for the particular production area. The lags are based on the average of equipment and buildings estimated from the 1981 and 1984 livestock surveys. Taxes include personal property tax on machinery and real estate taxes. Insurance is computed on the current values of machinery, equipment, buildings, and livestock and dairy. The taxes and insurance expense for machinery and land used to produce forage for the livestock and dairy enterprises are also included. State level data are used to update tax rates between surveys. Insurance costs associated with the machinery, equipment, and buildings are estimated using a procedure similar to that for taxes. Machinery and equipment prices are updated annually from data in *Agricultural Prices* (USDA, NASS).

Interest. Information on interest expenses for the whole-farm are obtained on the FCRS. These include finance charges and service fees for loans on machinery, the farm share of motor vehicles, purchases of inputs, land contracts, mortgages, and any other loans secured by real estate. ERS allocates interest expense to each commodity and updates during nonsurvey years in a manner similar to that of general overhead expenses.

Capital Replacement. Capital replacement is the portion of the value of machinery and equipment, in addition to repairs, that is used up in the production of a particular crop. Information is collected on the FCRS to determine machine use, including size, number of passes over a field, type of fuel used, equipment age when purchased, total hours of annual use. These data are combined with information on list and purchase price and engineering coefficients developed by the American Society of Agricultural Engineers. An hourly capital replacement cost per machine is calculated by dividing the machine's current purchase price less salvage value by the hours that the machine was used. Machinery and equipment prices are updated annually from *Agricultural Prices* (USDA).

Capital replacement charges for livestock enterprises include machinery, equipment, buildings, and purchased breeding stock. Capital replacement represents a charge sufficient to maintain production capacity through time. The value of purchased breeding animals depreciates to salvage value. The number of years which the animal's value depreciates varies by region and species. Livestock used for replacement but raised on the farm do not depreciate because the cost of raising these replacements is included in the budgets. The procedure to estimate the capital replacement for depreciable assets in the livestock and dairy enterprises are similar to crops.

Economic (Full-Ownership) Costs

Economic costs are long-term costs, designed to account for the value of all inputs used in production. An imputed cost is calculated for all nonexpenditure inputs in a consistent manner, whether owned, rented, or financed. Economic costs reflect the production situation as if the operation and landlord fully own the production inputs. Therefore, interest is ignored when deriving economic costs and returns. This full-ownership assumption of costs and returns allows comparisons across crops without regard to the actual ownership and debt positions of producers. Economic costs are defined as variable cash expenses, general farm overhead, taxes and insurance, capital replacement, allocated returns to the capital invested in the production process, unpaid labor, and land. Opportunity costs are imputed from values of capital, land, and unpaid labor in alternative uses, and are called allocated returns to owned inputs.

Operating Capital. The return to operating capital is the cost of carrying input expenses from the time they are used until harvest. The imputed cost of operating capital is calculated based on the assumption that the rational producer expects money invested in variable production inputs to earn at least as much as if the money were placed in a savings account or similar financial instrument. The 6-month U.S. Treasury bill (T-bill) rate is used to measure the cost of capital since they represent a relatively risk-free alternative investment.

Other Nonland Capital. The return to other nonland capital is the cost of capital invested in the nonland assets used in crop, livestock and dairy enterprises such as machinery, equipment and breeding stock. The calculation of this imputed cost is not independent of the method chosen for estimation of capital replacement (depreciation). Since capital replacement represents depreciation of the current purchase price of the asset over its expected life, the corresponding average amount of capital invested should be allocated a return based on a real rate of return. As a result, the imputed cost of other nonland capital is calculated by multiplying the current market value of machinery, equipment and breeding stock used in production by the rate of return to production assets from current income over the previous 10 years. This closely approximates a long term real rate of return in agriculture. The long term rate of return in agricultural production is used as the basis for calculating the opportunity costs of nonland capital because it is assumed that the next best alternative use for these fixed assets is in another agricultural enterprise.

Livestock, however, is valued at the current year's price. An operator could expect the capital invested in these inputs to receive an annual return equal to their opportunity costs, exclusive of asset valuation changes because of inflation or deflation. The rate of return must be calculated using earnings from the current production process. Earnings from price appreciation or price depreciation are not included here or in the cash receipts. The expected rate of return is estimated by taking the average of the previous 10-year total return to production assets in the agricultural sector, subtracting the value of operator's labor used each year, and dividing by the respective cash receipts assets. All these rates are obtained from *Economic Indicators of the Farm Sector: National Financial Summary* (USDA, ERS).

Land. In order to maintain comparability between costs and returns, the opportunity costs for land is based only on the contribution of land to the current year's production. The procedure for estimating the returns to land used in crop production (for accounts that omit the direct effects of government programs) is based on information about cash and share market rental rates. These two rental rates are weighted together based on the proportion of land rented in an area on a cash or share basis to yield one rental rate. This weighted rate, minus real estate taxes that are already included in the taxes and insurance costs, is then applied to all acres in production whether owned or rented. The cash rental rate is the rate actually paid by producers as reported on the FCRS. The share rental rate is a calculated rental rate based on the share of production producers agreed to provide landlords and the inputs landlords provided in the production process. A cost of co-op shares owned by sugarbeet growers in Minnesota and eastern North Dakota is included in the land charge for sugarbeets. The per-acre value of peanut quotas are considered when estimating land charges for peanut enterprises.

The procedures used to estimate land charges for livestock enterprises differs from those applied to crops. The allocated returns to land is the value of land used for the livestock enterprise (including land for livestock feed production) multiplied by the calculated 10-year rate of return to agricultural assets. Information on land used in the livestock enterprises are collected from FCRS. Land values are obtained from *Situation and Outlook Summary: Agricultural Resources* and *Agricultural Land Values* (USDA, ERS). Between the surveys, returns to land are updated by the percentage changes in the land value from the previous year.

Unpaid Labor. A large portion of the labor input in most farm operations is supplied by the operator and family. Like other capital, this labor has an opportunity cost. These costs cannot be determined during the production period through market transactions because there are none. Labor costs are divided between hired and unpaid labor depending upon whether a cash wage is paid. Additional value of unpaid labor, such as for management and entrepreneurial skill, is measured as a residual return.

There are two approaches used to estimate the amount and value of unpaid labor, depending on the data available for the commodity. Under the first approach, the value of unpaid labor is calculated using information collected from the survey on hours of work devoted to the enterprise multiplied by the wage rate for agricultural workers adjusted to include the operator's share of social security. The second approach measures the hours of labor on a crop in proportion to the machinery used plus an estimate of the hours of hand labor.

Conclusions

In large part, the credibility of statistics provided by USDA is determined by the amount, timeliness, and clarity of information provided to users about concepts and procedures used. As agriculture continues to change and adopt new technologies cost of production methodology will also need to change. Recognizing this dynamic environment, USDA has a rich history of introspection regarding methods used to develop national estimates of enterprise cost and returns. Much of the evolution in USDA methodology has been the result of interaction with economists from land grant universities, policy institutions and the National Agricultural Cost of Production Standards Review Board. Improvements tend to be incorporated gradually over time, although these improvements have often originated from major conferences and reviews such as this. For example:

1. In June 1978, USDA held a Cost of Production Workshop to review its cost computation procedures (USDA, ESCS). Some of the principal topics discussed included the method of calculating land charges, management and labor charges, farm overhead, and non-land depreciation. Although there was no consensus about whether imputed values should be used to determine a farmer's portion of land, labor, capital, and management, several recommendations were later incorporated in USDA accounts.

2. In 1980, USDA analysts participated in a conference on *Developing and Using Farm and Ranch Cost of Production and Return Data* sponsored by the Great Plains Agricultural Council (Helmers). Numerous papers were presented dealing with a broad range of issues related to cost of production.

3. A Cost of Production Task Force was established within USDA in 1981 to reexamine cost concepts and the format in which cost of production

information is conveyed to policymakers. The task force identified several aspects of USDA's cost of production program that were in need of modification, such as the treatment of charges for land and management. The inconsistency of published results with economic theory spurred most of the proposed changes. Many of the items identified for change, however, had specific estimation procedures outlined in prior legislation.

4. In 1983, USDA solicited wide-spread review of a new procedure for valuing and allocating the returns to resources used in production and a new format for reporting cost of production (Hoffman and Gustafson). Much of the impetus for making these adjustments in procedures was a direct result of changes in cost of production legislation that occurred in 1981.

5. More recently, USDA held conferences to examine methods used to develop estimates for wheat (1988) and rice (USDA, ERS, 1988). Extension economists, industry representatives, producers and others interested in cost estimation for these commodities were given an opportunity to review USDA procedures.

Differences between USDA and land grant university estimates of costs of production have captured people's attention at times. Most recently, the costs of producing rice became the focus of controversy during the debates of the 1990 farm legislation; a meeting of government and land grant agricultural economists was held to discuss the differences. This meeting resulted in reports by land grant and government economists (Land Grant Universities; USDA, ERS, 1990). A single cost of production estimate that is suitable for all purposes does not exist. The initial consideration in distinguishing between different estimates of a commodity's cost of production is the original purpose for which the estimates were constructed. Different purposes lead to different estimates. In general, the primary purpose of projected cost of production budgets developed by universities and other research institutions is to provide local farmers guidance in their management planning. Congressional mandate dictates the primary purpose for USDA cost of production estimates. USDA cost of production estimates are historical costs and could not be used as effectively for individual producer planning purposes. The advantage of USDA cost and return estimates are that they are based on the actual production practice of a statistically representative sample of farmers and that they are calculated using the same methods for all commodities and states.

Finally, recognizing the limitations of a single point estimate, USDA is establishing a broader analytical view. The development of farm level enterprise accounts has facilitated the analysis of cost distributions and allowed differences between low and high cost farms to be explored. This may lead to a more complete and useful information system and could reduce the confusion and misuse of single point cost estimates.

References

Ahearn, Mary, Mir Ali, Robert Dismukes, Hisham El-Osta, Dargan Glaze, Ken Mathews, Bill McBride, Robert Pelly, and Mike Salassi. 1990. "How Costs of Production Vary," Agriculture Information Bulletin Number 599, Economic Research Service, U.S. Department of Agriculture.

Commerce Clearing House. "State Tax Handbook." Washington, D.C. Published annually.

Ericksen, Milton H. 1983. "Plans for Cost Estimate Reports". Presented to the National Agriculture Cost of Production Standards Review Board, Feb. 11, Washington, D.C.

Federal Reserve System. "Federal Reserve Bulletin." Published monthly.

Glaze, Dargan. "A New Approach to Estimating COP Budgets." 1988. *Agricultural Income and Finance Situation and Outlook Report*, AFO-29, May, USDA-ERS.

Glaze Dargan and Mir Ali. 1988. "Distribution of Costs of Production for Wheat Farms." *Agricultural Income and Finance Situation and Outlook Report*, AFO-31. Dec., ERS-USDA.

Harrington, David H. 1983. "Costs and Returns Economic and Accounting Concepts." *Agricultural Economics Research* 34, No. 4: 1-8.

Helmers, Glenn. 1980. "Developing and Using Farm and Ranch Cost of Production and Return Data: An Appraisal." Great Plains Agricultural Council, Publication No. 95., University of Nebraska, Dept. of Ag. Economics Report No. 104.

Hoffman, George and Cole Gustafson. 1983. "A New Methodological Approach for Estimating Agricultural Costs of Production." Staff Report No. AGES830513. Economic Research Service, U.S. Department of Agriculture.

Krenz, Ronald D. 1975. "The Farm Enterprise Data System: Capabilities and Applications." *Southern Journal of Agricultural Economics*, July: 33-38.

Land Grant Universities. 1990. "Five State Land Grant University Report on Rice Cost and Returns Estimates." Unpublished paper prepared by Robert Coats, Jr., Leo Guedry, James Hamill, Arthur Heagler, Karen Klonsky, Brian McManus, and Edward Rister, February.

McBride, William D. 1991. "Characteristics and Production Costs of U.S. Soybean Farms, 1989." AIB No. 623, USDA-ERS.

Mighell, Ronald. 1973. "Farm Production Economics in ERS and Before." Unpublished CED Working Paper, Economic Research Service, USDA.

Salassi, Michael E., Mary Ahearn, Mir Ali, and Robert Dismukes. 1990. "Effects of Government Programs on Rice Production Costs and Returns, 1988." Agricultural and Rural Economy Division, ERS-USDA, Agricultural Information Bulletin No. 597.

U.S. Department of Agriculture, Economic Research Service. "Agricultural Resources: Agricultural Land Values and Markets Situation and Outlook Report." Published annually.

_____. "Agricultural Resources Situation and Outlook: Inputs." Published each April and August.

_____. "Economic Indicators of the Farm Sector: Costs of Producing Major Field Crops." Published annually.

_____. "Economic Indicators of the Farm Sector: Costs of Producing Livestock and Dairy." Published annually.

_____. "Economic Indicators of the Farm Sector: National Financial Summary." Published annually.

_____. 1990. "The Economic Research Service's Response to the Five State Land Grant Universities Report on Rice Cost and Returns Estimates." Unpublished paper prepared by Mary Ahearn, Mike Salassi, Jim Johnson, and Dave Harrington, June.

_____. 1988. "*U.S. Rice Production Costs: Workshop Proceedings.*" Agriculture and Rural Economy Division, Staff Report No.AGEE880509, July.

U.S. Department of Agriculture, Economics, Statistics and Cooperative Service. 1979. "Estimating Agricultural Costs of Production: Workshop Proceedings." Report No. ESCS-56.

U.S. Department of Agriculture, National Agricultural Statistics Service. "Agricultural Statistics." Published annually.

_____. "Crop Production." Published each January.

_____. "Small Grains." Published each January.

_____. "Agricultural Prices." Agricultural Statistics Board. June issue.

_____. "Commercial Fertilizers." November issue.

_____. "Farm Labor." February issue.

_____. "Field Crop Production: Disposition and Value." Published annually.

_____. "Milk Production." Published monthly.

_____. "Livestock Slaughter." Published monthly.

_____. "Cattle on Feed." Published monthly.

_____. "Hogs and Pigs." Published monthly.

_____. "Livestock Market News: Meat and Wool." Published weekly.

U.S. Department of Energy. "Natural Gas Monthly." Energy Information Administration. Published monthly.

Van, Harry, ed. "Chemical Marketing Reporter." New York: Schnell Publishing Co., Inc. Selected issues.

9

Rice Cost and Return Estimates: Similarities and Differences Among Predominant Methods

Leo J. Guedry

Introduction

The focus on cost of production estimates for a specific commodity, rice, arises from a substantial effort on the part of extension and research personnel at five land grant universities and the Economic Research Service of the U.S. Department of Agriculture, to reconcile differences between university and Economic Research Service cost of production estimates for rice.[1] This interaction resulted in the clarification of many differences existing between these estimates, but it also highlighted areas in which differences still remained (Land Grant Universities; USDA, 1990b). The detailed analysis for rice likely will be relevant for other commodities.

This chapter will attempt to identify the similarities and differences between the cost of production estimates for rice prepared by the five universities and the Economic Research Service (ERS). No attempt will be made to address whether these cost of production (COP) estimates are the most appropriate for the purposes for which they are being used. The focus will be on comparing the cost and return components included in developing estimates and estimation procedures, where appropriate. This comparison will be followed by some general observations (questions-debatable issues) with respect to these similarities and differences.

Informational Sources

Information used in developing this chapter was obtained from materials prepared by personnel at ERS; the University of Arkansas; the University of California, Davis; the Louisiana State University; the Mississippi State University; and Texas A&M University. Two sets of estimates prepared by the universities will be reviewed: (1) COP estimates published as a regular output from their extension and research programs and (2) COP estimates developed for input to the 1990 Farm Bill discussions. The use of rice cost of production estimates and the supporting data bases for research activities is an explicit or implicit purpose of COP programs at most of the universities. However, the degree of use in research programs is highly dependent on data collection procedures and accessibility of the data bases. Two of the universities were using probability-based sampling and maintaining budget generator data bases for use by other researchers. However, these uses were not central to the reconciliation discussions between the universities and ERS. Consequently, the COP estimates used for this purpose will not be further addressed in this chapter.

The estimates can be differentiated by the general purposes for which they were developed into the following three categories: firm level decision making and planning, research input, and policy analysis. The format used by the Economic Research Service to report COP estimates for rice will be utilized for comparison purposes, since it is a common frame of reference for users of COP information (USDA, 1990a, pp.42-46). Likewise, generalized ERS estimation procedures for the various items will serve as the common reference for university estimates (Table 9.1).

Firm Level Decision Making and Planning

Each university provides short run estimates of the cost of producing rice for farm level planning and decision-making purposes. This information commonly serves as the basis for educational programs in farm management carried out by state extension services. Most are presented in a format typical of cost and return enterprise budgets and are presented on a cost or net returns per acre basis. While all of the estimates are based on historical production practice information, four of the universities annually update costs and returns by using updated prices based on expectations for the production season, i.e., projected costs and/or returns for the upcoming year or production season are estimated for planning purposes. Components of these COP estimates for rice will be reviewed according to the following categories: gross value of production, variable cash expenses, fixed cash expenses, and capital replacement.

TABLE 9.1   Data Sources and Generalized Procedures for COP Estimates, USDA-ERS

Item	Source and Procedures
Gross value of production:	
Market value	HM
Cash expenses:	
Seed	SM
Fertilizer	SM
Chemicals	S
Custom operations	S
Fuel, lube and electricity	EE
Repairs	EE
Hired labor	EE or S
Purchased irrigation water	S
Drying	EE or S
Miscellaneous	S
Technical services	S
Total variable cash expenses	Σ
General farm overhead	TC
Taxes and insurance	TC
Interest on operating loans	TC
Interest on real estate	TC
Total fixed cash expenses	Σ
Capital replacement	EE

HM = The yield per planted acre comes from FCRS surveys or NASS, and the price is the harvest month price from NASS.

SM = Producer surveys on application rates and prices from surveys or published market sources.

S = Producer survey of total line item cost.

EE = Calculations using a labor supply or equipment complement developed from producer surveys and prices from other surveys or published market sources.

TC = Whole farm cost obtained from producer surveys and allocated by value of commodity.

Source: (USDA, 1990a).

Data Sources

The data sources were producers, extension service specialists, suppliers, researchers, and published sources for interest rates, labor costs and use, pesticide costs, equipment operation and repair. Two of the university estimates (Mississippi and Louisiana) were developed from statistically based samples of rice farmers surveyed every three and five years, respectively, for production practice information.

Gross Value of Production

Gross value of production or gross returns estimates provide some information on the general character of the estimates being provided (Table 9.2). Two of the states did not directly include gross return estimates; instead, they focused their estimates on cash expenses. One of these was a point-in-time estimate for a particular size farm for a region in the state. The other was a projection estimate for a representative operation by major-producing regions in the state. The other three state estimates provided revenue estimates--two use the loan rate as a price proxy and one a five-year average market price; the former two, obviously, assume participation in the rice program. In comparison, the traditional ERS estimate was developed using harvest month prices. However, ERS has begun to prepare COP estimates with and without the direct effects of Government payments (Salassi, *et al.*).

TABLE 9.2 Data Sources and Generalized Procedures for Revenue Estimates by USDA-ERS and States

Item		State					
	ERS	AR	CA	LA	MS	TX
		- Sources and Procedures -				
Gross value of production:						
Market value	HM	N/A	N/A	LR	SM	LR
Deficiency payment	N/A	N/A	N/A	ASCS	TP	ASCS

N/A = Not applicable.

HM = The yield per planted acre comes from FCRS surveys or NASS, and the price is the harvest month price from NASS.

LR = The yield per planted acre comes from producer surveys, and the price is the announced Farm Program loan rate.

ASCS = Area average base acres and proven yields information from USDA-ASCS along with the deficiency rate are used to calculate the deficiency payment.

SM = The yeild per planted acre comes from producer surveys, and the price is a five year average of market prices. If the price is below the Farm Program loan rate, the loan rate is used.

TP = The deficiency payment is the producer survey yield times the difference between target price and the greater of the loan rate or the market price.

Sources: (USDA, 1990a; Coats, *et al.*; Williams and Dubruille; McManus and Heagler; Mississippi State University; and Gerlow.

Variable Cash Expenses

A major difference between the state estimates in the handling of variable cash expenses occurs in the costing of irrigation water (Table 9.3). This difference appears to reflect state differences in the customary practices for obtaining water. In those areas where it is customary for the operator to obtain water from owned wells, the irrigation charges are reflected in other categories. However, in Louisiana, where a common practice is to purchase water on a crop share basis, crop share is treated as a reduction in returns. There are other differences that are not reflected in the table, since they are not included in ERS estimates but are included in some of the state budgets. Louisiana, Mississippi, and Texas include a storage charge, and Texas includes a crop insurance and marketing charge. Louisiana includes charges for tractor, fuel, and labor associated with set-aside maintenance resulting from program participation.

Fixed Cash Expenses and Capital Replacement

Three of the universities did not account for general farm overhead in their estimates and four did not include a charge for interest on real estate (Table 9.4). However, each included a charge for interest on operating capital, which was calculated using a nominal market rate of interest that was commonly being incurred by farmers within their areas. Each provided an estimated value for capital replacement needed to maintain an equipment bundle.

Summary of Commonalities and Differences

In general, while there were differences in the data sources and in those items where regional or state uniqueness existed, all of the COP estimates accounted for specified variable cash expenses. A major difference existed, however, as one of the state estimates (Louisiana) included set-aside maintenance charges. All estimates accounted for capital replacement. Only one included a charge for interest on real estate and three did not account for overhead.

Policy Analysis

Policy analysis is a very broad term used to encompass a multitude of different analyses associated with policy development and implementation. The analyses completed by the universities can best be described as inputs

TABLE 9.3 Data Sources and Generalized Procedures for Variable Cash Expense Estimates by USDA-ERS and States

Item	ERS	State				
		AR	CA	LA	MS	TX
		- Sources and Procedures -				
Cash Expenses:						
Seed	SM	EX	EX	SM	SM	SM
Fertilizer	SM	EX	EX	SM	SM	SM
Chemicals	S	EX	EX	SM	SM	SM
Custom operations	S	EX	EX	SM	SM	SM
Fuel, lube and electricity	EE	EEX	EX	EE	EE	EE
Repairs	EE	EEX	N/A	EE	EE	EE
Hired labor	EE/S	EX	EX	EE	EE	EE
Purchased irrigation water	S	N/A	S	N/A	N/A	SM
Drying	EE/S	EX	EX	SM	SM	SM
Miscellaneous	S	EX	EX	SM	SM	SM
Technical services	S	EX	N/A	N/A	SM	SM

N/A = Not applicable.

SM = Producer survey of application rates and prices from surveys or published market sources.

EX = Expert opinion (including farmer validation) and prices from surveys or published market sources.

S = Producer survey of total line item cost.

EE = Calculations using a labor supply or equipment complement developed from producer surveys and prices from surveys or published market sources.

EEX = Calculations using an equipment complement developed from expert opinion (EX) and prices from surveys or published market sources.

Sources: (USDA, 1990a; Coats, *et al.*; Williams and Dubruille; McManus and Heagler; Mississippi State University; and Gerlow).

to the implementation component of a policy. That is, university estimates were not designed to be used in determining whether or not there should be a farm program; instead, they were used in identifying the level of parameters defined in the 1990 legislation.

Since ERS estimates are considered benchmark values for commodity cost of production at the national level with respect to policy analysis, the ERS framework was again used as the basis of comparison. The basic question addressed with respect to the 1990 farm policy debate was: How did the cost of production estimates developed by ERS compare with those from the five land grant universities? In preparing their estimates, three of the universities used the firm level decision-making and planning estimates previously discussed and developed comparable estimates (Hamill and

TABLE 9.4 Data Sources and Procedures for Fixed Cash Expenses and Capital Replacement Estimates by USDA-ERS and States.

Item		State				
	ERS	AR	CA	LA	MS	TX
		- Sources and Procedures -				
Fixed cash expenses:						
General farm overhead	TC	N/A	EX	SM	N/A	N/A
Taxes and insurance	TC	TC	EX	S	N/A	SM
Interest on operating						
loans	TC	IC	IC	IC	IC	IC
Interest on real estate	TC	N/A	IR	N/A	N/A	N/A
Capital replacement	EE	EEX	EE	EE	EE	EE

N/A = Not applicable.

TC = Whole farm cost obtained from producer surveys and allocated by value of commodity.

S = Producer survey of total line item cost.

EE = Calculations using a labor supply or equipment complement developed from producer surveys and prices from surveys or published market sources.

IC = Operating expenses as incurred times a market rate of interest.

EX = Expert opinion (including farmer validation) and prices from surveys or published market sources.

EEX = Calculations using a labor supply or equipment complement developed from expert opinion (EX) and prices from surveys or published market sources.

IR = Current amount of loan times a market rate of interest.

SM = Producer survey of application rates and prices from surveys or published market sources.

Sources: (USDA, 1990a; Coats, *et al.*; Williams and Dubruille; McManus and Heagler; Mississippi State University; and Gerlow)

Spurlock; McManus; Wick). Two of the universities used a whole farm analysis of representative size farms and production regions for rice in their states. One used an economic engineering approach complemented with information from the firm level decision-making and planning estimates to arrive at the estimates (Clark, *et al.*); and the other used a modified delphi approach in obtaining data from a producer panel and complemented this with information from firm level decision-making and planning estimates, existing whole farm data bases and other sources (Rister, *et al.*). For purposes of this discussion, the same categories used previously are used, including estimates of Economic Costs as defined by ERS, to compare the state estimates developed as an input to policy deliberations.

Gross Value of Production

Four of the universities assumed participation in the rice program (Table 9.5). The estimate that was not based on program participation used a harvest month price on a comparative basis with the estimates prepared by ERS.

Variable Cash Expenses

The differences from the configuration of costs used in the short run estimates discussed previously are those associated with the incorporation

TABLE 9.5   Data Sources and Generalized Procedures for Revenue Estimates
Used in Policy Analysis by USDA-ERS and States

Item	State					
	ERS	AR	CA	LA	MS	TX
			- Sources and Procedures -			
Gross value of production:						
Market value	HM	EX	HM	LR	SM	PP
Deficiency payment	N/A	N/A	N/A	ASCS	TP	PP

HM = The yield per planted acre comes from FCRS producer surveys or NASS, and the price is the harvest month price from NASS. For the policy study, California used the USDA-ERS number.

EX = The gross returns were calculated by multiplying average yields obtained from verification trials times the Farm Program target price.

LR = The yield per planted acre comes from surveys, and the price is the announced Farm Program loan rate.

ASCS = Area average base acres and proven yield information from USDA-ASCS along with the deficiency rate is used to calculate the deficiency payment.

SM = The yield per planted acre comes from producer surveys, and the price is a five year average of market prices. If the price is below the Farm Program loan rate, the loan rate is used.

TP = The deficiency payment is the survey yield times the difference between target price and the greater of the loan rate or the market price.

PP = The market value is calculated by multiplying the expected yield obtained from producer panels times a market price. The deficiency payment is calculated by multiplying the proven yields as reported by the producer panel times the difference between the target price and the market price.

Sources: USDA, 1990a; Clark, *et al.*; Wick; McManus; Hamill and Spurlock; and Rister, *et al.*)

TABLE 9.6   Data Sources and Generalized Procedures for Variable Cash
Expenses Estimates Used in Policy Analysis by USDA-ERS and States

Item	ERS	State				
		AR	CA	LA	MS	TX
		- Sources and Procedures -				
Cash expenses:						
Seed	SM	EX	EX	SM	SM	PP
Fertilizer	SM	EX	EX	SM	SM	PP
Chemicals	S	EX	EX	SM	SM	PP
Custom operations	S	EX	EX	SM	SM	PP
Fuel, lube & electricity	EE	EE	EX	EE	EE	PP
Repairs	EE	EE	N/A	EE	EE	PP
Hired labor	EE/S	EX	EX	EE	EE	PP
Purchased irrigation water	S	N/A	S	N/A	N/A	PP
Drying	EE/S	EX	EX	SM	SM	PP
Miscellaneous	S	EX	EX	SM	SM	PP
Technical services	S	EX	EX	N/A	SM	PP

N/A = Not applicable.

SM = Producer survey of application rates and prices from surveys or published market sources.

S = Producer survey of total line item cost.

EE = Calculations using a labor supply or equipment complement developed from producer surveys and prices from surveys or published market sources.

EX = Expert opinion (including farmer validation) and prices from surveys or published market sources.

PP = Line item cost obtained from producer panel.

Sources: (USDA, 1990a; Clark, *et al.*; Wick; McManus; Hamill and Spurlock; and Rister, *et al.*)

of participation in the government program, which gave rise to the inclusion of set-aside maintenance (Table 9.6).

Fixed Cash Expenses and Capital Replacement

Capital replacement was calculated using basically the same procedure in all estimates with the exception that one university used a percentage of total market value of investment as a means of estimating replacement (Table 9.7).

Not all estimates published by the universities for firm level decision-making and planning included all categories of fixed cash expenses as defined by ERS, estimates were developed for all categories for policy analysis. Interest on operating loans was estimated by the universities using

TABLE 9.7   Data Sources and Generalized Procedures for Fixed Cash Expenses and Capital Replacement Estimates Used in Policy Analysis by USDA-ERS and States

Item	State					
	ERS	AR	CA	LA	MS	TX
	- Sources and Procedures -					
Fixed cash expenses:						
General farm overhead	TC	EC	EX	SM	VC	PP
Taxes and insurance	TC	EC	EX	S	EX	PP
Interest on operating loans	TC	IC	IC	IC	IC	IC
Interest on real estate	TC	IR	IR	AA	AA	IR
Capital replacement	EE	EE	EE	EE	EE	M

N/A = Not applicable.

TC = Whole farm cost obtained from producer surveys and allocated by value of commodity.

S = Producer survey of total line item cost.

EE = Calculations using a labor supply or equipment complement developed from producer surveys and prices from surveys or published market sources.

EC = Whole farm cost obtained from expert opinion (including farmer validation) and allocated by value of commodity.

IC = Operating expenses as incurred times a market rate of interest.

IR = Current amount of loan times a market rate of interest.

EX = Expert opinion (including farmer validation) and prices from surveys or published market sources.

SM = Producer survey of application rates and prices from surveys or published market sources.

AA = Average annual investment times a market rate of interest.

VC = Calculated as a percentage of total variable cost.

M = Calculated as a percentage of total market value of investment.

PP = Line item cost obtained from producer panel.

Sources: (USDA, 1990a; Clark, *et al.*; Wick; McManus; Hamill and Spurlock; and Rister, *et al.*).

the same procedures employed in the firm level decision-making estimates (operating expenses times a market rate of interest). ERS used survey results of whole farm interest paid proportioned by the value of rice production on the farm. Two of the universities used the same procedures as used in the firm level planning estimates, to provide estimates for general farm overhead, while the other three used the following different procedures: whole farm costs allocated by value of the commodity, a percent of total variable cost; and cost obtained from a producer panel. Interest on real estate was included by the four universities not including this cost in their planning budgets, two used the current amount of the loan times a market rate of interest, and the other two used an average annual investment times a current rate of interest.

Economic or Ownership Costs

All of the universities, except one, followed the same general approach used by ERS in developing estimates for economic costs (Table 9.8). The university that did not follow this general approach did not develop estimates for economic costs. In general, equilibrium was assumed, that is, all returns were included. While the same generalized approach was used, the estimates were not all derived in the same manner. All calculated return to operating capital in the same manner, operating capital times a market rate of interest (the latter considered to be the opportunity costs for capital). While returns to other nonland capital was estimated by one of the universities as average annual investment times a real rate of interest, the other three used a market rate of interest. Net land rent was estimated using the same procedures as ERS by three of the universities, while one used responses from a producer panel. Only one of the universities provided an estimate for unpaid labor and that was developed from producer panel data.

Summary of Commonalities and Differences

Two of the universities used a whole farm analysis as a basis for developing the COP estimates, based on specified sizes and locations within their state's rice producing areas. The other estimates were developed through the usual procedures of an enterprise budget. Other areas of differences included the incorporation of marketing charges by two universities, while storage charges and set-aside maintenance were included in other estimates. In general, the universities used market rates of interest (opportunity costs) in estimating the returns to other nonland capital, while ERS and one of the universities used a real rate of return to assets in agriculture, based on the assumption that the next best use was in some other agricultural production activity.

Observations-Questions-Debatable Issues

The comparison of the cost of production estimates for rice prepared by the universities and ERS suggests that purpose affects the procedures in developing the estimates. However, differences were observed even in the

TABLE 9.8 Data Sources and Generalized Procedures for Estimates of Allocated Returns to Owned Inputs Used in Policy Analysis by USDA-ERS and States

Item	ERS	State				
		AR	CA	LA	MS	TX
		- Sources and Procedures -				
Allocated returns to owned inputs:						
Return to operating capital	OC	N/A	OC	OC	OC	OC
Return to other nonland capital	SR	N/A	AR	AA	AA	AA
Net land rent	R	N/A	R	R	R	PP
Unpaid labor	S	N/A	N/A	N/A	N/A	PP

N/A = Not applicable.

OC = Operating capital as incurred times a market of interest.

SR = Market value of investment from surveys times a real rate of interest.

R = A composite value of cash and share rents less any costs already account for. Cash and share rent values came from surveys.

S = Producer survey of total line item hours times the state average hired farm worker wage rate.

AR = Average annual investment times a real rate of interest.

AA = Average annual investment times a market rate of interest.

PP = Line item cost obtained from producer panel.

Sources: (USDA, 1990a; Clark, *et al.*; Wick; McManus; Hamill and Spurlock; and Rister, *et al.*)

COP estimates developed for the same purpose. These differences and those associated with the purpose for COP estimates indicate a need for professional dialogue. Some of the observations, questions, and/or debatable issues that should be a part of the dialogue are briefly addressed below.

Include or Exclude Government Program Participation

The estimates suggest that the inclusion or exclusion of government program participation will depend on the use for which the COP estimate is derived. The provision of annual or projected costs and returns for use in farm level decision-making with and without the inclusion of costs and returns associated with participation in government programs for program commodities is clear cut with respect to this purpose. Such estimates provide the producer with information needed to decide whether or not to produce outside the government program for program commodities or if there is an advantage to the farmer in producing non-program commodities.

In the case of policy analysis, the "with" or "without" scenario depends on what policy issue is being addressed. If the decision before policy makers is whether there should or should not be a program; then the issue is to determine the expected profit and loss situation for commodity producers if there was no program, then COP without government programs would be useful. It is not possible to remove the effects of government programs which get built into land values, prices, etc. However, as a first approximation, the direct cash costs of program participation can be removed. If, on the other hand, this political decision has been decided in the affirmative and the policy question has to do with implementation of a program, then the "with" scenario would be appropriate. In the situation where the policy instruments being used are price support and/or acreage control programs that are voluntary, which require the participant to incur a cost for participation, the "with" scenario would be appropriate. That is, the acreage reduction or the price support decision should be based on factors affecting firm level decisions, if the intent is to generate the desired reaction on the part of the producer with respect to the program. Otherwise, a distortion in the competitive position of the program commodity relative to non-program commodities may exist with respect to firm level decisions and the aggregate supply response may be affected. Harrington has pointed out that there are cost spiral problems if government payments are included. However, these are real costs to producers and policy in general is designed to generate a desired producer response.

Long Run vs. Short Run

Firm level decision-making and planning COP estimates are short run economic estimates. However, when policy analysis is conducted, there is, at least with respect to the length of time involved, a time period that should not be classified as short run from an economic perspective. Practically, these estimates should relate to the length of time that the policy or program will be in effect.[2] Given the manner in which the COP estimates for rice have been developed, it is questionable whether any COP estimates being developed for rice are truly long run economic estimates. Given the structure of the most recent farm programs, where support levels and acreage reduction levels are set annually, once the program has been designed, the decisions become a series of short run or year to year decisions.

Tenancy

If policy is designed to be voluntary and incentive driven, as has generally been characteristic of U.S. farm policy, then the desired reaction on the part of farmers to some extent will depend on the tenancy characteristics of farmers. Do owners and tenants allocate resources in the same manner? Again, if there is to be a desired outcome to policy in the short run, should any differences be reflected?

Cost of Firm Level Marketing

Some estimates include a cost for storage and marketing. Should these be included as part of COP estimates for policy analysis purposes? Generally, such costs would not be included as part of COP but would be part of managing price risk, especially storage costs. The question really relates to what the purpose of policy is. Is it to affect a commodity, such as rice, or is it to affect the farmer? If it is designed to represent an income safety net for the farmer, then it is his cost of doing business that is appropriate. This observation raises a fundamental question as to what is cost of production in the situation where it is common practice to use on-farm storage to secure market price.

Firm vs. Regional or National Estimate

COP estimates for firm level planning and decision making are typically developed on a per animal or per acre basis. The information contained within these estimates are for the average farm situation or for what some refer to as a typical or representative farm situation (more of a median than an average measure). Each cost item that is included in variable costs represents a particular item or set of items, such as chemicals. Given that aggregation problems exist when the estimates being developed are regional or national, requiring some items to be more broadly defined than they would be in state level estimations, the aggregated items represent basically the same general category of inputs. That is, for the same purpose the estimation model used at the national level should, if applied at the state level yield the same results as the estimation model used at the state level. It would seem that the same should hold for items included in estimates of economic cost, as well. Should the items used in estimating COP for the same purpose differ at the state, regional, and national level? A good example would be returns to non-land capital (machinery and equipment). In general, this item is estimated as an opportunity cost for capital. ERS assumes the next best use in agriculture and uses a real rate of return to production assets in the farm sector (including land) over the past 10 years. Some of the universities assume a market rate of interest. The actual rate of interest would not be the same in each state, but it should represent the farmer's opportunity cost for those funds.

Commodity Cost of Production and the Profession

Cost accounting and production economics principles are the basic disciplinary foundation upon which commodity cost of production estimates are based and have been basic elements in the tool kit of agricultural economists since the turn of the century. Their role in applied economic analysis has been fundamental and was reinforced during the 1950s and early 1960s in publications such as the regional research bulletin edited by Irving F. Fellows entitled "Budgeting: Tool of Research and Extension in Agricultural Economics." The developments in the area of production economics were used to refine the essentials of enterprise budgeting. However, it would appear that, since that time, little effort has been made by the profession to use the conceptual developments of multi-output production, duality theory, and risk analysis to further refine the old standby of COP estimation. Has the profession forsaken the 'ole' standby for something better. If so, what?

Program vs. Non-Program Crops

The interrelationship between program and non-program crops, is it such that the resource allocative efficiency on farms is being adversely affected? For areas of the country in which program crops are produced in rotation with other crops, would the concept of a rotational acre be a more appropriate measure to use than a crop acre? Given the evolving interest in sustainable agriculture, the idea of a rotational acre may prove to be of value in COP estimation. In those areas without a rotational strategy, would a whole farm analysis provide a more appropriate set of COP estimates for measuring the farm level cost-income situation?

Summary Comments

This review of COP estimates for rice suggests that purpose does influence the manner in which estimates are developed. Estimates developed for policy analysis more fully accounted for all cost items and are designed to reflect more than one production season. Those developed for annual planning purposes are designed to provide useable estimates within the resource constraints of those preparing the estimates. Information that is developed annually is, in general, part of an on-going educational program that has been developed over the years. Such programs have the advantage of being validated annually by users, and their formats have been designed according to what experience has identified as most effective within a given state. Therefore, the differences that exist in data and procedures used in developing estimates should be the focus, not the format in which the estimates are presented.

It is important to note that even though there were differences in data collection methods and general estimation procedures between ERS and the universities, there was a great deal of similarity with respect to variable cash expenses for comparable years. The primary differences occurred in the fixed cash expenses and economic cost categories, i.e., those categories of costs that depend on assumptions with respect to imputed returns (and/or opportunity costs). This cursory review of rice cost of production estimates indicates that it is in this area that additional attention is needed by those working in the area of cost of production estimation. Such a focus seems appropriate since it is the incorporation of "opportunity costs" in the estimation of cost of production that to a great extent differentiates the perspective of the agricultural economist from that of the cost accountant. Given the various purposes for which COP estimates are developed, it is important to recognize that someone will use these published estimates for a purpose for which they were not designed. All that can be done is to provide the best documentation possible for the intended purpose and not assume the responsibility for misuses. It is incumbent, however, that we ask the following question: Are we and others asking more of COP estimates than these estimates are capable of answering?

Notes

Assistance in preparation of the tables contained throughout the chapter was provided by Brian McManus, Research Assistant, Department of Agricultural Economics and Agribusiness, Louisiana State University. Sole responsibility for their use rests with the author.

1. COP estimates as used in this chapter should be interpreted to include returns from production as well as costs.

2. The exception may be the firm level policy simulator model being used at one of the universities, which incorporates aggregate economic forecasts in the basic model from which rice cost of production estimates were derived.

References

Clark, Jr., Jerry D., Robert E. Coats, Jr., Billy E. Herrington, Jr., Clyde A. Stuart, and Tony E. Windham. 1989. "Economic Analysis of Two Rice-Soybean-Wheat Production Farms Located in the Grand Prairie Region of Arkansas." University of Arkansas Cooperative Extension Service. Unpublished paper presented at the Conference on the Costs of Producing Rice, November, Economic Research Service, Washington, D.C.

Coats, Jr., Robert, Billy Herrington, Clyde Stuart, and Tony Windham. 1990. "An Economic Summary of Extension Technical Bulletin No. 126." University of Arkansas Cooperative Extension Service, November.

Fellows, Irving, Jr, ed. 1960. *Budgeting: Tool of Research and Extension in Agricultural Economics.* Northeast Farm Management Research Committee. Bulletin 357, Storrs Agricultural Experiment Station, Storrs, Connecticut.

Gerlow, Arthur R. 1991. Personal communication. Texas Agricultural Extension Service, Texas A&M University, College Station.

Hamill, James G., and Stan R. Spurlock. 1989. "Estimated Costs and Returns of Rice Production in the Delta Area of Mississippi." Department of Agriculture Economics, Mississippi State University. Unpublished paper presented at the Conference on the Costs of Producing Rice, November, Economic Research Service, Washington, D.C.

Harrington, David H. 1983. "Costs and Returns: Economic and Accounting Concepts." *Agricultural Economics Research* 35, No. 4: 1-8.

Land Grant Universities. 1990. "Five State Land Grant University Report on Rice Cost and Returns Estimates." Unpublished paper prepared by Robert Coats, Jr.; Leo Guedry; James Hamill; Arthur Heagler; Karen Klonsky; Brian McManus; and Edward Rister. February.

McManus, Brian, and Arthur Heagler. 1990. "Projected Costs and Returns-1990, Rice, Louisiana and Soybeans, Corn, Milo, Wheat, Wheat-Soybean Double Crop, and Rice-Crawfish Double Crop, Southwest Louisiana." A.E.A. Research Report No. 80, Department of Agricultural Economics and Agribusiness, Louisiana Agricultural Experiment Station, Louisiana State University Agricultural Center.

McManus, Brian. 1989. "Estimated Returns and Costs of Production, Louisiana Rice." Unpublished paper presented to a combined meeting of the Louisiana Rice Research Board, Louisiana Rice Promotion Board, and Louisiana Farm Bureau Rice Advisory Committee, July, LSU Rice Research Station, Crowley, Louisiana.

Mississippi State University. 1988. "1989 Costs and Returns: Rice." Agricultural Economics Report 31, Mississippi Agricultural and Forestry Experiment Station, Mississippi Cooperative Extension Service.

Rister, M. Edward, Edward G. Smith, James W. Richardson, Ronald D. Knutson, Warren R. Grant, and Troy N. Thompson. 1989. "Texas Rice Production Costs." Agricultural and Food Policy Center, Department of Agricultural Economics, Texas Agricultural Experiment Station, Texas Agricultural Extension Service, Texas A&M University.

Salassi, Michael, Mary Ahearn, Mir Ali, and Robert Dismukes. 1990. "Effects of Government Programs on Rice Production Costs and Returns, 1988." Agricultural Information Bulletin Number 597, Economic Research Service, U.S.D.A.

Gerlow, Arthur. 1991. Personal communication. Texas Agricultural Extension Service. Texas Agricultural Extension Service, Texas A&M University.

U.S. Department of Agriculture. 1990a. "Economic Indicators of the Farm Sector: Costs of Production--Major Field Crops, 1988." Agricultural and Rural Economy Division, Economic Research Service, ECIFS 8-4.

_____. 1990b. "The Economic Research Service's Response to the Five State Land Grant Universities Report on Rice Cost and Returns Estimates." Unpublished paper prepared by Mary Ahearn, Mike Salassi, Jim Johnson, and Dave Harrington. June.

Wick, Carl. 1989. Personal communication with Karen Klonsky, Univ. of California, Davis. Farm Advisor, Butte County, California, July 27.

Williams, Jack and Jim Dubruille. 1989. "1989 Sample Costs to Produce Rice, Sutter, Yuba, Placer, and Sacramento Counties, California." University of California Cooperative Extension Service. January.

10

Results of a National Survey on Data and Methods

Karen Klonsky

Introduction

Most land grant universities develop production estimates for the state's major commodities that are available to the general public. Each institution's estimates are constructed following an internally standardized set of methods. However, the methods used to estimate costs vary from state to state. Conceptual and empirical problems that arise in estimating cost of production make it difficult to assign costs to certain production inputs. Differences in estimation methods are the result of alternative solutions to these problems. Methods differ in several ways:

- the cost factors included in the budget
- the computational procedure or mathematical model used to calculate costs
- data sources
- format of the budget reports

The primary purpose of this chapter is to describe the different methods actually used by university researchers to construct enterprise budgets and report the predominance of each method on a national level. The results of a formal survey are presented. The study also identifies the primary users of the university budgets and discusses the use of USDA cost of production estimates by university researchers.

Methods

A written survey was sent to the USDA Cooperative Extension mailing list of all university researchers with farm management assignments. The survey was also sent to the AAEA mailing list of department chairs. Departments of rural sociology and Canadian departments were excluded from the mailing. Each survey was accompanied by a letter requesting that the survey be given to the appropriate person in that institution. Usable surveys were returned from 41 states. Three states (Montana, Rhode Island and Vermont) do not produce enterprise budgets except for specific research projects. The remaining six states (Alabama, Alaska, Massachusetts, New Hampshire, North Dakota, and West Virginia) are not represented in this analysis.

In five cases two surveys were returned from the same state because two mailing lists were used. These were reconciled so that there is only one survey response from each of the 41 states. Follow-up telephone interviews were also conducted for 23 states.

The survey questions pertained to crop budgets only. Livestock budgets were not included in the survey.

Enterprise Budget
Development Methods in Practice

The following discussion pertains to crop budgets generated by land grant universities that are available to the general public in a published form. The terms "cost of production estimates" and "enterprise budgets" will be used interchangeably. In both cases reference is to crop and not livestock enterprises. The statistics presented are for the 41 states included in the survey.

The discussion is divided into four sections. First the general types of budgets and the distribution systems used will be presented. Next the actual development of the budgets will be described for the three major sections of a budget report: (1) receipts, (2) operating expenses and (3) investment costs. Each of the budget sections will be discussed in terms of the costs (or returns) included, the computation method used to calculate costs (or returns) and data sources.

A problem of semantics exists in discussing these sections of a budget format. For clarification, the term receipts used here refers to gross income or gross receipts. Operating expenses are cash variable expenses plus cash fixed expenses. Investment costs are also referred to as fixed costs or ownership costs.

Budget Development and Distribution

One fundamental difference among enterprise budgets is that some are intended to reflect historical events and are a summary for the current year. Others are intended to be projections of costs for the coming year. Twenty eight of the 41 states responding to the survey reported that they develop projected enterprise budgets (Table 10.1). About half of the states produce enterprise budgets for the current year. Seven states produce both projected and current year budgets.

Whole farm budgets, as opposed to enterprise budgets, include all enterprises for a sample farm. They are a summary of all costs for the farm. Only 6 states develop whole farm budgets for the current year and 2 states develop projected whole farm budgets.

Survey respondents were also asked how often they update their budgets. Most states (61%) update their budgets annually, while about one quarter have no set schedule (Table 10.2). The budgets are distributed through the Cooperative Extension or Agricultural Department office on campus and county Extension offices. Several respondents mentioned that they maintain mailing lists for distribution in addition to answering direct requests. Other outlets include farm lenders, grower meetings, and the Soil Conservation Service of USDA.

Receipts Section

Farm receipts include the income generated from the sale of the production from the enterprise. Six states do not include a receipts section in their enterprise budgets. Government payments (when relevant) are included in 16 (39%) of the state budgets. In-kind payments are not included in any of the state reports.

In all cases the receipts are calculated as the multiple of the yield times price. Yield and price information usually are from different sources. Many states use more than one source for both price and yield.

Yield information most commonly comes from Extension agents or specialists (90%) followed by farmers (64%) (Table 10.3). Information from farmers are from interviews in a group setting or one-on-one. The individual interviews may be either written or oral. One third of the states rely on yield data reported by the state Department of Agriculture.

Price data also come primarily from Extension agents and specialists (72%). Farmers are relied on much less for price information than for yield information. Only 30% of the states ascertain price information directly from farmers. Other major sources used about equally include the National Agricultural Statistical Service of USDA, state Department of Agriculture,

TABLE 10.1 Types of Crop Budgets Produced

Crop Budget Type	Number of States	Percent of States [a]
Enterprise budgets for current year	21	51.2
Whole-farm budgets including more than one crop for current year	6	14.6
Projected enterprise budgets	28	68.3
Projected whole-farm budgets	2	4.9

[a] Total is greater than 100% because more than one response was possible.

TABLE 10.2 Frequency of Budget Updates

Frequency of Updates	Number of States	Percent of States
Anually	25	61.0
Less frequently than annually	6	14.6
No set schedule	10	24.4
Total	41	100.0

the futures market and the loan price or target price. Industry figures are used to a lesser extent.

Operating Costs

Operating costs include variable cash costs and fixed cash costs. Variable cash costs are the costs directly associated with production of the enterprise, and vary with the level of production. They include fertilizer, seed, pesticides, repairs, fuel and labor. Fixed cash costs are the annual overhead costs of running the farm business which do not vary greatly with the level of production. They include road maintenance, office expenses, property taxes and insurance.

The variable operating costs included in the budgets do not vary from state to state. All states include the cost of fuel, repairs, machine labor, input materials (seed, fertilizer, pesticides, and water), and hauling costs for moving the product out of the field (Table 10.4). However, the fixed operating costs included in budgets do vary significantly from state to state. Insurance is included by all but three states. Most states include property taxes on equipment or land (75%) but a significant number do not. From the telephone interviews it became clear that property taxes in many states

TABLE 10.3 Sources of Information: Receipts Section of Budgets

Sources Used by Respondents	Yield Number of States	Yield Percent of States [a]	Price Number of States	Price Percent of States [a]
National Agriculture Statistical Service	7	17.9	11	28.2
State Department of Agriculture	13	33.3	13	33.3
County Agriculture Commissioner	3	7.7	0	0
Industry figures	2	5.1	9	23.1
Farmer panel	20	51.3	10	25.6
Statistical survey of farmers	5	12.8	2	5.1
Extension agents/specialists	35	89.7	28	71.8
Future markets	0	0	15	38.5
Loan price or target price	0	0	14	35.9
No receipts included in budgets	6	14.6	6	14.6

[a] Total is greater than 100% because more than one response was possible.

TABLE 10.4 Operating Costs Included in Crop Enterprise Budgets

Cost Factor	Number of States	Percent of States
Fuel	41	100.0
Lube	19	46.3
Repairs	41	100.0
Machine labor	41	100.0
Management (explicit cost)	15	36.6
Management (residual return)	10	24.4
Input material: (fertilizer, seed, pesticides)	41	100.0
Processing (drying)	34	87.2
Hauling	41	100.0
Interest on operating capital	40	97.6
Property	31	75.6
Insurance on property	38	92.7
Office expenses	22	57.9
Attorney's fees	22	57.9

are extremely low, particularly on land, and may not be an important cost item. Business overhead costs such as office expenses and attorney's fees are included in slightly more than half of the budgets. Fixed operating costs are sometimes collected from farmer surveys. Alternatively, they are estimated as a percent of variable operating costs or as a percentage of the value of equipment.

Farming Operations. Most of the states include a list of farming operations. These are the production practices followed for the enterprise. The operations included are important because the cost of each operation is calculated and then added together to generate the total operating cost. The operations included in the budget calculation are developed from information gathered from Extension agents and specialists (78%) and farmers (83%) (Table 10.5). Farmers may be sampled as a panel, through individual interviews or through farm records maintained by farm business associations.

Input Materials. Materials for an enterprise include the seed, fertilizer, pesticides and water inputs to production. Expenditures on materials for individual enterprises can be easily extracted from a farm enterprise accounting records when available. Alternatively, the cost of input materials can be estimated by determining the application rate per acre and multiplying it by the cost per unit of material. Information concerning these two factors are usually collected separately. Also, more than one source is usually consulted for both the rate and the price.

TABLE 10.5 Source of Information: Operations and Input Materials

Sources Used by States	List of Operations		Rates for Input Materials		Price of Materials	
	N	% [a]	N	% [a]	N	% [a]
Panel of farmers	18	43.9	19	46.3	13	31.7
Statistical survey of farmers	7	17.1	6	14.6	4	9.8
Farm business association	9	22.0	2	4.9	2	4.9
Extension agents/specialists	32	78.1	32	78.1	22	53.7
Local suppliers	N/A	N/A	5	12.2	35	85.3

[a] Total is greater than 100% because more than one response was possible.

The most common source of information for input rates is a combination of Extension agents or specialists (78%) and farmers (83%) (Table 10.5). Extension personnel and farmers are relied on less for price information than for application rate information (53% and 41%, respectively). Local suppliers are asked for rates in only 5 states (12%) but provide price information to 35 of the 41 states (85%). The numbers total to more than 100 percent because most states obtain this information from more than one source.

For all budgets including irrigation, water costs are calculated per acre inch and then charged back to the individual enterprise depending on the amount used. The operating cost per acre inch includes the cost of electricity for pumping, labor, repairs and any fees paid to irrigation districts.

Machinery Operating Costs. In general, the operating costs of equipment are estimated by constructing an equipment complement and calculating the cost per hour to run each piece of equipment. In two states the custom rates for operations are used, so there is no need to calculate the cost to operate machinery. The time per acre and the equipment requirements are specified for each operation. The cost for each piece of equipment used can then be easily calculated as the number of hours per acre times the hourly rate.

The most commonly used budget generator is the Oklahoma State Budget Generator or a modified version of this program. The Mississippi State Budget Generator, Budget Planner, originally developed by North Carolina State, and MBMS developed by Texas A & M are also used by more than one state. Spreadsheets are also commonly used.

Performance rate is calculated from agricultural engineering equations by three quarters of the states (Table 10.6). About one third of the states consult Extension specialists and half of the states consult farmers. Quite often these are a double check of the engineering equation estimates.

Usually the hours per acre for tractors exceeds the machine hours. One commonly used standard is to increase the time for tractors by ten percent over machine time to account for the time moving in and out of the field and the time the tractor is running idle for fill up and adjustment of implements. The annual hours of use for tractors reported by respondents ranges from 500 hours per year to 1200 hours per year. The median usage is close to 500 hours per year. These differences reflect the different growing seasons and cropping intensities experienced in different states. The numbers are obtained from farmers, university agricultural engineers, field trials and ASAE standards.

The hourly equipment cost always includes fuel and repair costs. Most states (83%) rely on engineering equations to estimate fuel consumption and repair costs (Table 10.6). Farmers are asked about fuel consumption by 13 states and about repair costs by 10 states. Equipment suppliers are

TABLE 10.6   Sources of Information: Cost of Estimates for Equipment

Sources of Information	Fuel		Repairs		Performance Rate		Value of Equipment	
	N	% [a]	N	% [a]	N	% [a]	N	% [a]
Agri. engineering equations	34	82.9	34	82.9	30	73.2	0	0
Panel of farmers	8	19.5	6	14.6	15	36.6	8	19.5
Statistical survey of farmers	5	12.2	4	9.8	4	9.8	2	4.9
Extension agents specialists	8	19.6	9	22.0	13	31.7	11	26.8
Equipment suppliers	14	34.1	9	22.0	1	2.4	34	82.9

[a] Total is greater than 100% because more than one response was possible.

consulted about fuel consumption by 14 states (34%) and repair costs by 9 states (22%). The cost of lube is either estimated as 10 percent of the total fuel cost or considered to be part of repair costs.

Almost all states (83%) base repair costs on agricultural engineering equations (Table 10.6). The engineering equations are a function of the type of equipment, annual hours of use, total hours of life, and purchase price.

Only four states (10%) said they do not use agricultural engineering equations in any way. Of those states using engineering equations, three quarters use equations published in the American Society of Agricultural Engineering (ASAE) Yearbook and the rest use equations developed by their university's Agricultural Engineering Department. Of course, these equations may or may not be directly from the yearbook.

Respondents reported using ASAE yearbook information from years ranging from 1971 to 1990. Since the categories for equipment, the functional form and/or the coefficients used to estimate repair costs are periodically revised, equations from different years may result in different repair cost estimates.

Hourly equipment costs may also include property taxes, insurance, and housing. In many budgets these costs are considered equipment overhead. In such cases they are calculated as a percentage of the value of the equipment and then charged to the enterprise based on the number of hours the equipment is used.

Labor. The cost of labor is calculated by multiplying the hours per acre of machine time by the hourly labor rate. Typically the hours of labor are increased by between 10 to 21 percent over machine time to account for downtime, the time used to move equipment in and out of the field, and set-up time. One standard guideline is to increase tractor time by 10 percent over machine time and labor by ten percent over tractor time. This results in labor time of 21 percent over machine time. The other common approach is to increase labor by 10 percent over machine time and set tractor time equal to machine time.

None of the budgets using a simulation, or budget generator, approach make a distinction between hired labor and owner's labor. All machine operator labor is charged at the same rate and there is no accounting for unpaid owner or family labor. Labor for irrigation, pruning, hand hoeing or other non-machine labor is charged to individual operations on an hourly basis. Roughly two-thirds of the states include benefits in the hourly labor rate while one-third do not. Benefits include workers' compensation insurance, social security and other fringe benefits.

Management. Fifteen states (roughly one-third) include an explicit charge for management (Table 10.4). Of these, eight calculate the management charge as a percentage of gross receipts, four as a percentage of costs, two by multiplying the number of hours by a wage rate, and one as the going wage rate for hired managers. Another ten states include returns to management as part of the residual return above total costs. The remaining 16 states do not include management in any way.

Interest on Operating Loans. All of the states except one include interest on operating loans in their budgets (Table 10.7). Almost three-quarters (30) of the states use the Production Credit Association rate or another local bank rate. Another nine states use some other nominal interest rate. Only one state uses a real rate of interest to calculate the cost of operating capital.

Investment Costs

Investment costs are the fixed costs of owning assets used in production. The costs associated with these assets do not vary directly with production levels from year to year and are long term investments. They include machinery, buildings and land.

Machinery. In all but four states, the ownership costs for equipment are estimated as the sum of depreciation and interest of the portion of the equipment charged to the enterprise (Table 10.8). Two states calculate a

TABLE 10.7   Interest Rates Used for Cost of Capital

Interest Rate Used by Respondent	Operating Capital		Equipment Capital	
	N	%	N	%
PCA, FLB, or local banks	30	73.1	26	63.4
Other nominal interest rate	9	22.0	8	19.5
Real interest rate	1	2.4	4	9.8
Interest not included	1	2.4	3	7.3
Total	41	100.0	41	100.0

TABLE 10.8   Fixed Costs Included in Enterprise Budgets

Cost Factor	Number of States	Percent of States
Depreciation on equipment	37	90.2
Interest on equipment loan	37	90.2
Replacement cost for equipment	2	4.8
Fixed costs included in custom rates	2	4.8
Buildings	22	53.7
Irrigation system	34	82.9
Land charge (explicit)	27	65.9
Land charge (residual return)	4	9.8

replacement cost that does not separate depreciation from interest on investment. Two states use all custom rates and do not include equipment ownership costs.

Both depreciation and interest can be calculated in a number of different ways. The vast majority of states (90%) calculate depreciation using the straight-line method. However, salvage values range from zero to 30 percent of new costs. The years of life also vary. Three states use the ASAE remaining value equations, two use the ACRS method from the Internal Revenue Service, and one state uses 8 percent of the value of the machinery.

The equipment value used to calculate depreciation, interest on investment and replacement costs also varies across states. The equipment value used most commonly (40 percent, 16 states) is new prices (Table 10.9). A mix of new and used prices is used by one-quarter of the respondents (10 states) while a percentage of new prices is used by another quarter (10 states). Three states use the average value over the life of the equipment. Two states use custom rates and, therefore, do not include a value for equipment.

TABLE 10.9 Equipment Values Used to Estimate Ownership Costs

Equipment Value	Number of States	Percent of States
New prices	16	39.0
Mix of new and used prices	10	24.4
Percentage of new prices	10	24.4
Average value	3	7.3
Custom rates used	2	4.9
Total	41	100.0

Most states (83%) rely on equipment suppliers for machinery values (Table 10.6). Ten states ask farmers for equipment values and 11 states also query Extension agents and/or specialists.

The interest rate used to calculate the cost of equipment capital comes predominantly (63%) from the Federal Land Bank or other local bank (Table 10.7). This should be interpreted as the cost of borrowing capital, implying interest only loans and no equity financing. Eight states use some other nominal interest rate, which can be interpreted as full equity financing with the opportunity cost being the next best investment including investments outside of agriculture. Four states use a real interest rate in contrast to only one state using a real interest rate for return to operating capital.

Land. A charge for land may or may not be included. One argument is that land is a residual claimant of return in agriculture. This follows if agricultural land is viewed as having no opportunity cost. In other words, land has no alternative use in the very short run. Another approach is to use the going rental rate as the opportunity cost and include that as the land charge. The logic underlying this approach is that land can be rented for agricultural use and is the only alternative use in the short run. A third approach is to calculate the cost of owning land as the sum of the interest payment on the real estate loan and the opportunity cost of the equity tied up in the land. This is approximated by multiplying the market value of the land by a selected interest rate.

Two-thirds of the states (27 states) include an explicit charge for land in their cost estimates (Table 10.8). Of these, 16 states charge land at the going rental rate, 4 as the cost of owning land and 7 as a combination of owning and renting land (Table 10.10). Another four states do not include an explicit cost but include land in the residual return above all costs.

TABLE 10.10 Land Charge Estimation Methods

Method	Number of States	Percent of States
Cost of renting land	16	39.0
Cost of owning land	4	9.8
Combined cost of rented and		
owned land	7	17.1
Residual return	4	9.8
Land not included	10	24.4
Total	41	100.0

Buildings and Irrigation Systems. Although 22 states include buildings in their cost estimates, these buildings generally include only those that can be charged entirely to the enterprise (Table 10.8). These are specialized buildings such as those used for drying tobacco. Buildings for housing equipment and/or shop buildings are almost never included in ownership costs. Many states include a housing cost as part of the hourly cost for equipment operation.

Over three-fourths of the states (78%) include ownership costs for irrigation systems when appropriate. Only 7 states never include irrigation systems.

Report Formats and Supplemental Tables

The formats of the published enterprise budgets vary in level of detail presented, explanation of assumptions and major section headings. The budgets derived using a budget generator usually include a list of cultural operations used to produce the commodity being studied. The costs for each operation are usually broken out into machine costs, labor and material. The performance rate for each operation may also be given. Alternatively, a summary by category of expenses may be provided. In this case, quantities of inputs may or may not be specified.

In comparing budgets produced by different sources, it is important not to confuse differences in format with differences in estimation procedures. On the other hand, it is equally important not to assume that two budgets using similar formats were constructed using the same estimation procedures. For example, what one budget calls variable costs may not include equipment overhead (taxes, insurance and housing) while another budget might include these costs as variable costs. The treatment of interest on operating capital and investment capital is another example of cost items that should be analyzed carefully when comparing budgets produced by different sources.

Many respondents commented that they produce several budgets for the same commodity to reflect differences in production practices, costs and yields found in different parts of the state, as well as different soil types and different farm sizes. They indicated it is not meaningful to develop a single representative budget for a commodity grown in the state.

Almost all of the budget reports include some sort of supplemental table or information. Most of the reports (85%) include a general introduction and explanation of assumptions (Table 10.11). Three-quarters of the budgets include a list of the machinery required for the enterprise and the associated hourly costs.

Over half of the budgets include some type of sensitivity analysis. Twenty eight states (68%) present a breakeven analysis and eight states (20%) calculate returns at alternative prices. These are the only two ways that any of the budgets analyze risk factors related to production. Finally, 18 states (44%) generate a monthly cash flow or seasonal distribution of costs and returns.

Use of USDA Cost of Production Estimates

Cost of production estimates are published annually by the Economic Research Service of USDA on a state-by-state basis for selected commodities. As with the university-generated reports, a number of sources are used to construct budgets.

TABLE 10.11 Supplemental Tables Included

Supplemental Table	Number of States	Percent of States
General introduction, explanation of assumptions	35	85.4
Machinery list and costs	30	73.2
Breakeven analysis	28	68.3
Seasonal distribution of inputs and/or costs	18	43.9
Returns at alternative prices	8	19.5

TABLE 10.12 Use of ERS Cost of Production Estimates

Use	Number of States	Percent of States
Don't use	20	50.0
Research	12	30.0
Extract information for use in state budgets	11	27.5
Distribute to clientele	3	7.5
Policy analysis	3	7.5
Compare to state budgets	2	5.0
Look at trends	1	2.5
Speech material	1	2.5

Half of the survey respondents reported that they do not use ERS estimates in any way (Table 10.12). The remaining respondents indicated that the ERS estimates are most often used as a source of data for university generated budgets and in research. Only 3 states distribute ERS estimates directly to clientele. Other uses include policy analysis, trend analysis, comparison to university budgets and speech material.

End Users

Respondents were asked to rank the end users of the enterprise budgets according to the number of budgets distributed. The greatest number of budgets are distributed to farmers. Ranked next are agricultural lenders, followed by researchers and potential farmers. Other users include private consultants, real estate appraisers, policy makers, lawyers, and tax assessors in that order. Out-of-state requests far outweigh international requests.

Summary and Recommendations

Enterprise budgets are published by all but 3 state universities and made available to the general public. Researchers were surveyed to determine the differences in methods used to develop budgets. Usable responses were received from 41 states. The survey questions pertained to crop budgets and did not include questions about livestock budgets.

Respondents were asked about the cost factors included in their budgets, the computational procedures used to interpret data and data sources. The questions were asked for each of the three major sections of an enterprise budget: receipts, operating costs and investments costs.

Six states do not include a receipts section in their budgets. Government payments are included as gross income by 16 states (39%). In-kind payments are never part of income. In all cases receipts are the multiple of price times yield.

Enterprise budgets vary significantly from state to state with respect to the operating costs included in the budgets. All of the budgets include fuel, repairs, interest on operating capital and labor. However, only 15 states include a management charge, 22 states include business overhead costs, and 31 states include property taxes on equipment.

The investment costs included in the budgets also varies. All states include a charge for equipment in some way. Only 22 states include an explicit cost for buildings. Usually these buildings only include specialized facilities directly pertaining to the enterprise such as drying facilities. Housing for equipment and shop space are usually not included unless they are part of a general equipment overhead expense.

Fourteen states do not calculate a land charge. Of the states that do include an explicit land charge, the majority use the cost of renting land. Others use the cost of owning land or a combined cost of owning and renting land.

Although all states include a charge for all labor, about two-thirds of the states include fringe benefits in the labor rate and one-third do not. The hours of labor are tied to the hours of machinery operation. The hours for labor are increased from between zero and 21 percent over machinery hours to account for down time, fill up and implement adjustments.

Two states use all custom rates and there is no owned equipment. Therefore there are no operating or ownership costs for equipment. All but two of the remaining states estimate operating costs for equipment based on agricultural engineering equations.

Although all but two states include an ownership cost for equipment in some manner, the method of calculation differs. Ninety percent of the states value the ownership cost of equipment as the sum of depreciation and interest. Two states use a replacement value. Two states use all custom charges and do not include ownership costs. The equipment value used to calculate ownership costs can be new prices (16 states), a mix of new and used prices (10 states), a percentage of new prices (10 states) or the average value over the life of the equipment (3 states).

Every state except one uses a nominal interest rate to estimate the cost of operating capital. Only one state uses a real interest rate. The interest rate used to calculate the cost of owned inputs is a nominal rate for all but four states that use a real interest rate.

Estimating machinery ownership costs from a mix of new and used prices and a nominal interest rate is the closest to an accounting procedure used for an income statement or tax purposes. However, the most common approach to estimating ownership costs is new prices and a nominal interest rate. This results in a larger value for depreciation and interest than a mix of new and used prices or a percentage of new prices.

New prices more closely reflect the current cost of providing the service of equipment assets to the enterprise. Taxes, insurance, housing and repairs will be higher when new prices are used to calculate ownership costs. Fuel, lube and labor costs are not changed by the equipment value.

The nominal interest rate includes inflation and takes into account increasing asset prices. Using new prices and a nominal interest rate may be double-counting asset appreciation. Therefore, it is recommended that either the nominal interest rate be used with an estimate of historic list prices (a mix of new and used equipment, a percentage of new prices or the average value of the life of the equipment) or the real interest rate be used with new prices to estimate the ownership costs of equipment.

Many states start out the budget process by specifying equipment complements for specific crop rotations and farm sizes. Others create a large data file of equipment and select equipment from this list to use in the enterprise budget. In either case the annual hours of use for each piece of equipment must be specified in order to apply the engineering equations and to allocate ownership expenses to the enterprise. The annual hours of use for tractors ranged from 500 hours to 1200 hours.

Most states use more than one source of information for yield and price data. Extension agents and/or specialists are the most important source of yield information (35 states, 90% of respondents). About two thirds of the states rely on farmers for yield information. Extension agents and specialists are also the primary source for price data, although to a lesser extent than for yield data (28 states, 72% of respondents). The National Agricultural Statistical Service and farmers are each sources of price information for about 30% of the states.

The major source of information concerning the application rates used for inputs such as fertilizer, seed and pesticides is Extension personnel and farmers while many states rely on both sources. Price information is primarily from local suppliers.

Half of the respondents indicated that they do not use ERS cost of production studies in their work. The primary uses are in research and as sources of information for state enterprise budgets.

It is clear from these results that there is no standard set of methods used to develop enterprise budgets. The budget generator approach is used by almost all states but the methods differ with respect to the underlying mathematical model, the costs included and the data sources. Also, the terminology used varies from state to state. There appears to be a problem with jargon.

The cost factors included in the enterprise budgets vary primarily with respect to overhead and ownership costs. This difference can be explained by assuming that some budgets are short-run and only include costs that are variable within one growing season while others are long-run and include all economic costs. However, there seems to be a mixing of short-run and long-run considerations in the budgets that should be corrected.

The inclusion of government payments and nominal interest rates is introducing macro effects into the budgets. This may be the intent but again these factors should be used consistently.

All but four states use agricultural engineering equations in some way. There is a need for agricultural economists to work more closely with agricultural engineers to develop coefficients that are appropriate with local production conditions. This effort is already underway in the North Central region.

There needs to be a step added to the estimation of equipment costs that assures that the appropriate number of pieces of equipment are included in the equipment complement to accomplish the given operations in the time frame required. For example, the annual hours of use for the tractors in the complement may not be exceeded but there still may not be enough tractors to complete cultural practices during the busiest part of the growing season. While most farmers are probably over equipped, most budget generator complements are probably under equipped.

Although it is undoubtedly impossible to get a consensus on the exact methods that are used, it is possible to improve the explanations and assumptions that accompany the budgets in printed form. In fact there are 6 states that do not currently include any assumptions or explanations with their studies. In particular, these assumptions should list the inputs that are not included in the budgets, calculations used for ownership costs, and sources of prices.

Finally an inclusion of some sensitivity analysis with respect to variation in yields and prices received should be included with the enterprise budgets. In this way the reader can be introduced to the concept of risk and risk management.

Notes

Financial support for the survey described in this chapter was provided by the Federal Extension Service, U.S. Department of Agriculture.

References

Boehlje, M. and V. Eidman. 1984. *Farm Management.* New York: John Wiley and Sons.

Eck, D. 1990. "A National Survey of Enterprise Budget Development and Use by The Extension Service." Unpublished Masters Thesis, Utah State University.

Harrington, D. 1983. "Costs and Returns: Economic and Accounting Concepts." *Agricultural Economics Research* 35, No.4: 1-8.

Hoffman, G. and C. Gustafson. 1983. "A New Approach to Estimating Agricultural Costs of Production." *Agricultural Economics Research* 35, No.4: 9-14.

Klonsky, Karen. 1989. "Enterprise Accounting" in D. Smith, ed. *Farm Management: How to Achieve Your Farm Business Goals.* U.S. Government Printing Office.

Libbin, J. and A. Torell. 1990. "A Comparison of State and USDA Cost and Return Estimates." *Western Journal of Agricultural Economics* 15: 300-310.

McElroy, Robert. 1987. "Major Statistical Series of the U.S. Department of Agriculture: Costs of Production, Volume 12." U.S. Department of Agriculture, Economic Research Service. Agricultural Handbook No. 671. Washington, D.C.

Walker, O. and R. Krenz. 1980. "Conceptual and Empirical Problems in Cost and Return Estimation." Paper presented at GPC-10 Workshop on Farm and Ranch Production Costs and Returns. May 21-22. Lincoln, Nebraska.

Comparability of Predominant Methods: Discussion

DeeVon Bailey

I am pleased to offer my comments regarding these three informative chapters. They offer a sharp contrast between the goals of the USDA, ERS and land grant institutions relative to the information and use of cost of production (COP). It also reveals that substantial diversity exists between methods used by the land grant institutions.

Obviously, COP information is being generated for different clientele groups at the federal and state level. For example, ERS COP information is highly aggregated and also very uniform to facilitate comparisons between states, regions, and commodities. Conversely, COP information generated by the land grant institutions is directed toward farmers and extension personnel who desire to have this type of information on a disaggregated basis such as a sub-state region or a county.

Since the purposes for generating COP information differs between federal and state agencies one would expect that differences in methods and, consequently, results will occur since sampling methods and calculation techniques vary. The recent discussions in the literature surrounding the differences in COP estimates by USDA, ERS and land grant institutions suggests that some reconciliation between methods is appropriate (Libbin and Torell and Rister *et al.*).

One can not argue with the sampling techniques used by USDA, ERS to collect COP information from farmers. The information is obtained from a stratified probability sample that would be difficult for any university to replicate because of resource constraints. Consequently, many universities rely on much smaller and, admittedly, nonprobability samples to obtain COP information. This suggests that USDA, ERS data should be unbiased and should be good point estimates for costs and returns for particular locations.

The methods and procedures used by USDA, ERS are well developed in the chapter by Morehart *et al.* The questions that can be raised, I believe, relate to the broad assumptions USDA, ERS must make to accomplish their task of developing state and regional budgets. While some of these assumptions are necessary and probably can not be improved on at a reasonable cost, others appear to ignore some local and regional peculiarities that may influence production costs but are not accounted for in ERS's COP estimates. The fact that over 50% of the land grant institutions do not use ERS budgets at all raises legitimate concerns about the validity of the USDA budgets as they relate to information for public policy decisions.

One source of differences in the COP estimates produced by the land grant institutions and ERS is the method of periodic updating used by ERS. Klonsky reports that 61% of the states update input costs and coefficients for their COP information annually. ERS updates their technical coefficients approximately every four years but then uses indices to update input costs and USDA, NASS information to update yields annually. While this should not create a problem relative to gross returns, it may underestimate costs associated with obtaining higher yields through investment in new technologies. The information in Guedry's chapter points out that ERS calculation methods are meant to facilitate this type of updating.

Klonsky points out that most land grant institutions, as well as ERS, use engineering equations to estimate per-acre costs of production for equipment use. This method does help to remove subjectivity from the calculations. However, costs can vary substantially depending on what assumptions are used in the ASAE and other engineering equations. For example, if one looks at the range of possible costs for plowing in Utah based on the level of field efficiency of the equipment it can be seen that costs per acre could vary from about \$31.00/acre, if field efficiency is assumed to be 0.5, to about \$17.00/acre, if field efficiency is assumed to be 1.0.

This implies that costs will be affected by soil type, size of field, etc. As a result, the assumptions regarding average field efficiency and other engineering parameters need to be clearly identified to facilitate comparisons between USDA, ERS and land grant institution estimates. In addition, this method calculates some minimum level of input use given a set of assumptions about the size and relative efficiency of machinery. It can ignore that producers may have capital assets beyond a minimum machinery complement that are held in reserve for peak planning and harvesting periods. I agree with Klonsky that: "There is a need for agricultural economists to work more closely with agricultural engineers to develop coefficients that are appropriate with local production conditions."

Labor costs need to be tied to the level of skill required to perform certain operations. Labor costs also should include housing and other considerations for some commodities (especially dairy operations).

One problem cited in the literature is the length of the COP questionnaire administered by USDA, ERS and NASS. While most of the surveys may be administered during the winter months, clearly a 30-40 page instrument must surely tax the patience of even the most forebearing farmer. As a result, the accuracy of the estimates may be influenced. In addition, farmers may tend to underestimate production costs for financial creditors. This suggests that the Farm Cost and Returns Survey (FCRS) would be expected to have costs somewhat lower than those obtained by other methods. The method used by some land grant institutions of interviewing farmer panels with face-to-face interaction is one way to arrive at a consensus about what costs really are. However, this method may not be desirable or practicable for ERS.

The procedure of calculating individual farm budgets (Farm Level Budget Models (FLBM)) is an excellent move on the part of ERS. However, this information should be verified by a sample of survey participants to validate the calculations. This would help to minimize errors associated with administering the survey instrument and would also increase producer confidence about the accuracy of the COP information.

While concerns can be raised about USDA's methods, the magnitude of the undertaking must allow for some patience on the part of the land grant institutions as some reconciliation occurs. Over time, ERS has clearly made significant progress in aligning their COP estimates with economic theory and empirical considerations. The fact that different clientele are served by the ERS and the land grant institutions is obvious as one considers Klonsky's numbers showing that less than half use the ERS budgets in the states. Aside from this and other concerns, I believe that more attention needs to be paid to actual costs versus the engineering relationships used to calculate costs. Also, some validation of the farm level results obtained in the FLBM is essential to increasing the confidence of the public in the ERS numbers.

References

Libbin, J.D. and L.A. Torell. 1990. "A Comparison of State and USDA Cost and Return Estimates." *Western Journal of Agricultural Economics* 15: 300-310.

Rister, M.E., E.G. Smith, J.W. Richardson, R.D. Knutson, W.R. Grant and T.N. Thompson. 1989. "Texas Rice Production Costs." AFPC Policy Working Paper 89-8. Agricultural and Food Policy Center, Department of Agricultural Economics, Texas A&M University, College Station, Texas.

Comparability of Predominant Methods: Discussion

John E. Ikerd

Budgets of cost and return estimates are developed for a variety of different purposes. Consequently, no single budget format or cost and return estimation procedure can be expected to be adequate, or even appropriate, for all purposes. Morehart, Johnson, and Shapouri state that cost information is of "general interest to farmers and farm-management specialists on the one hand, and to policy makers on the other. These two groups use information on costs for different purposes." Guedry identifies researchers as a third group who use cost of production estimates for yet another purpose.

Klonksy found significant differences among budgets prepared by extension specialists in different states even though the primary purpose of extension budget development is to support farm-level decision making. "Conceptual and empirical problems that arise in estimating costs of production make it difficult to assign costs to certain production inputs. Differences in estimation methods are the result of alternative solutions to these problems."

Each of these three chapters outlines, in varying detail, the conceptual approaches and empirical methods used by ERS and the land grant universities in addressing the issues that arise from differences in purposes and problems associated with cost and returns budgets. It is interesting to note from Klonsky's survey that less than half of the university economists make use of ERS cost of production estimates. Likewise, the primary source of ERS cost and return data is their own survey of farmers, the Farm Cost and Returns Survey. There is no indication that ERS makes significant use of cost and return information generated by university economists. The rice budgeting case study by Guedry suggests there may be significant differences between ERS- and university-developed cost of production estimates even for the same cost components.

Taxpayers might legitimately ask whether we should be basing government programs on cost and returns estimates that are different from the estimates being provided to the farmers who must respond to those policies. Our continuing failure to resolve these differences raises legitimate questions regarding duplication of efforts, at best, and the potential for counterproductive conflicts, at worst. Cost and return data needs of all government agencies are not the same. At the very least, budget data must be presented in different formats to achieve different purposes. However, the basic data collected by one government agency should be freely available and fully utilized by other agencies. The Morehart, *et al.* chapter gives a historical perspective on how ERS has responded to changing policy mandates with changes in cost and return estimation procedures and changing budget formats. These same types of changes have occurred at the farm decision-making level, although they are less apparent because they are reflected by the needs of farmers rather than by law.

For example, during the 1970s, an accounting for appreciation in value of farm assets and equity was critically important in farm planning. Asset appreciation was the primary source of increases in farm net worth during this period of double-digit inflation rates. In many cases, farms with negative cash flows increased in net worth. Thus, cash budgets, alone, were of little value. During the 1980s, however, budgets of cash flow and farm debt-to-asset ratios became critical factors in farm planning. In many cases, farms with positive net returns were unable to generate the cash flow needed to service previously incurred debts. Many university economists changed data collection and budget formats between the 1970s and 1980s to reflect the changing needs of farm decision-makers.

U.S. agriculture is now faced with another set of issues that will require development of yet another set of farm budgeting concepts, data, and formats. The fundamental issue confronting policy-makers and farmers in the 1990s is that of the sustainability of U.S. agriculture. Farm policies related to this issue fall primarily under the headings of resource conservation and water quality. At the farm level, however, the objectives of productivity and profitability must be balanced against those of conservation and environmental protection.

Until now, the conservation and environmental aspects of sustainable agriculture have dominated the issue. The economic aspects of sustainable farming systems, for the most part, have been given only superficial treatment, primarily by non-economists. Agricultural economics, as a profession, has not yet made the professional commitment that ultimately will be necessary to adequately address the economics of a sustainable agriculture.

Our current budgeting concepts and formats do not adequately address the economic issues associated with sustainable agriculture in three important areas:

- costs of external inputs versus internal resources
- evaluation of production and financial risks
- costs, returns, and risks for systems versus enterprises

Agricultural economists should be preparing for significant changes in cost and return budgeting as we enter the 1990s.

The question of external versus internal costs relates to the basic definition of low-input farming systems. Lower input farming systems may be defined as systems that rely less on external purchased inputs, such as commercial pesticides and fertilizers, and, thus, must rely more on management of internal resources, such as operator, or family labor, and owned land (Rodale). The motive typically stated for moving to lower input systems is to reduce the use of agricultural chemicals associated with environmental risks. However, many farmers interested in low input systems may be more concerned with reducing the use of hired labor and borrowed capital, both of which are external inputs, than with reducing environmental risks. These external inputs are associated with economic rather than environmental risks.

Traditionally, economists have not distinguished between input and resource costs. All factors of production are treated as inputs. However, the economic risks associated with external inputs may be quite different from the risks associated with internal resources. For example, assuming other things equal, an enterprise using a lower proportion of external inputs, relative to internal resources, will face lower short run financial risks. In simple terms, the lower input system can absorb more of a revenue short-fall in any given year and still meet its cash commitments.

The input-resource risk concept is quite similar to that of debt-equity risk implications in financial risk. Other things equal, a farm with a low debt-to-equity ratio is less vulnerable to economic adversity than is a more highly leveraged firm. A input-to-resource ratio is a more general measure in that it is not limited to capital but includes all factors of production. The budgeting process for external inputs and internal resources may be similar, if not identical, to budgeting cash and no-cash costs. The challenge is to relate the old concept of cash budgeting to the issue of sustainability.

The second critical factor in evaluating the sustainability of a farming operation is a probabilistic assessment of risks. The production or yield risks associated with use of external inputs and internal resources, for example, may be quite different. Lower input farming methods are not necessarily more sustainable because sustainable systems must be both economically viable and ecologically sound. Less reliance on commercial pesticides, for example, could increase yield variability. An increase in yield variance could result in a greater probability or risk of failing to cover external or cash costs, even if external costs were lower for the low input system. There is no way of determining whether one system is more or less economically viable than another without a probabilistic assessment of the relevant economic risks. However, none of the states in Klonsky's survey indicated that they made probabilistic risk assessments in their budgeting processes.

Several procedures are available for calculating production, market, and financial risks from basic estimates of yield and price predictability (Anderson and Ikerd). Apparently, few economists have considered the probabilistic assessment of risks to be worth the added complexity it implies for budget development and the associated instruction regarding the use of budgets by farmers.

The issue of sustainability should justify a reexamination of the worth of risk rated budgeting. Finally, sustainability cannot be evaluated through analysis of individual enterprises. The implicit strategy of enterprise analysis is specialization as a means of achieving economies of scale. Analysis, by its very nature, implies the separation of wholes into their various component parts. However, many of the problems encountered in allocating machinery and overhead costs in the farm budgeting process, for example, result from trying to divide farming operations that are essentially wholes into collections of components.

A fundamental philosophy of sustainable agriculture is that farmers must find ways to work with nature rather than to rely solely on new technologies designed to remove nature's constraints. To achieve this goal, they must make more efficient use of natural production processes through a holistic approach to farm resource management. This strategy, conceivably, could reduce environmental and resource risks without increasing economic risks.

The challenge of sustainability is to develop integrated, diversified farming systems that can compete economically with conventional, specialized systems of farming. If this challenge is to be met, farmers will have to find synergistic gains from integration of the diversified farming systems that will more than offset any future gains from further specialization. Without such gains, diversified systems will not be economically viable even if they are more ecologically sound.

Some of these gains may come from diversification as a means of risk management. The variance of whole-farm net returns will be less than the sum of the variance of net returns of the individual enterprises, assuming that net returns from the individual enterprises are not perfectly, positively correlated. The larger the number of enterprises and the lower the positive or higher the negative correlations among enterprises, the lower the resulting whole-farm variance relative to the sum of the individual enterprise variances. The reduction in variance, and reduction of risks for a given whole-farm net return, is a characteristic of the whole system and is not an inherent characteristic of its components or enterprises.

Sustainability is a systems concept. Consequently, budgets that support a systems approach to farm management should be budgets of systems, or at least subsystems, rather than budgets of individual enterprises. For example, corn in a corn-soybean rotation is different from corn in a continuous corn operation, which is different, still, from corn in a corn-soybean-wheat rotation. At a minimum; pesticides, fertilizer, soil erosion, expected yields, and yield variability may be different for corn in each rotation. A single corn budget cannot accurately reflect the economic or ecologic contribution of corn to different rotations. However, Klonsky reports that current economic budgets are designed almost exclusively for enterprise analysis rather than whole-farm synthesis.

In addition, costs of internal resources will be different for each farming system. The typical practice of calculating opportunity costs of a owned land, operator or family labor, and management ability at market values, for example, is not adequate. The real opportunity cost of cultivation labor may be practically zero if cultivation comes at time when there is no other productive use of the farmer's time. Off-farm, two-or-three day employment opportunities for farmers are rare. On the other hand, cultivation labor may be extremely expensive if cultivation comes at a time when everything else on the farm also needs to be done. Any simple process for assigning costs to internal resources is likely to be more misleading than helpful in developing individual whole-farm systems.

Agricultural economists have responded to changes in past policy and farm decision-making demand for cost and returns information by changing data collection and budgeting procedures. There is no data collection or budget format that is best for all purposes over all times. The issue of agricultural sustainability presents a new challenge to those of us who are responsible for providing information for farm policy-makers and decision-makers.

There is no reason to expect that we will not respond to this new challenge, as well. Hopefully, we can respond to this issue by working together, so that the basic budget information used in research and policy development will be consistent with the information used to support decisions made by the farmers who must implement new technologies and respond to new policies.

References

Anderson, Kim B. and John E. Ikerd. 1985. "Whole Farm Risk-Rating Micro-computer Model." *Southern Journal of Agricultural Economics* 17, No.1: 183-188.

Rodale, Robert. 1988. "Agricultural Systems: The Importance of Sustainability." Paper presented at the National Forum, Honorary Society of Phi Kappa Phi, Louisiana State University, Baton Rouge, LA.

PART FOUR

Treatment of the Effects of Government Programs

11

Recognizing the Effects of Government Programs in Developing Cost and Returns Statements

B. F. Stanton

Introduction

Government programs have important, often unrecognized effects on both costs and returns for many commodities. These effects range from obvious ones like deficiency payments for wheat and corn, to more subtle ones such as implicit subsidies for irrigation water or agricultural assessments that reduce real estate taxes. It is clearly difficult to disentangle and correctly identify these effects, but as analysts we should surely make some effort to recognize them. These efforts should be quantitative whenever possible, but at least qualitative for the important ones which seem to defy consistent procedures for measurement. This is particularly important in making comparisons across production regions where different cost structures may apply.

At the outset, it is also important to salute the consistent, careful work of economists at ERS in their annual efforts since 1973 to meet the mandates of Congress in preparing annual estimates of production costs for key commodities by region across the United States. These have provided a national frame of reference for nearly all discussions of production costs for individual commodities. Methods used have been clearly identified; we have all benefitted from ERS' efforts on these basic annual series.

Effects of Government Programs on Returns

Government commodity programs clearly have important effects on farmers' planting and harvest decisions. These programs influence both production and prices, and hence both gross and net farm incomes. The aggregate level of direct government payments to farmers over the last two decades is indicated in Table 11.1. Payments are a relatively small proportion of total cash farm receipts in the U.S., in most years less than 10 percent. But for major program commodities they are much more significant. The cover illustration of the December 1990 *Agricultural Income*

TABLE 11.1 Direct Government Payments to Farmers, Major Programs, United States, 1970-89

Year	Feed Grains	Wheat	All Cotton	Other	Total
		- Million dollars -			
1970	1,504	871	919	423	3,717
1971	1,054	878	822	391	3,145
1972	1,845	856	813	448	3,962
1973	1,142	474	718	273	2,607
1974	101	70	42	317	530
1975	279	77	138	313	807
1976	196	135	108	295	734
1977	187	887	130	614	1,818
1978	1,172	963	127	768	3,030
1979	494	114	185	583	1,376
1980	382	211	172	520	1,285
1981	243	625	222	843	1,933
1982	713	652	800	1,327	3,492
1983	1,346	864	662	6,424[a]	9,296
1984	367	1,795	275	5,994[a]	8,431
1985	2,861	1,950	1,106	1,788	7,705
1986	5,158	3,500	1,042	2,114	11,814
1987	8,490	2,931	1,204	4,122	16,747
1988	7,219	1,842	924	4,495	14,480
1989	3,140	603	1,184	5,947	10,874

[a] PIK payments not distributed to individual crops.
Source: USDA, 1989b.

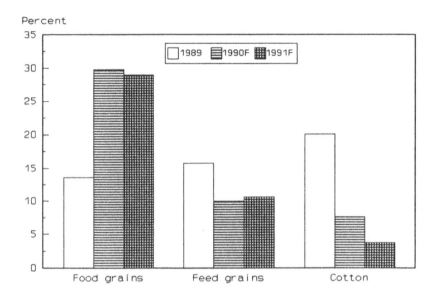

Figure 11.1 Direct payments as a percent of sales plus payments. *Source:* USDA, 1990.

and Finance Situation and Outlook Report, Figure 11.1, provides current perspective for food grains, feed grains, and cotton where as much as 30% of the total come from government payments.

In a state such as North Dakota, where government programs are particularly important, the impact of direct payments is more readily evident. The 1987 Census of Agriculture collected data on government payments and market sales and summarized the effects of including payments with sales on size distributions in Volume 2, Part 5 (USDC). The effect of including government payments along with market sales on changes in the size distributions is presented in Tables 11.2 and 11.3. On farms where deficiency payments were received for most of the crops produced, the effects on gross income were important. Often they turned what would have been negative net incomes into positive ones.

Consistent Treatment in Cost and Returns Statements

Cost and returns statements for wheat and peanuts are both included in the annual publication, *Economic Indicators of the Farm Sector: Costs of Production* (USDA). In the case of wheat, the gross value of production is

TABLE 11.2 Impact of Adding Government Payments to Market Sales on Size
Distribution of Farms, North Dakota, 1987

Size Class	Market Sales Only	Sales Plus Government Payment	Difference
		- Farm numbers -	
$1,000,000 or more	46	53	+ 7
500,000 - 999,999	200	264	+ 64
250,000 - 499,999	846	1,270	+ 424
100,000 - 249,999	4,855	6,646	+ 1,791
50,000 - 99,999	7,808	8,414	+ 606
25,000 - 49,999	7,725	7,120	- 605
10,000 - 24,999	6,817	5,810	- 1,007
5,000 - 9,999	2,982	2,444	- 538
2,500 - 4,999	1,750	1,376	- 374
Less than 2,500	2,260	1,892	- 368
Total farms	35,289	35,289	

Source: U.S. Dept. of Commerce, 1990.

estimated based on harvest month market prices; the value of secondary products, such as straw, is included as well. Since 1986, the annual publication has explicitly noted that direct government payments are excluded from the estimate of gross value of production.

In the case of peanuts, procedures followed are the same as those for all of the other grains and oilseeds. Yet, there is one strikingly different result in the final economic calculation, "residual returns to management and risk" (Table 11.4). There is a large positive value per acre for peanuts, unlike all those calculated for the other grain crops. This is because the effects of government programs through the imposition of marketing quotas are included in the case of peanuts, where a two price system for *quota* peanuts and *additional* peanuts is in place. The price of quota peanuts is about four times the support rate for additional peanuts. This government program has worked well in the interest of both producers and processors. Interestingly, the annual rental value of quota is reported by Schaub to be about $150 per ton, about the same as the average of the residual returns to risk and management per acre reported in Table 11.4.

The reason for dwelling on this comparison, which could be explained to the general public by most agricultural economists with a little effort, is that

TABLE 11.3 Impact of Adding Government Payments to Market Sales on Total Values by Size Class, North Dakota, 1987

Size Class	Market Sales Only	Sales Plus Government Payments	Difference
	- Million of dollars -		
$1,000,000 or more	79.2	89.8	+ 10.6
500,000 - 999,999	134.0	177.1	+ 43.1
250,000 - 499,999	279.4	420.2	+140.8
100,000 - 249,999	714.6	993.1	+278.5
50,000 - 99,999	554.1	605.4	+ 51.3
25,000 - 49,999	280.8	262.6	-18.2
10,000 - 24,999	115.4	99.4	-16.0
5,000 - 9,999	21.8	18.1	- 3.7
2,500 - 4,999	6.5	5.1	- 1.4
Less than 2,500	2.3	1.9	- 0.4
Total value	2,188.2	2,672.8	+484.6

Source: U.S. Dept. of Commerce, 1989 and 1990.

TABLE 11.4 Residual Returns to Management and Risk on ERS Cost of Production Statements, 1985-87

Crop	Residual Returns to Management and Risk		
	1985	1986	1987
	- Dollars per planted acre -		
Corn	- 24	- 73	- 51
Grain sorghum	- 23	- 42	- 39
Barley	- 42	- 47	- 40
Oats	- 33	- 50	- 30
Wheat	- 29	- 44	- 37
Rice	- 42	- 189	- 149
Soybeans	- 4	- 8	+ 4
Sunflower	- 10	- 16	- 17
Peanuts	+ 148	+ 177	+ 124

Source: USDA, 1989a.

the effects of government programs on returns is included in one estimate of gross value of production (peanuts) but not the other (wheat). If one turns a few more pages in ECIFS 7-3 to sugar beet production costs for the same years, one finds residual returns to risk and management per acre of $100 (1985), $157 (1986), and $236 (1987). It is no wonder that stories appeared in the *Wall Street Journal*, June 26, 1990, with the headline, "Range War: Small Minnesota Town Is Divided by Rancor Over Sugar Policies". While the programs for sugar beets, which compete with corn and soybeans in some locations, are different from peanuts, gross returns per acre are even more strongly influenced by government programs although direct government payments are not involved.

The single, strongest recommendation in this chapter is that ERS include direct government payments and other quantifiable additions to income resulting from government programs when estimating gross value of production in their annual cost of production reports. Two sets of estimates were developed for 1988 rice production costs and returns excluding and including the effects of government programs (Salassi, *et al.*). Variable cash expenses were the same in both estimates. Gross value of production per acre was $370 in one case and $596 in the second. Fixed costs differed in the two sets of estimates, primarily reflecting differences in the way land was valued. Residual returns to risk and management increased from $-105 to $+49 per acre.

Included in gross value of production when government programs were included were (1) market value, (2) deficiency payments, (3) marketing loan proceeds, and (4) haying/grazing on ACR acreage. The third item is one of the "additional" benefits which may occur when marketing loans are part of government programs. All such quantifiable benefits resulting from participation in a commodity program should be identified insofar as possible as part of the returns statement. Thus, if corn producers consistently benefitted from "PIK and roll" such an estimate of the addition to gross value would be warranted. At the same time if land is forced to be idled and no crop can be harvested because of ARP, the costs associated should be recognized in a separate line in the summary statement to recognize this effect of government programs as well.

Other Effects on Returns by Government Programs

Government programs have a direct effect on returns through control over supply and hence in market prices (peanuts, sugar beets, sugar cane) or through direct payments (food and feed grains). In the case of marketing loans there is also the difference between the loan rate and the world market price which can be received which adds to revenue in years when this occurs. With appropriate "...adjustments made for program yields, the basis on which producers are actually paid, program participation rates, and payments under the 50/92 program..." (Salassi, *et al.*) most of the quantifiable effects of government programs are considered.

Other effects also exist which are much less easily identified in numerical form. Among those commonly cited are the depressing effect on market prices of Commodity Credit Corporation stocks and those in the Farmer Owned Reserve when the market views them as burdensome (Knutson, Gardner). Conversely, quotas and controls on imports for commodities like beef and dairy products can have a positive effect on prices even when the quantity of imports is small relative to domestic production. The major efforts devoted to estimating producer subsidy equivalents as part of the data assembled for the GATT Negotiations, have provided evidence for both exporting and importing countries of the numerous ways in which governments influence both costs and returns in each domestic market (Mabbs-Zeno). In an international framework quantitative estimates of these separate effects have been made using a consistent methodology.

In my view it would be useful in national estimates of costs and returns for individual commodities to recognize the more important of these effects. These can be identified in a footnote to the existing tables even if a quantitative estimate of their implicit effect on returns cannot be calculated systematically.

Estimates of Fixed and Variable Costs

Before examining the ways in which government programs have direct and indirect effects on costs of production, the following assertions about the treatment of fixed and variable costs provide my perspective: (1) Valid comparisons can be made across regions with respect to variable costs per unit of production.   (2) Most problems in developing "acceptable" cost statements are associated with the assumptions required in estimating and allocating fixed costs.   (3) Recognition of the influence of government programs on capital requirements, land values, and production rights, like marketing quotas, must be addressed directly if a "full" cost of production statement is desired.

Variable Costs

In the view of this analyst, it is variable costs that deserve the most care and attention in collecting information on costs of production. These are the out-of-pocket costs over which short-run production decisions are made. It is possible to get comparable information from producers in different locations and situations for a given technology. It is even possible to make comparisons on variable costs across national boundaries. For many producers, the difference between gross value of production and variable costs, excluding cash rent, is the number that deserves special study. This is what is left over to cover fixed costs, overhead, depreciation, and pay for the use of family labor, capital, land, and management. If consistent estimates of gross margin, or alternatively value of production less cash expenses (interest payments on real estate may confound these results), useful comparisons between years and regions can be made.

Fixed Costs

Despite 100 years of making estimates of production costs in Europe, North America and Asia, agricultural economists have not yet agreed on a standard way to calculate and then allocate fixed costs to individual enterprises. The approaches to this problem from both accounting and economics are commonly mixed together. The opportunity cost principle and market prices get priority in most calculations. Yet, accounting rules and conventions remain important parts of most estimation procedures because better alternatives have not been developed to handle depreciation, charges for management, and interest on family contributions of capital.

Expected future benefits associated with the rights to produce, such as an acreage allotment or a marketing quota, are quickly capitalized into a value either associated with land or the production right itself. Thus, annual rent for one ton of quota peanuts is about $150, roughly 25% of the sale price. Marketing quotas, when sold, can range in value between $1200 and $2000 per ton. In a somewhat more complex, but similar manner, expected future benefits from current government programs involving target prices, deficiency payments, and loan rates are capitalized into land values. Separating the effects of government programs on land values from other effects like distance from markets, productivity of soils, risk from weather related hazards, etc. is difficult or impossible if one seeks agreement on how to allocate the contributions to value. Nevertheless, most will agree that government programs do have an influence on land values even though quantifying that component from others is difficult to achieve.

Special Issues Concerning Costs Resulting
from Government Programs Subsidies

Within a national context, if all competing producers have equal access to the same production subsidy then it can be argued that recognition of that subsidy in cost of production statements is unnecessary except for international comparisons. If, on the other hand, some producers benefit substantially while others do not, there is a greater need to recognize these differences. Federal and state water projects are examples of this kind of issue. As the competing demands for water become increasingly strong in the West this will necessarily be reflected directly in variable costs. Land values and their associated rental rates already reflect the capitalized value of "free" access to irrigation water.

The question posed here is whether water costs, both cash and subsidized, should not be recognized as one of the key items in variable cost statements for crops where irrigation is required. Such action might require a compensating reduction in "net land return" or cash rent in the national accounts prepared by ERS. Such action would recognize more directly the difference in variable costs associated with new and old sources of irrigation water.

Acreage Reduction Program

Some cropland must be idled as a result of participation in commodity programs. Recognition of this cost item should be made explicit for a typical or representative producer in the cost statement. Because the ACR varies by crop and by year the associated "cost", or income foregone from the use of the land resource, can be incorporated as a component of the land charge. In principle, the percentage of set-aside, for example, 15%, could be reflected as 1.15 acres of land charged per acre of crop actually produced. This complication, while an additional burden in a standard cost of production statement, will provide a mechanism to recognize some of the costs as well as the addition to income from deficiency payments.

How far should the effort to account for effects of government programs go? With the triple base option in 1991, deficiency payments are now denied on an additional percentage of the acreage of a program crop. This will be reflected automatically in cash deficiency payments received. Likewise there will be some cash costs of meeting the conservation compliance requirements of the ARP. The rules on payment limitations may also have an effect on some of the largest operations. Some capital costs may be necessary to meet the requirement for having an approved conservation plan in place by 1995. Insofar as it is feasible to determine the actual cash, or their equivalent effects of government programs in a cost and returns statement it will make the final product more realistic. Pointing out these several influences may also be beneficial to both analysts, producers, and policy makers.

TABLE 11.5 Production Costs and Returns: With and Without Effects of Government Programs United States, 1988

Description	Without Effects	Including Effects
	- Dollars per planted acre -	
Gross value of production:		
Market value	$ 370.33	$ 370.33
Deficiency payments	0	204.03
Marketing loan proceeds	0	21.30
Other	0	.05
Total value	370.33	595.71
Cash expenses:		
Variable cash expenses	$ 296.03	297.46
General farm overhead	21.50	25.05
Taxes and insurance	12.13	14.28
Interest on operating loans	14.04	16.33
Interest on real estate	9.42	11.18
Total cash expenses	353.18	364.30
Returns less cash expenses	17.15	231.41
Capital replacement	46.34	47.49
Returns less capital and replacement	-29.19	183.92
Economic costs:		
Variable cash expenses:	296.09	297.46
General farm overhead	21.50	25.05
Taxes and insurance	12.13	14.28
Capital replacement	46.34	47.49
Subtotal	376.06	384.28
Allocated returns to owned inputs:		
Operating capital	6.68	6.74
Other non-land capital	10.86	11.15
Net land rent	57.84	121.15
Unpaid labor	23.70	23.84
Total economic costs	475.14	547.16
Residual returns to management and risk	-104.81	348.55

Source: Salassi, Ahearn, Ali, and Dismukes.

Effects on the Cost and Returns Statement

One important effort to compare cost and returns statements for a crop, rice, where the effects of government programs are important, was published by ERS, USDA in 1990 (Table 11.5). All of the detail is not reproduced but the key differences are suggested. Value of production is increased primarily by deficiency payments and net proceeds from marketing loans. Cash expenses differ only by the additional costs of maintaining conserving acres. The most important difference among all the cost calculations is the calculation for "net land rent" under the heading, allocated returns to owned inputs. The rental rate of $57.84 contrasts with $121.15 per acre when the effects of government programs are included (Salassi, *et al.*).

From my perspective the statement which includes the effects of government programs provides the more realistic assessment of conditions facing producers. One could argue that the controversial land charge should be handled like capital replacement and moved to the first half of the table. Moreover, some report of cash rental rates could be provided directly or in a footnote. The important point is that the statement including the effects of current programs more nearly represents the combination of incentives and disincentives facing producers in the field. A cost statement which consciously excludes deficiency payments and the costs of conservation compliance, leaves out an important part of the cost and returns picture facing the industry.

Summary Comments

The effects of government programs on both costs and returns should be identified and assessed wherever possible in the statements prepared annually by ERS. Because the government programs change from year to year, the annual statements for corn or wheat may not be "comparable" in all respects. But many other things of importance change as well including technology, the weather, and world supply-demand conditions. Recognizing the influences of government programs provides the public and policy makers a more accurate and realistic picture of the industry.

A full assessment of all the ways in which government programs affect costs and returns for a given commodity may not be possible. But a large share of them can be put in quantitative terms just like fertilizer and seed. Certainly deficiency payments should be included as part of cash returns for food and feed grains if sugar beets and peanuts are evaluated at "market" prices established by the intervention of government controlling supply. An effort should be made to evaluate as many of the items of costs and returns as possible. If a subsidy or cost can only be recognized in qualitative terms it is worthy of a footnote. In the long run legislators, industry personnel and the general public will be served best if these cost and returns statements reflect actual conditions as clearly as possible.

References

Gardner, Bruce L. 1978. "Public Policy and the Control of Agricultural Production." *Amer. J. Agr. Econ.* 60: 836-43.

Ingersoll, Bruce. 1990. "Range War: Small Minnesota Town Is Divided by Rancor over Sugar Policies." *Wall Street Journal.* June 26.

Knutson, Ronald, J. B. Penn, and W. T. Boehm. 1990. *Agricultural and Food Policy.* New York: Prentice Hall, 2nd edition.

Mabbs-Zeno, Carl. 1988. "Estimates of Producer and Consumer Subsidy Equivalents." ATAD, ERS, USDA, Staff Report AGES880127, April.

Salassi, Michael, Mary Ahearn, Mir Ali, and Robert Dismukes. 1990. "Effects of Government Programs on Rice Production Costs and Returns, 1988." U.S. Dept. of Agriculture, Econ. Res. Serv., Agr. Info. Bull. 597.

Schaub, James. 1990. "The Peanut Program and Its Effects." *National Food Review* 13 No. 1: 37-40.

Stanton, B. F. 1986. "Comparative Statements on Production Costs and Competitiveness in Agricultural Commodities." Cornell Univ. A.E. Staff Paper 86-27, October.

U.S. Dept. of Agriculture. 1990. *Agricultural Income and Finance Situation and Outlook Report.* Economic Research Service. AFO-39. December.

_____. 1989a. *Economic Indicators of the Farm Sector: Costs of Production, 1987.* Economic Research Service. ECIFS 7-3. February.

_____. 1989b. *Economic Indicators of the Farm Sector: National Financial Summary, 1988.* ECIFS 8-1. September.

U.S. Dept. of Commerce. 1989. 1987 Census of Agriculture. *North Dakota State and County Data.* Vol. 1, Part 34, AC87-A-34. June.

_____. 1990. *1987 Census of Agriculture. Government Payments and Market Value of Agricultural Products Sold.* Vol. 2, Part 5, AC87-S-5. September.

12

Measuring Cost and Returns: Do We Include Government Program Impacts?

Otto Doering

Introduction

Since John Peek and Hugh Johnson developed and began campaigning for government transfers to the farm sector to be based on their notion of parity, economists have derided the measure as inaccurate, inappropriate, or both. Much of the concern of economists centered on the 'inaccuracy' of parity. Economists have found flaws in its formulation and application so that it is discredited for helping to measure effectively the economic health or financial position of farming, let alone the general welfare of farmers and the agricultural sector.

There has been, however, a lemming-like desire on the part of economists to have some uniform and accepted measure or standard that would assist in looking at the operative economics of producing crops. This has included a desire to have something that would help in farm management, help in the policy decisions about the appropriate levels of income transfers to agriculture, and something that would be helpful in looking at broader concerns about regional or international differences. If we had hoped that parity would be very helpful in these respects we were soon disabused of that notion. But, in a sense, cost of production (COP) has been endowed with whatever hopes might have been held out in these respects for parity.

A Background to Concerns About Cost of Production Numbers

Cost of production was borne on a political wind that stressed equity and fairness in the late 1970s to become part of the formula for setting target prices under the 1977 act. The unusually high prices of 1973-75 involved some very real increases in cost of production. Some were due to more expensive production at the margin and some to a numeraire increase in prices with the development of inflation and the impact of the initially successful petroleum cartel. Increasing costs were seen as a wedge that was cutting farm profits to zero--a wedge beyond the control of farmers A decade later, we tend to forget the hold that the high inflation rates of the late 1970s had on our thinking about agriculture, what policies might be appropriate, and what was 'fair' to the sector. Inflation and its impacts were a major policy concern even though the formal COP/target price linkage was broken in 1981 (Conner).

Having had cost of production put on them by the political system, economists tried to develop as accurate and useful a set of measures as possible, but also tried to avoid its use for anything where it might be misused. As the then Administrator of the Economic, Statistics, and Cooperative Services (ESCS) put it: "We are concerned about how cost of production estimates are used by noneconomists as well as by economists and wonder about our responsibility in presenting COP estimates so that they will be used correctly, to the advantage of society. We are continually faced with both conceptual and practical issues concerning what we can and should do in developing COP estimates." (Farrell)

The most important battle fought about the use of COP in determining target prices in the 1977 Act was the battle over potential self-generating spirals. That concern remains today in the debate over the inclusion of the impacts of government programs. Then, as now, a key factor was the debate over how land prices were treated and whether they were included in the formula for determining the level of support prices. The public resolution of this debate was that the cost of land would be included in the setting of the initial target prices, but that only nonland costs of production would be used in subsequent years to adjust target prices. With the 1981 Act, most of the linkage of government price supports to cost of production estimates was severed, and nothing as formal remains today.

Writing about economic and accounting concepts for costs and returns in 1983, Dave Harrington stressed the potential circularity problem pointing out that "if one uses the same assumptions in valuing assets as in calculating costs of production, then costs of production will always exactly equal the price received for whatever period of time one uses" (Harrington).

If one looks at the flow of prices received, program prices, and costs of production over time, both Farrell's and Harrington's comments are tested.

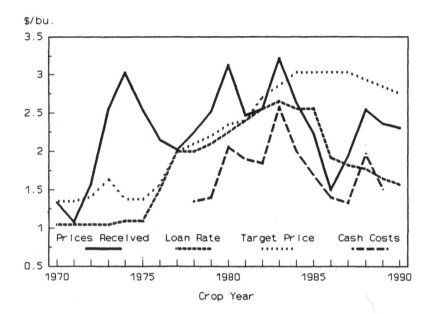

FIGURE 12.1 Trends in corn program provisions and total cash costs. *Source*:
Agricultural Statistics and USDA, 1990.

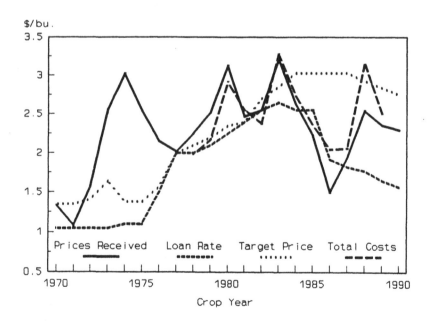

FIGURE 12.2 Trends in corn program provisions and total economic costs. *Source*:
Agricultural Statistics and USDA, 1990.

In Figures 12.1 and 12.2, showing corn program provisions, note that both total cash expenses and total economic costs follow the same pattern and also follow basic price movements. The corn price peaks of 1980, 1983, and 1988 were weather-induced, yield-reduction price peaks that (by whichever accounting measure) resulted in like movement of both costs and prices. It is also interesting to note that while target prices climbed from 1977 to 1981 along with the total cash expenses, the 1981 decoupling of cost of production from corn target price setting did not halt the upward swing. Nor did the decline in costs from 1983 to 1987 result in much of a lowering of target prices or the cost increase in 1988 stimulate a target price increase.

Having been through the debate in Congress and USDA in 1976/77 about the use of cost of production and having watched what followed, there is a broad array of factors that influence congress to raise or lower target prices and loan rates. Costs of production are only one of these factors in the array and appear to be seized upon primarily when the rest of the train is also moving in the same direction. In that sense, "correct" use of cost of production is likely to be a rare event, if one means by "correct" use an explicit role for COP in setting support prices that would override other concerns such as fairness or budget constraints.

An Early Analysis Including Government
Program and Other Costs

One early attempt to broaden the notion of costs of production and apply it to something beyond a 'fair' return to farmers was that by Schmitz, Miranowski and myself (Schmitz, Miranowski, and Doering; Doering, Schmitz, and Miranowski). Our attempt was to account for the full costs of production; the private costs of the farm firm, the costs of government programs, and some of the social externality costs. We included in this accounting the private costs of the farm firm that were then taken into account in the USDA cost of production studies (U.S. Senate). As an example of an input subsidy we included transportation subsidies and research expenditures. We included erosion costs as an example of an important externality. We then included direct government program costs in terms of the costs of surplus purchases and managing reserves programs. Finally, we included an estimate of the cost to other tax payers of the special tax advantages given to agricultural production as an indirect government benefit to agriculture. Table 12.1 indicates the costs we accounted for. Note, that we did not account for the very real private costs to the firm of participation in government programs when acreage set aside would have been required, nor benefits like government credit for the agricultural sector, let alone direct payments.

TABLE 12.1 Total Per-Bushel Costs and Farm Level Prices, 1978-1980

	Corn	Wheat
Private costs	$2.39	$3.94
Input subsidies		
Transportation	.03	.03
Research	.002	.006
Social costs		
Erosion	.10	.14
Tax advantages		
In a profitable year	.60 - .90	.30 - .50
In a unprofitable year	.15 - .25	.07 - .12
Program costs		
Surplus purchases	.26 - .52	.61 - 1.35
Managing reserves	.04	.06
Total costs		
Assuming profitable year		
and managing reserves	$3.16 - 3.46	$4.48 - 4.68
Assuming unprofitable year		
and surplus pruchases	$2.93 - 3.29	$4.80 - 5.59
1978 - 1980 farm level		
prices	$2.68	$3.57

Source: Doering, Schmitz, and Miranowski, 1982.

Our concern was with what we believed to be distorted market signals caused by input subsidies, output price supports and externalities beyond the farm firm. We believed that these distorted prices caused the flow of excess resources into agricultural production and export promotion in the late 1970s. Our attack through cost of production was to try and demonstrate the extent to which private costs did not represent the total cost of such production, justify the level of resource use, or indicate the ultimate level of export benefits. Hoffman and Gustafson stated parallel concerns: "In recent years farmers have expanded production for some commodities even when USDA estimates have shown that production costs are not being covered. If producers are rational, this suggests USDA cost estimates may have been too high, returns may have been underestimated, or a combination of both may have occurred."

Few have argued since then with our numbers or the concern about the flow of excess resources into the sector. Many have argued with our conclusions about whether the extremely high full costs of production we postulated resulted in a negation of gains from trade (Schmitz, Sigurdson, and Doering). What we also should have pointed out, and did not, was that the upward trend of private costs through inflation and production expansion was increasingly unsustainable on international grain markets as prices receded--irrespective of the gains from trade argument.

The Current Debate About Including
Government Program Impacts

Part of the current debate about including government program impacts involves the concern about the inappropriate use of the numbers for assessing the consequences of policy. This brings us back to the three concerns that Dave Harrington had in 1983. He viewed the economic COP framework as problematic for policy formulation on three grounds: "the circularity of the calculations, the potential for price spirals, and the incidence of any price spirals primarily on land values (regardless of their source)." Hoffman and Gustafson responded to this by a suggested change in the frame and content of the COP accounting system to try and capture longer term forces and impacts related to such things as land costs.

One question is whether you have useful cost of production information if you do not include the costs of the major productive asset, land. The first attempt to publish the "Full Costs of Farm Exports" was derailed by reviewers who believed it was inappropriate from the standpoint of economic theory to include land as a cost of production. I agree with the concern about circularity and the potential for price spirals--especially with respect to land (as impacted by government programs). However, to be useful, cost of production data has to include as many of the costs as possible that are actually part of the resource use and management decisions of the farm firm if they are to be realistic or useful to the farm firm or useful to an understanding of its behavior. This is a Pandora's box situation where economists only want to let some COP components out of the box. I am not sure that is possible in spite of our fears of trouble. If the box remains open long enough, complications will emerge.

After going back over some of the more recent literature on value to land I am convinced again that returns to land are critical determinants of its value (Alston; Burt; Scott). I am also persuaded that such things as real interest rates also have had an impact on land values since 1972 (Gertel). The lack of relationship for the several decades prior to 1972 should be of little surprise given the low and stable interest rates that prevailed. Insofar as government programs increase returns to land they are going to influence the value of land--be that influence cyclical or counter cyclical.

What uses of cost of production numbers might a policy economist or an administrator of ERS think appropriate? The COP numbers should give an indication of costs that aids in farm management and farm resource use planning. The COP numbers should give an indication about the appropriateness of a new or old farm program--from the standpoint of resource use, returns or subsidies to the agricultural sector, and distribution within the agricultural sector. The COP numbers also should be of some help in making interregional or international comparisons. However, in all these cases, they will only be of help if they represent some accurate picture of those costs and returns that actually drive firm decisions, represent the results from program decisions, and actually represent critical resource use, farm size, income distribution and other regional or international characteristics. The greatest analytical challenge here is the diversity of firm and the economic decision-making/benefit-conferring reality that is to be adequately represented. What we are also dealing with is a dilemma between micro and macro analysis. Farm level cost of production analysis does not easily fit into the frame of agricultural sector modeling and analysis. How do we ensure an appropriate and useful linkage between the two?

Conclusion

How does the inclusion, or non-inclusion, of government program costs and benefits impact the usefulness of the COP numbers for these purposes? My feeling at this point is that the inclusion of government program costs and benefits can be extremely valuable for utilizing COP numbers for these purposes. I also see great value in reporting the figures both with and without the impacts of the program. If nothing else, the recent report on government program impacts on rice production costs and returns certainly emphasizes the importance of government payments on both the benefits and costs side (Salassi, *et al.*). It also reflects important issues concerning the distribution of farm returns among owners, operators, and providers of essential inputs like water that are critical to both farm firm decisions and to understanding the distributional impacts of farm programs in different regions that may have different institutional production arrangements. However, as the number of options for participation expands under more flexible farm programs in the. future the computations may become unreasonably complex. In addition, the options may become too multi-faceted for useful comparisons.

What concerned those of us who were involved in the 1976/77 cost of production debate, what concerned Ken Farrell in 1979, and what should concern ERS Administrators and others today is the representative accuracy of the cost of production work and the status and use of the numbers themselves. Are the cost of production numbers to be the reflective honest mirror to agricultural production or are they to be part of the direct mechanical linkage that sets specific policy parameters?

It is primarily when we think of cost of production numbers as specific program drivers that we rightfully flinch. However, we must remember that the dividing line between the two different roles for cost of production numbers may be something of a continuum rather than a Chinese wall. The attempts by congressional staff working on the 1990 Farm Bill to look at the COP numbers for guidance about which commodities and which regions might be most or least harmed by differential reductions in target prices has got to be considered an appropriate use of cost of production numbers as a mirror to the sector and to the producers of various commodities. The problem was that while the use may have been appropriate, some of the numbers themselves may not have been. However, cost of production numbers are going to be used for comparing different policy options and their consequences. What that does is leave us with the continuing job of improving the numbers for just such purposes. The box has been left open.

References

Alston, Julian, M. 1986. "An Analysis of Growth of U.S. Farmland Prices." *American Journal of Agricultural Economics* 68: 1-9.

Burt, Oscar R. 1986. "Econometric Modeling of the Capitalization Formula for Farmland Prices." *American Journal of Agricultural Economics* 68: 10-26.

Conner, Richard J. 1981. "The New Food and Agriculture Bill - Where Is It Headed? Potential Impacts on Southern Agriculture." *Southern Journal of Agricultural Economics* 13: 21-23.

Doering, Otto, Andrew Schmitz, and John Miranowski. 1984. "Farm Costs and Exports," in Farm Foundation, *Increasing Understanding of Public Problems and Policies,* pp. 117-127.

_____. 1982. "The Full Costs of Farm Exports." Giannini Foundation Working Paper No. 206, Giannini Foundation.

Farrell, Kenneth, R. 1979. "Opening Comments" in U.S. Dept. of Agriculture, *Estimating Agricultural Costs of Production-Workshop Proceedings.* ESCS Report No. 56, June, p. 1.

Gertel, Karl. 1990. "Farmland Prices and the Real Rate of Interest on Farm Loans." *Agricultural Economics Research* 42, No.1: 8-15.

Harrington, David. 1983. "Costs and Returns: Economic and Accounting Concepts." *Agricultural Economics Research* 35, No.4: 1-8.

Hoffman, George and Gustafson, Cole. 1983. "A New Approach to Estimating Agricultural Costs of Production," *Agricultural Economics Research* 35, No.4: 9-14.

Salassi, Michael, Mary Ahearn, Mir Ali and Robert Dismukes. 1990. "Effects of Government Programs on Rice Production Costs and Returns, 1988." U.S. Dept. of Agric., Econ. Res. Serv. Agric. Info. Bull. No. 597.

Schmitz, Andrew, Dale Sigurdson, and Otto Doering. 1986. "Domestic Farm Policy and the Gains from Trade." *American Journal of Agricultural Economics* 68: 819-27.

Scott, John T. Jr. 1990. "Midwest Farmland Values Affected by Government Subsidies to Agriculture." Unpublished paper. University of Illinois.

U.S. Department of Agriculture. 1990. *Economic Indicators of the Farm Sector: Costs of Production--Major Field Crops, 1988.* Agriculture and Rural Economy Division, Economic Research Service. ECIFS 8-4.

U.S. Senate. 1981. "Costs of Producing Selected Crops in the United States - 1978, 1979, 1980, and Projections for 1981." Committee on Agriculture Nutrition and Forestry, 97th Congress, 1st Session, Committee Print. Estimates prepared by the Economic, Statistics, and Cooperative Services, USDA.

Treatment of the Effects of Government Programs: Discussion

Patrick Westhoff

Introduction

Government programs greatly complicate computations of farm production costs and returns. Stanton and Doering argue that a proper accounting of the costs and benefits associated with government farm programs is important to developing estimates of costs and returns that will be useful to both producers and policymakers. Rather than addressing all the issues raised by Stanton and Doering, I will focus my remarks on three particular areas of concern to a user of USDA cost and return estimates:

1. Costs and returns for program participants and nonparticipants.
2. Distinguishing private and social costs and returns.
3. Government programs and comparisons of costs and returns.

Costs, Returns and Program Participation

Both Stanton and Doering applaud the efforts of Salassi, *et al.*, to develop estimates of costs and returns for rice producers that directly account for the effects of government farm programs. The USDA report methodically accounts for the various features of government farm programs that affect average national and regional costs and returns. The estimates appear reasonable and useful, and I look forward to seeing future reports in the series that will examine other commodities.

As a user of USDA cost and return estimates, I would like to suggest a way the information could be made more useful to analysts like myself, and perhaps to producers and policymakers as well. Models utilized by the Food and Agricultural Policy Research Institute (FAPRI) use estimates of both participant and nonparticipant returns over variable production costs to determine participation rates and planted acreage for major program crops. To construct the national average estimates of costs and returns for rice producers, Salassi *et al.* utilize information about both participants and nonparticipants, weighting the data by the proportion of farmers participating in the rice program. It would be very useful to us if USDA were to report the costs and returns for participants and nonparticipants separately.

Participant costs and returns can be reported on a per-planted-acre basis (as done by USDA), or on a per-base-acre basis (as done by FAPRI). As a first approximation, it is reasonable to assume that the variable costs on acres actually planted would be similar for participants and nonparticipants, although acres idled by government programs will result in important differences in total variable costs. Land costs and other fixed costs will, of course, differ significantly between program participants and nonparticipants. It would be useful to know if there are systematic differences in input usage and yields between participants and nonparticipants (Of course, if differences are found, it would be difficult to know if they are to be explained by the effects of the government program, or by some factor like land quality that affects both the participation decision and production technologies).

The triple-base program makes it particularly important to separate participant and nonparticipant net returns. Under the triple base, deficiency payments do affect the participation decision and total participant returns. However, on the 15 percent of acreage not eligible for program payments, deficiency payments should have no affect whatsoever on planting decisions. On the mandatory flexible acres (MFA), producers will receive only the market price for their production (plus, in some cases, returns from commodity loan programs). Thus, planting decisions on the MFA will depend on a producer's comparisons of expected market (nonparticipant) net returns over variable production costs. Thus, estimates of both participant and nonparticipant costs and returns are necessary to understand the sequence of decisions made by producers of program crops under the new legislation.

Private and Social Costs and Returns

Stanton and Doering both point out that an accounting of the "true" costs and returns associated with government farm programs must include a variety of effects typically omitted. For example, Stanton mentions the price-depressing effects of government stocks, and Doering reports estimates of total production costs that incorporate input subsidies, erosion, and tax advantages. There is some confusion in both speakers' discussions of these nontraditional measures of costs and returns, and Stanton in particular fails to distinguish private and social costs and returns associated with farm programs.

Consider the case of Commodity Credit Corporation (CCC) stocks. While it may be true that high levels of CCC stocks tend to depress market prices, that "cost" of the program is already incorporated in the market price used to estimate returns--if market prices are lower than they otherwise would have been, then producer returns are also lower than they otherwise would have been. Private costs and returns are properly measured using the observed market price. To assess the true social cost of CCC stocks, one would need a lot of information that cannot be measured directly, but that might be estimated with a model like that maintained by FAPRI--the price-depressing effect of holding high stock levels, the price-supporting effect of building government stocks, the effects on the livestock sector and consumers of holding buffer stocks, the trade-offs with respect to government costs, etc.

A measure of total social costs and returns might contribute to a proper assessment of the true value of government farm programs, but the need to identify and estimate many unobservable factors will make it difficult for the profession to agree on any single measure. Producer subsidy equivalents (PSEs) and aggregate measures of support (AMS) have proven useful concepts for conducting international trade negotiations, making it possible to estimate the total amount of support received by the agricultural sector. PSEs and AMS do not measure total social costs and returns, however, since they do not distinguish transfers from deadweight losses.

Government Programs and
Comparisons of Costs and Returns

Production costs are not determined independently of government programs. This point is made forcefully by Doering when he discusses the history of linking government support levels to cost of production measures. Stanton points out that expected future benefits of government programs are capitalized into the cost of land and other fixed inputs. However, both Doering and Stanton suggest that carefully constructed estimates of variable production costs can be compared across countries. Such comparisons are possible, but it is very important to understand why they may be misleading.

Suppose one wanted to compare international production costs for corn. Government policies mean that the prices paid to farmers for corn vary greatly across countries. For example, corn prices in the European Community average the equivalent of five dollars per bushel, as import barriers and export subsidies isolate the domestic market from world market forces. In the United States corn sells for slightly more than two dollars per bushel (deficiency payments increase total U.S. producer returns, but frozen program yields mean that the marginal unit of production is produced at the market price). Producer prices in Argentina may be even lower, as export taxes reduce internal prices relative to world prices.

Policy-driven differences in output prices will result in corresponding differences in total production costs, as acknowledged by Doering and Stanton. They are also likely to result in significant differences in variable production costs, contrary to the contention of Stanton. Basic economics suggest that producers should equate marginal costs and marginal revenues in making production decisions. Because marginal revenues are higher in the European Community, producers would be expected to use inputs more intensively than producers in the United States or Argentina in an effort to maximize yields. Conversely, corn prices are so low in Argentina that producers are likely to use less fertilizer and other purchased inputs than are producers in the United States and the European Community.

Because of the differences in market prices, marginal production costs should be highest in the European Community and lowest in Argentina, all else equal, if producers are making rational economic decisions. Higher marginal costs may not necessarily translate directly into higher average variable production costs, but marginal and average costs are likely to be correlated. Thus, it would not be surprising if a carefully constructed survey of corn production costs found the highest variable production costs in the European Community and the lowest in Argentina. This would not imply that Argentina is more "efficient" than the United States or the European Community, but only that Argentina can produce the amount of corn it produces at a lower cost than the other countries can produce the amount of corn they produce. If government policies changed and Argentine market prices increased, production levels and marginal (and average) production costs would also be likely to increase. Likewise, if a change in the Common Agricultural Policy reduced marginal prices received by European producers, production levels and marginal (and average) production costs would be likely to fall. Simply estimating current variable production costs in different countries does little to establish which country is the least-cost or most efficient producer.

Concluding Comment

The papers by Doering and Stanton have raised a variety of issues concerning the relationship between government programs and measures of production costs and returns. Properly accounting for government programs is critical if one hopes to provide realistic estimates of costs and returns that will be of use to producers, policymakers, and analysts. The efforts of USDA to incorporate government programs into its estimates are to be applauded and encouraged.

References

Salassi, Michael, Mary Ahearn, Mir Ali, and Robert Dismukes. 1990. "Effects of Government Programs on Rice Production Costs and Returns, 1988." U.S. Dept. of Agric., Econ. Res. Serv. Agric. Info. Bull. No. 597.

PART FIVE

Estimating Costs of Land and Water Services

13

Market Value Versus Agricultural Use Value of Farmland

Lindon J. Robison and Steven R. Koenig

Introduction

The value of farmland is an important component of the farm sector's balance sheet. Moreover, opportunity costs of farmland investments are used in cost of returns studies. Since the farm sector's balance sheet and estimates of costs and returns are important information, farmland value data used in these estimates must also be important. Therefore, if there are problems in the way farmland values are estimated or used, these problems should be addressed. In fact, there are problems with these data, at least in the way they are frequently used.

One problem with the use of these data is the frequent assumption that the reported value of farmland is equal to the discounted present value of cash returns from farmland's use in agriculture, its agricultural use value. A related assumption is that agricultural cash rents reflect the value of the services generated by farmland. The conclusion reached in this chapter is that these assumptions are likely incorrect.

We assume that knowing the agricultural use value of farmland is important. Unfortunately, the current U.S. Department of Agriculture (USDA) data series provide only market value of farmland data. How to find the agricultural value of farmland is the focus of this chapter. In what follows, the origins of commonly used farmland value and agricultural rent data are discussed as well as other data series and issues. Then, we review the literature that has estimated the market value of farmland and deduce a model that will permit us to obtain estimates of the importance of agricultural income relative to farmland's market value. In the last section, our conclusions are summarized.

Divergence Between Agricultural Use and Market Value

Causes

Several factors may contribute to the divergence between farmland's agricultural use value and its market value. One factor may be that farmland's most profitable use in the future may not be agriculture. Using farmland for housing developments, recreational activities, roads, and land fill sites may in the future earn returns greater than those available from farmland's use in agriculture; when this is true, farmland's current value will not be reflected accurately by its agricultural use value.

Government payments associated with the control of farmland may also contribute to the divergence of farmland's market value from its agricultural use value. Farmland retirement and conservation payments, and subsidized credit are examples of government payments tied to the control of farmland but not necessarily associated with the production of agricultural products. In many cases, farmland subsidies require eligible participants to have established a record of agricultural production involving farmland resources. In these cases, farmland with an established record of agricultural use may have a capitalized value independent of what is produced on the farmland.

Farmland's market value may also diverge from its agricultural use value because it simultaneously earns returns (and suffers costs) in addition to those associated with agricultural production. Sources of income from farmland besides agricultural production may include recreational payments (Pope) or mineral extraction payments. Costs independent of agricultural production may include pollution abatement payments, insurance, maintenance, and the costs of meeting specific municipal ordinances. Tax policies of state and local governments may also affect the value of farmland. In addition, natural features of the farmland such as location relative to markets, topography, fertility, and rainfall in the area, affect the amount of other resources required to produce a crop and consequently how much tenants can afford to pay to acquire the use of farmland. Population density will also affect farmland's market value because the demand for farmland for nonfarm use has been found to be related to population density (Peterson).

Farmland's market value may diverge from its agricultural use value because of what are now called speculative bubbles. Simply stated, speculative bubbles may increase the value of farmland because farmland owners and prospective buyers incorrectly infer from past experiences the future earnings stream from farmland and consequently, farmland's future value. In the presence of speculative bubbles farmland may be priced differently than its agriculture use value simply because the future is hard to predict.

Finally, farmland's market value may diverge from its agricultural use value because of the expectation of future earnings that depart from current patterns. When these expectations exist, the relationship between market value and capitalization of current income based on past rates of growth may seem inconsistent or may even lead us to believe in speculative bubbles.

Importance of Separation

There are several reasons why it is important to separate farmland's agricultural use value from its market value. Many states have enacted laws that encourage the use of farmland for agricultural purposes near urban centers. To encourage the continued use of farmland in agricultural production, farmland near urban centers is assessed for tax purposes at its agricultural use value rather than its market value. In these instances, there is a need to distinguish between the agricultural use value of farmland and the market value of farmland.

Another reason for separating farmland's agricultural use value from its market value is to better understand and interpret the relationships between the market value of farmland and agricultural rents from farmland. Without such information, the farmland value to agricultural cash rent ratio of 94 in New Jersey and 13 in North Dakota in 1977 seem out of line (Robison, Lins, and VenKataraman). Finally, separating farmland's agricultural use value from its market value is important because it allows us to examine the influence that factors such as governments payments have on the market value of farmland. Making such a separation has the added benefit of encouraging economists to refine the capitalization theories traditionally used to explain farmland's market value.

Agricultural Farmland Value and Rent Data

An important data series frequently used in farmland value studies is the U.S. Department of Agriculture data series for average farmland and buildings values and average cash rent values for farmland and buildings. These values are benchmarked to the Census of Agriculture (which is conducted every five years) and moved on an annual basis with data from the Agricultural Land Values Survey, which is a voluntary survey of farm operators' opinions about farmland values (Barnard and Hexem). This means that the USDA data represent farmer operators' perceived values rather than an identifiable transaction value. Unless survey respondents have access to complete information, their opinions and the values they provide on surveys will be biased with respect to actual transaction values, especially during periods of rapidly changing values.

Since the farmland values represent state averages, an important factor in determining these values is the survey used and the weighting schemes employed to create the series (Kuchler and Burt). In addition, assumptions made about the portion of farmland's value attributed to the value of building and other capital investments tied to the land are important as well.

To illustrate the importance of survey design, contrast the way University of Minnesota and USDA collect their farmland value data. The University of Minnesota has collected rural farmland value data for over 80 years. Data for their farmland value estimates are obtained from brokers, appraisers, farm managers, insurance agents, bank officers, country officials and others familiar with their respective areas (Govindan and Raup). In contrast, the USDA sample design ignores nonfarmer owners who purchase roughly 30 percent of all farmland each year. The value of farmland for nonagricultural purposes, such as recreation, which may be important to nonfarmer owners, is thus systematically ignored. Moreover, because USDA's sample design tends to select large farmers located in rural areas there is likely an underreporting of higher farmland values reported by small and hobby-sized farmers near urban centers.

The difference in farmland value estimates resulting from different sample designs and weighting schemes is evident when the two estimates are compared. For example, USDA values are consistently below estimates obtained at the University of Minnesota from 1972 until the early 1980s and then were consistently higher than University estimates after the early 1980s. In 1988, the USDA estimated the value of Minnesota farmland to be $700. The University's estimate was $523 in 1988.

Some bias in USDA's estimate of farmland values could result from efforts to calibrate the USDA survey data with Census of Agriculture benchmark surveys. Farmland values from USDA Agricultural Land Value Survey are benchmarked every five years to Census of Agriculture survey data. Intercensus years are extrapolated forward and are then recalibrated when new Census benchmark values become available. The recalibration methodology has recently changed, raising the concern that the earlier data series using the old methodology are not valid. Regardless of which calibration method can be best defended, the data series are no longer consistently estimated.[1] Also, Kuchler and Burt suggest that intercensus estimates derived from USDA's survey typically underestimate Census values when nonfarm demands for farmland are relatively strong because of the bias in the sampling.

There are other concerns about the calibration of USDA data to Census of Agriculture data. One concern is that the calibration of the surveys is influenced by the differences in sample design. The Census targets a much larger owner population (including urban areas and small farmers) and is statistically reliable to the county level. Conversely, the USDA farmland survey targets large farm operators in rural locations, is voluntary, and is designed to produce aggregate state-level estimates. Moreover, the timing of the USDA survey has been changing, recently from April 1 to February 1 and now to January 1. Changing the surveys' timing raises additional concerns about the consistency of the data.

Using USDA's per-acre cash rent of farms and farmland value series to model agricultural returns from farmland also raises concerns. Cash rent values reflect an opportunity cost to the farm operator, but not necessarily equal to opportunity costs of nonfarmer owners. Another concern is that cash rents may not be the only source of income. Nonfarm owners may attach site value to the farmland or may receive income for allowing part of their farmland to be used for recreational purposes. Finally, it is not clear if part of the rents are payments for the use of buildings or if these payments in some cases are separate from the rents paid for the use of bare farmland. Nor is it clear how government payments and property taxes influence rents. In addition, costs such as applications of lime and tile, road, building, and fence maintenance may also influence rents.

Changes have also occurred in the methods used to estimate cash rents, raising concerns about the consistency of this data series. Before 1984, a panel of farmers were surveyed annually to obtain their opinion about the average cash rent and average value of rented farmland in their "locality." Now a stratified random sample is selected from the same overall population used to estimate the farmland and building value series. Beginning in 1989, the ambiguity of "locality" was resolved by asking for average cash rent values within a county. A difference that remains between methods used to estimate rents and farmland values is that rent data are not benchmarked to Census data.

Models employing state generated data for farmland values may provide alternatives to USDA supplied data. Studies by Burt and Scott are examples of studies that use state data. But not all states collect separate data and even among these will be methodological differences in data series generation making interstate comparisons using state data difficult.

Other Data Series

Proxies for variables used to separate agricultural farmland values and nonagricultural value from market value are not readily available. Estimates of nonfarm income influences can be proxied by a state level series available from USDA, but they come with caveats. For example, the nonfarm income series is an estimate of off-farm cash income of the farm operator and his or her household and, therefore, does not capture the nonfarm investor's income. Per capita personal income measures are available but may be too broad a measure for this application.

A number of alternative series can be used to proxy the opportunity cost of capital. An annual series of Federal Land Bank interest rates compiled by the Farm Credit Administration until 1989 provides one option. The weakness of this series is that it is tied to average bond costs that causes it to not reflect some changes in the marginal opportunity costs of capital. The prime rate, or other interest rates more accurately reflect changes in the opportunity cost of capital.

In addition to the data series issues already discussed are concerns about structural factors that influence farmland valuation and farmland returns and for which better data are needed. Preferential property tax assessments of farmland is an example. At the end of 1985, 49 states allowed preferential tax assessments of farmland, 27 states had these coupled with deferred taxes which are imposed when the farmland is converted to a nonqualifying use, and 6 offered programs providing tax concessions in exchange for restrictions on farmland use (Grillo and Seid). Local suburban governments in some states have begun to offer their own tax concession programs, often referred to as Greenbelt laws.

Another example of structural factors influencing farmland valuation is the limits on the ownership of farmland. These limits may affect the demand for farmland and hence its value. There is little empirical evidence of the effect on the market value of farmland of these limits. At the end of 1986, Schian and Seid report that 29 states had limits on foreign ownership of farmland and 15 states restricted business entities from owning farmland or engaging in the business of farming.

Having listed some concerns with USDA farmland values and rent series, it still is the best data generally available. Moreover, we believe their accuracy is sufficient to measure the influence of agricultural and nonagricultural influence on farmland's market value.

Finding the Market Value of Farmland:
A Review of Literature

The importance of farmland in the farm sector's balance sheet has made it a popular focus of many research efforts. The most remarkable feature about these studies, however, is their lack of consensus. Perhaps inaccurate data series have contributed to the lack of consensus. An alternative explanation is that the lack of consensus can be attributed to researchers having failed to distinguish between farmland's market value and farmland's agricultural use value. Many of the studies reviewed here attempt to estimate farmland's market value, sometimes failing to recognize that agricultural income may not be adequate to explain agricultural farmland's market value. A sample of such studies include the following.

Reynolds and Timmons, Tweeten and Martin, and Herdt and Cochrane use simultaneous equation models of supply and demand for farmland. Factors affecting farmland values suggested from these studies included: expected capital gains, government payments for farmland diversion, conservation payments, farm enlargements, and technological progress. Klinefelter used a simple linear model emphasizing the importance of capitalized agricultural rents in determining agricultural farmland's value. For these models, however, Pope *et al.* found that, "when recent data were added to the sample, numerous changes in signs of coefficients occurred for all of the simultaneous equations models. Further, most of the estimated coefficients were not statistically significant from zero."

Robison, Lins, and VenKataraman concluded that most farmland model specifications do not reflect accurately enough the relevant structural changes occurring in the farmland market. In an effort to model structural changes in the farmland market, they deduced a model in which farmland values were determined by the capitalized value of agricultural rents and nonagricultural returns to farmland in a two-sector farmland model. While they found some support for the importance of nonagricultural returns on farmland values, they also found considerable difference between states.'

Castle and Hoch concluded that expected capitalized rent and expected capital gains are the critical components determining farmland values. Featherstone and Baker criticized the Castle and Hoch study because capital gains can only result from capitalizing a growing income stream. Phipps, on the other hand, concluded from his study that farm-based returns unidirectionally change farmland values. Meanwhile, Burt demonstrated the link between rents and farmland values in Illinois approximating a capitalization formula using a second-order rational distributed lag on farmland rents.

In another study, Scott lists factors affecting the decline in farmland values, emphasizing the increasing ratio of debt servicing required relative to cash rents. Shalit and Schmitz concluded savings and accumulated real estate debt are important determinants of farmland prices. Meanwhile, Reinsel and Reinsel reaffirm the importance of farm income in the determination of farmland values.

Recently, modelers have discovered still another factor influencing farmland values about which they can disagree, speculative or rational bubbles. A speculative bubble is essentially an overreaction to current price information. Bubbles suggest that because of overreaction to current price information, market corrections will be required later and the market participants will find their expectations unfulfilled. The findings of DeBondt and Thaler, and Grossman and Shiller, are that the stock market has greater price variance than one would expect without price bubbles. Featherstone and Baker reached a similar conclusion regarding farmland values. Moreover, high returns in 1973 and 1974 set off a boom-bust cycle in part caused by an overreaction to prices. Falk, on the other hand, found no evidence of speculative bubbles in Iowa farmland prices. But a working paper by Baffes and Chambers supports the conclusions of Featherstone and Baker using a different methodology. In contrast, Tegene and Kuchler concluded that there is no evidence to support the hypothesis that speculative bubbles contribute to farmland prices.

These conflicting results, unfortunately, do not help us decide the question raised earlier: What is the agricultural use value of farmland? Pope concluded that only 22 percent of the total market value of rural farmland in Texas can be explained by its productive value. Peterson concluded that nearly two-thirds of the differences in land prices among states is attributed to nonagricultural uses. Particularly useful in Peterson's study was his land quality indexes calculated showing relative differences in the agricultural productivity of U.S. farmland. In one sense, our analysis can be viewed as an effort to examine more closely Peterson's useful results.

Present Value Models

Assume that farmland produces a stream of cash flows from agricultural activities of R_t, R_{t+1}, R_{t+2} ... in periods t, $t+1$, $t+2$ ... Moreover, assume that the opportunity costs of capital for periods t, $t+1$, $t+2$, ... equal r_t, r_{t+1}, r_{t+2} ... The present value of the future returns from agricultural activities generated from the farmland, V^a_{t-1}, can be written as:

$$(1) \quad V^a_{t-1} = \frac{R_t}{(1+r_t)} + \frac{1}{(1+r_t)} \left[\frac{R_{t+1}}{(1+r_{t+1})} + \frac{R_{t+2}}{(1+r_{t+1})(1+r_{t+2})} + ... \right]$$

and V^a_t can be written as:

$$(2) \quad V^a_t = \frac{R_{t+1}}{(1+r_{t+1})} + \frac{R_{t+2}}{(1+r_{t+1})(1+r_{t+2})} + ...$$

Substituting V^a_t for the bracketed expression in equation (1), we can solve for V^a_t and write:

$$(3) \quad V^a_t = (1+r_t) V^a_{t-1} - R_t$$

Suppose there is another stream of cash income or expenses Q_t, Q_{t+1}, Q_{t+2} ... associated with farmland for periods t, $t+1$, $t+2$, ... whose present value equals:

(4) $\quad V^0_{t-1} = \dfrac{Q_t}{(1+r_t)} + \dfrac{1}{(1+r_t)}\left[\dfrac{Q_{t+1}}{(1+r_{T+1})} + \dfrac{Q_{t+2}}{(1+r_{t+1})(1+r_{t+2})} + \ldots\right]$

Then following the same procedures used to find equation (3), equation (4) is rewritten as:

(5) $\qquad\qquad V^0_t = (1 + r_t)V^0_{t-1} - Q_t$

Then the market value of farmland V_t can be expressed as:

(6)
$$V_t = V^a_t + V^0_t$$
$$= V_{t-1}(1+r_t) - R_t - Q_t$$

and Q_t is found equal to:

(7) $\qquad\qquad Q_t = V_{t-1}(1+r_t) - R_t - V_t$

We assume that the relationship described in (6) between the opportunity cost of owning farmland and returns from farmland (including capital gains) exists in the farmland market.[2] In order to find V^a_t and V^0_t requires additional assumptions. These are: that R_{t-1} grows at rate g_t and Q_{t-1} grows at rate h_t. Given these assumptions, V_{t-1} can be written as:

$V_{t-1} = V^a_{t-1} + V^0_{t-1}$

$= R_{t-1}\left[\dfrac{(1+g_t)}{(1+r_t)} + \dfrac{(1+g_t)(1+g_{t+1})}{(1+r_t)(1+r_{t+1})} + \ldots\right] + Q_{t-1}\left[\dfrac{(1+h_t)}{(1+r_t)} + \dfrac{(1-}{(1+}\right.$

(8)

Substituting α_{t-1} and β_{t-1} for the bracketed expressions in (8) allows us to write:

(9) $\qquad\qquad V_{t-1} = \alpha_{t-1}R_{t-1} + \beta_{t-1}Q_{t-1}$

and

(10) $$V_t = \alpha_t R_t + \beta_t Q_t$$

After exploring many different ways to approximate α_t and β_t, estimates $\hat{\alpha}_t$ and $\hat{\beta}_t$ were found equal to:

(11) $$\hat{\beta}_t = \beta_0$$

and

(12) $$\hat{\alpha}_t = \alpha_0 + \alpha_1 t^2$$

Finally, we substitute the right-hand side of (7) for Q_t in (10) and solve for V_t. The result is:

(13) $$V_t = \left(\frac{\alpha_t - \beta_t}{1 + \beta_t} \right) R_t + \left(\frac{\beta_t}{1 + \beta_t} \right)(1 + r_t) V_{t-1}$$

Speculative Influences on Farmland's Market Value

Recent attention has been given to the influence of speculative values on farmland. To introduce these influences into our model, suppose that n periods from year t, an additional source of income or expense is expected. This income or expense may be associated with changes in government programs, different uses of land, new costs associated with pollution abatement, or new restrictions on zoning ordinances that alter earnings potential for land.

Whatever the source let its value in period (n) equal w_0 so that its value in period t is:

(14) $$V_t^w = w_0(1+r)^{t-n} \gamma_0(1+r)^t$$

Empirical Tests and Interpretations

Our generalized model for explaining farmland's market value can now be expressed as:

$$V_t = \alpha_0 R_t + \alpha_1 R_t t^2 + \beta_0 Q_t + \gamma_0 (1 + r_t)^t$$

Then substituting for Q_t and solving for V_t we write:

$$(15) \quad V_t = \frac{(\alpha_0 - \beta_0) R_t}{(1 + \beta_0)} + \frac{\alpha_1 R_t t^2}{(1 + \beta_0)} + \frac{\beta_0 1 + r_t) V_{t-1}}{(1 + \beta_0)} + \frac{\gamma_0 (1 + r_t)^t}{(1 + \beta_0)}$$

$$= c(1) R_t + C(2) R_t t^2 + C(3)(1 + r_t) V_{t-1} + C(4)(1 + r_t)^t$$

Equation (15) can be estimated using ordinary least squares. When error terms are found to be autocorrelated they are corrected using the Orcutt-Cochrane method. Since (15) is just identified the structural coefficients can be recovered.[3]

If nonagricultural sources of income are significant, the hypothesis that $C(3) = 0$ must be rejected. If speculative influences affect farmland value then the hypothesis $\gamma_0 = 0$ must be rejected. Finally, if agricultural rents are important then the hypothesis that $C(2) = C(1) = 0$ must be rejected.

To estimate equation (14) requires estimates of V_t, r_t, and R_t. We use as our estimate for the market value of land V_t USDA's value of farmland series. The R_t series is estimated using USDA's farmland rent series adjusted for property taxes. The opportunity cost r_t is the prime rate plus 3% to adjust for risk and liquidity differences between the prime rate and rates of return in agriculture.[4]

Estimating the Agricultural Use of Farmland

Having obtained estimates of α_0 and α_1, it is a small step to next calculate the percentage of farmland's market value attributed to agricultural returns. This ratio ra_t is simply

$$(16) \quad ra_t = \frac{(\alpha_0 + \alpha_1 t^2) R_t}{V_t}$$

Moreover, the ratio of returns attainable to current nonagricultural influence equals

(17)
$$rq_t = \frac{\beta_o Q_t}{V_t}$$

Finally, future income streams not part of R_t or Q_t have an influence on V_t estimated by the formula:

$$rf_t = \frac{\gamma_0 (1 + r_t)^t}{V_t}$$

Table 13.1 reports the estimates of C(1), C(2), C(3) and C(4) and the corresponding t statistics. Only statistically reliable values are included in the model and used to calculate values for α_0, α_1, β_0, and γ_0. Finally R^2 values and d. w. statistics are reported for each equation along with the autoregressive coefficient used to correct for autocorrelation. In Tables 13.2, 13.3, and 13.4, our estimates of ra_t, rq_t, and rf_t for 1970 to 1989 are reported. The sum of ra_t, rq_t, and rf_t is reported in Table 13.5.

If the farmland market behaves similar to the model developed in this chapter, then the most important message of the empirical model is the difference between states. This difference is illustrated by the size and significance of coefficients. Table 13.1 results show that Ohio, Indiana, Illinois, Iowa and Missouri all have similar market structures. Based on similar evidence, it appears that Maryland, Michigan, Wisconsin and Minnesota have land markets with similar characteristics. Other similarities, such as between North and South Dakota, could be pointed out as well. Clearly Delaware, Virginia, and Mississippi are exceptions with insignificant C(1) coefficients.

Expectation about future growth rates of income are reflected by the magnitude of the γ_0 coefficients. In descending rank are: New Jersey, Pennsylvania, Ohio, Michigan, Iowa, Minnesota, North Carolina, Tennessee, Kentucky, South Carolina, Illinois, and Wisconsin. In all states but Minnesota, Ohio, Indiana, Illinois, Iowa and Missouri, nonfarm income (and expense) was a significant influence on farmland's market value. Important to note, however, is the changing magnitude and signs of this influence reported in Table 13.3.

The significance of the γ_0 term can be considered evidence of speculative influence on farmland's market value. The significance of speculative influences are reported in Table 13.4. What this study shows is that their influence is small but still significant in 11 of the 22 states examined. Finally, our results show that land prices were under stress in the late 70's and early 80's and that land prices during this period in many of the midwestern states were in part supported by these speculative influences.

TABLE 13.1 Estimates of Coefficients for Equation (14) and Capitalization Factors

	NJ	DE	MD	PA	MI	WI	MN	OH	IN	IL	IA
C(1)	20.27		11.12	16.29	11.50	8.99	15.62	21.67	15.95	16.24	20.10
(t stat)	4.3		2.3	3.1	6.3	5.5	9.1	3.2	18.4	14.0	8.1
C(2)				.005				-.002			
(t stat)				1.9				1.68			
C(3)	.31	.34	.17	.39	.42	.40	.21				
(t stat)	3.0	2.7	1.7	2.4	5.8	4.4	3.4				
C(4)			.85		.17	.20	.32	.43	.48	.56	.44
(t stat)			4.8		2.7	3.8	6.2	2.9	3.9	3.6	3.2
AR(1)	1.08	1.05	1.04	.55	.78	.89	1.03	.61	.83	.86	1.04
(t stat)	42.5	47.0	45.6	2.5	6.1	7.8	35.1	3.4	8.3	9.4	28.6
R^2	99	98	99	99	99	99	99	99	99	99	98
dw	1.83	2.02	2.56	2.03	2.11	1.70	1.47	1.77	1.14	1.00	1.51
α_0	29.8	.51	13.67	27.36	20.70	15.80	20.08	21.67	15.95	16.24	20.10
α_1				.009				-.002			
β_0	.45	.51	.21	.64	.73	.68	.27	.43	.48	.56	.44
y_0			1.02		2.30	.33	.41				

Continued—

TABLE 13.1 Estimates of Coefficients for Equation (14) and Capitalization Factors—Continued

	MO	ND	SD	VA	NC	KY	TN	SC	GA	MS	AR
C(1)	15.06	7.90	10.81		2.42	6.21	4.46	8.49	4.90		9.30
(t stat)	18.8	5.9	5.7		2.0	4.6	2.9	3.7	4.74		3.8
C(2)								.007	.009	.005	
(t stat)								3.4	4.5	3.4	
C(3)		.42	.22	.33	.84	.61	.72	.46	.44	.35	.45
(t stat)		5.7	2.0	2.8	21.5	9.4	10.9	4.8	5.2	5.1	4.0
C(4)	.18			.12						.39	.19
(t stat)	3.1			1.8						8.0	3.0
AR(1)	.83	.86	.79	1.05	.44	.82	.71	.66	.56	.99	.79
(t stat)	8.2	8.0	6.1	77.1	2.8	7.3	5.0	4.0	3.4	29.5	5.4
R^2	99	99	99	99	99	99	99	99	99	99	99
dw	2.09	1.73	2.08	2.02	1.89	1.98	1.76	2.26	1.85	1.71	2.02
α_0	15.06	14.42	14.22		19.90	17.51	18.37	16.69	9.54	.52	17.78
α_1								.014	.016	.008	0
β_0		.73	.29		5.11	1.57	2.55	.86	.79	.53	.82
γ_0	.18		1.02							.60	.35

TABLE 13.2 Percent of Value Attributed to Capitalized Agricultural Rents

Year	NJ	DE	MD	PA	MI	WI	MN	OH	IN	IL	IA
1970	56	2	32	100	98	124	189	106	115	116	164
1971	59	2	32	108	122	119	188	110	117	119	168
1972	45	2	33	127	107	108	166	119	120	112	169
1973	44	2	30	118	99	98	155	109	115	111	163
1974	52	2	28	101	100	94	164	103	111	109	173
1975	43	2	26	97	103	100	163	107	124	107	163
1976	32	1	25	97	103	95	161	107	116	103	149
1977	32	1	27	89	94	95	141	95	104	90	124
1978	34	2	23	87	87	86	127	95	94	84	123
1979	35	1	28	82	84	77	119	91	85	80	115
1980	36	1	24	77	85	70	110	81	80	78	104
1981	36	2	24	82	81	67	99	82	79	78	102
1982	41	2	27	92	80	73	107	94	87	90	112
1983	48	2	34	98	86	80	117	90	94	90	123
1984	47	2	33	97	91	83	124	97	95	100	139
1985	42	2	36	98	84	88	133	110	109	121	180
1986	44	2	35	104	88	81	152	106	113	131	189
1987	46	2	33	104	90	85	163	97	111	120	191
1988	38	2	35	114	81	94	148	93	101	106	172
1989	40	1	30	99	87	94	145	94	99	101	165

Year	MO	ND	SD	VA	NC	KY	TN	SC	GA	MS	AR
1970	112	129	134	37	116	131	134	115	109	30	106
1971	114	128	136	44	110	137	134	122	120	31	135
1972	123	139	130	32	109	135	127	116	120	35	135
1973	109	133	128	30	89	126	115	111	110	36	117
1974	113	144	124	25	85	118	112	95	94	34	108
1975	108	122	116	22	84	112	105	93	102	36	112
1976	101	121	120	26	76	102	105	95	104	37	100
1977	95	105	114	32	83	111	96	92	107	39	95
1978	94	94	103	27	68	93	94	91	98	35	90
1979	92	93	95	20	65	81	79	89	90	34	77
1980	84	85	93	19	54	82	77	81	85	34	77
1981	80	84	90	20	56	81	75	85	87	31	69
1982	84	86	86	24	60	86	79	83	87	36	74
1983	87	88	93	22	61	83	73	84	92	37	75
1984	88	92	88	18	60	84	72	96	98	42	79
1985	102	99	99	19	68	77	69	95	96	47	85
1986	98	115	110	18	56	85	81	90	93	41	90
1987	96	111	109	19	47	86	68	91	91	43	84
1988	105	114	99	17	45	84	63	98	98	54	83
1989	105	106	101	16	51	79	71	100	100	58	91

TABLE 13.3 Percent of Value Attributed to Capitalized Nonfarm Income

	NJ	DE	MD	PA	MI	WI	MN	OH	IN	IL	IA
1970	-2	-2	0	-4	2	-4	-1	0	0	0	0
1971	2	-3	0	0	1	-6	-1	0	0	0	0
1972	-1	1	0	-2	-6	-4	-1	0	0	0	0
1973	0	-3	-1	-6	-9	-9	-2	0	0	0	0
1974	-4	-6	-1	-8	-6	-7	-5	0	0	0	0
1975	-3	-5	0	-6	0	-5	-6	0	0	0	0
1976	-4	-4	-2	-3	-4	-7	-5	0	0	0	0
1977	2	-3	0	-8	-14	-10	-6	0	0	0	0
1978	2	0	-1	-2	-4	-8	-2	0	0	0	0
1979	1	1	0	-1	0	-5	-2	0	0	0	0
1980	5	-2	-1	0	0	-2	-2	0	0	0	0
1981	11	5	1	7	1	1	0	0	0	0	0
1982	7	12	5	13	11	10	4	0	0	0	0
1983	9	4	5	7	11	8	5	0	0	0	0
1984	13	6	0	5	6	7	3	0	0	0	0
1985	6	13	4	15	17	18	10	0	0	0	0
1986	5	1	4	11	13	14	10	0	0	0	0
1987	-8	4	2	-4	13	10	6	0	0	0	0
1988	3	2	-1	4	2	0	-4	0	0	0	0
1989	-1	-2	1	-4	6	9	0	0	0	0	0

	MO	ND	SD	VA	NC	KY	TN	SC	GA	MS	AR
1970	0	-1	0	-12	33	-5	-8	4	-4	-1	-1
1971	0	-1	-1	13	-41	-8	-5	-2	-6	0	3
1972	0	-3	-1	-17	-21	-15	-19	-8	-9	0	-12
1973	0	-6	-2	-14	-46	-11	-24	-1	-6	-4	-7
1974	0	-18	-5	-34	-46	-16	-29	-11	-13	-8	-10
1975	0	-19	-5	-8	-4	-10	-18	-4	-5	-4	1
1976	0	-13	-3	-16	-41	-23	-22	-3	-2	-2	-7
1977	0	-9	-5	-12	-33	-24	-29	-8	-7	-5	-7
1978	0	-3	-3	-5	-5	-13	28	0	-7	-7	-4
1979	0	-5	-1	-15	61	-13	-13	-5	-1	-4	-11
1980	0	-3	-1	10	-4	-1	0	-1	-1	-3	-4
1981	0	5	1	21	41	16	18	9	7	-4	2
1982	0	5	1	39	96	16	43	12	16	11	8
1983	0	9	2	18	47	16	32	13	8	11	20
1984	0	5	1	29	14	19	25	13	10	2	10
1985	0	21	10	27	91	28	48	12	11	11	13
1986	0	12	4	16	78	13	20	11	10	10	20
1987	0	11	5	25	43	22	19	17	3	12	12
1988	0	-1	-2	13	50	8	4	0	4	3	2
1989	0	3	-1	0	34	12	25	2	2	4	5

TABLE 13.4 Percent of Value Attributed to Capitalized Future Income

Year	NJ	DE	MD	PA	MI	WI	MN	OH	IN	IL	IA
1970	0	0	1	0	1	1	2	1	1	1	1
1971	0	0	1	0	1	1	1	1	1	1	1
1972	0	0	1	0	0	1	1	1	1	1	1
1973	0	0	1	0	1	1	2	1	1	1	1
1974	0	0	3	0	1	2	3	2	2	2	2
1975	0	0	1	0	1	1	1	1	1	1	1
1976	0	0	1	0	1	1	1	1	1	1	1
1977	0	0	1	0	1	1	1	1	1	1	1
1978	0	0	2	0	1	1	1	1	1	1	1
1979	0	0	4	0	2	3	4	2	2	2	2
1980	0	0	8	0	5	6	7	5	5	5	5
1981	0	0	23	0	13	16	18	13	13	14	12
1982	0	0	10	0	5	7	7	6	6	6	5
1983	0	0	4	0	2	2	3	2	2	2	2
1984	0	0	6	0	3	4	5	4	4	4	4
1985	0	0	4	0	2	3	4	3	3	3	3
1986	0	0	3	0	2	2	4	2	2	2	2
1987	0	0	3	0	2	2	4	2	3	3	3
1988	0	0	4	0	3	4	5	3	4	4	4
1989	0	0	8	0	5	7	10	6	7	7	8

Year	MO	ND	SD	VA	NC	KY	TN	SC	GA	MS	AR
1970	1	0	0	0	0	0	0	0	0	2	1
1971	0	0	0	0	0	0	0	0	0	2	1
1972	0	0	0	0	0	0	0	0	0	2	1
1973	1	0	0	0	0	0	0	0	0	3	1
1974	1	0	0	0	0	0	0	0	0	4	2
1975	1	0	0	0	0	0	0	0	0	2	1
1976	0	0	0	0	0	0	0	0	0	2	1
1977	0	0	0	0	0	0	0	0	0	2	1
1978	1	0	0	0	0	0	0	0	0	3	2
1979	2	0	0	0	0	0	0	0	0	7	4
1980	4	0	0	0	0	0	0	0	0	13	7
1981	10	0	0	0	0	0	0	0	0	33	18
1982	4	0	0	0	0	0	0	0	0	14	7
1983	2	0	0	0	0	0	0	0	0	5	3
1984	3	0	0	0	0	0	0	0	0	9	5
1985	2	0	0	0	0	0	0	0	0	6	3
1986	1	0	0	0	0	0	0	0	0	4	2
1987	2	0	0	0	0	0	0	0	0	5	3
1988	3	0	0	0	0	0	0	0	0	8	4
1989	5	0	0	0	0	0	0	0	0	15	8

TABLE 13.5 Percent of Actual Market Value of Farmland Explained

Year	NJ	DE	MD	PA	MI	WI	MN	OH	IN	IL	IA
1970	54	1	33	96	100	121	190	107	116	117	165
1971	61	-1	32	108	123	114	188	111	117	120	169
1972	44	3	34	125	101	105	166	120	121	113	169
1973	45	-1	30	112	91	90	155	110	116	112	164
1974	48	-5	30	92	95	90	162	105	113	110	175
1975	40	-3	28	91	103	96	159	108	125	108	164
1976	28	-2	24	94	100	89	157	108	117	104	149
1977	35	-1	28	81	81	86	136	96	105	90	124
1978	36	2	24	86	84	79	127	96	95	85	124
1979	35	2	32	81	86	74	121	93	87	82	117
1980	41	-1	31	77	90	73	114	85	85	83	109
1981	47	7	47	70	94	84	116	96	92	92	114
1982	48	14	42	105	96	89	118	100	93	96	118
1983	57	6	43	105	99	90	125	100	96	100	128
1984	61	7	40	102	99	94	132	101	99	104	143
1985	49	15	43	113	104	109	147	113	112	124	183
1986	49	3	42	114	102	97	165	108	115	133	192
1987	38	6	38	99	105	97	173	99	114	123	194
1988	41	3	39	110	86	98	150	96	105	118	177
1989	38	-1	38	95	98	104	154	100	106	108	172

Year	MO	ND	SD	VA	NC	KY	TN	SC	GA	MS	AR
1970	112	128	134	25	149	126	126	119	105	31	106
1971	115	127	135	32	69	130	129	120	114	32	138
1972	123	136	129	15	88	119	108	108	11	36	123
1973	109	127	127	16	42	115	91	110	104	35	110
1974	114	126	119	-9	39	102	83	84	81	30	100
1975	108	103	111	14	81	102	87	89	98	35	114
1976	101	108	117	11	35	79	83	91	102	37	93
1977	96	96	109	20	51	87	67	84	100	36	88
1978	95	91	99	22	63	80	66	91	90	31	88
1979	93	88	94	5	4	68	65	84	89	37	69
1980	88	82	92	29	50	81	77	80	84	45	79
1981	90	89	90	41	97	98	93	94	94	60	89
1982	88	91	88	63	156	102	122	95	103	61	88
1983	89	97	95	40	109	99	106	97	100	54	97
1984	91	96	90	47	74	102	97	108	108	53	94
1985	104	120	109	46	160	105	116	106	108	63	101
1986	99	127	114	24	135	98	101	100	103	55	113
1987	98	122	114	44	90	108	87	108	95	60	99
1988	107	113	97	31	94	92	67	92	103	66	89
1989	110	109	101	16	85	91	97	104	101	77	104

A major concern of this chapter has been to describe the influence of agricultural use on farmland's market values. Table 13.2 reports these results. The story that emerges is that the influence of capitalized agricultural rents changes over time. In many states, capitalized values of rents more than equaled the farmland value. Later on they accounted for less than the market value of farmland. The main message, however, is that in those states with significant agricultural production, farmland use reflected by agricultural rents are still the dominant force explaining farmland's market value.

Finally, we add that our results must be interpreted with caution. We know of no better way than using Table 13.5 to warn researchers about the limitations of this study and the need to make inferences carefully. In every state autocorrelated disturbance required correction. To show how the resulting model values related to the unadjusted market value of farmland, we totaled ra_t, rq_t, and rf_t. The total is reported in Table 13.5. Clearly in states such as New Jersey, Delaware, Maryland, and Mississippi, the model explained a small portion of land's market value.

Finally, the presence of autocorrelated disturbance suggests the model used in this chapter fails to fully capture the influence of all factors determining farmland's market value.

Summary and Conclusions

This paper has suggested that one difficulty with our use of farmland value and rent data is that we fail to separate agricultural use value and market value of farmland. Moreover, most empirical studies attempt to explain the market value of agricultural farmland by considering only agricultural sources of income. This paper has obtained an instrument for rents and costs that affect farmland's market value and that are not included in the agricultural rent series by assuming that perfect market conditions prevail. In most cases it is shown that nonagricultural rents (and costs) are important. But perhaps the more significant message is that the structure of farmland markets are complex and differ considerably by state. Yet in the important agricultural states, agricultural rents are still the main factor in determining farmland's market value.

Finally, it goes without saying, that the USDA agricultural farmland values and agricultural rents data may have biases that limit the data's usefulness in conducting farmland market studies. We believe, however, they are still adequate for many uses. Perhaps one solution to this difficulty is to supplement USDA data with university farmland value data series such as those available in Minnesota, Illinois, Nebraska, and Iowa.

Notes

The authors thank Steven D. Hanson, Robert J. Myers, and James F. Oehmke for helpful comments made on an earlier draft of this paper. The Michigan Agricultural Experiment Station gratefully is acknowledged for its support of this research.

1. Beginning with 1984 intercensus, values are determined using a geometric interpolation method instead of an additive interpolation method.

2. To see this relationship more clearly (6) is rearranged to show the relationship between opportunity costs $r_t V_{t-1}$, capital gains $V_t - V_{t-1}$, and agricultural returns R_t and nonagricultural returns Q_t. This relationship equals $r_t V_{t-1} = (V_t - V_{t-1} + R_t + Q_t$.

3. $\beta_0 = C(3)/(1-C(3))$; $\alpha_0 = (1 + \beta_0)C(1) + \beta_0$; $\alpha_1 = (1 + \beta_0)C(2)$; and $\alpha_0 = (1 + \beta_0)C(4)$.

4. The interest rate on FLB loans is not used because it reflects average, rather than the marginal, costs of bonds sold to finance the loan.

References

Baffes, J. and R.G. Chambers. 1989. "Rational Bubbles and Land Prices." Working Paper No. 89-39. Department of Agricultural and Resource Economics, University of Maryland, November.

Barnard, C.H. and R. Hexem. 1988. "Major Statistical Series of the U.S. Department of Agriculture, Volume 6: Land Values and Land Use." Washington, D.C.: U.S. Department of Agriculture, Economic Research Service, Agricultural Handbook No. 671.

Burt, O.R. 1986. "Econometric Modeling of the Capitalization Formula for Farmland Prices." *American Journal of Agricultural Economics* 68: 10-26.

Castle, E.N. and I. Hoch. 1982. "Farm Real Estate Price Components, 1920-78." *American Journal of Agricultural Economics* 64: 8-18.

DeBondt, W.F.M. and R. Thaler. 1985. "Does the Stock Market Overreact?" *Journal of Finance* 49: 793-805.

Dickey, D. and W. Fuller. 1979. "Distribution of the Estimators for Autoregressive Time Series With a Unit Root." *Journal of American Statistics Association* 74: 427-431.

Falk, B. 1988. "A Search for Speculative Bubbles in Farmland Prices." Working Paper. Department of Economics, Iowa State University.

Featherstone, A.M. and T.G. Baker. 1987. "An Examination of Farm Sector Real Asset Dynamics: 1910-85." *American Journal of Agricultural Economics* 69: 532-546.

Govindan, K. and P.M. Raup. 1990. "The Minnesota Rural Real Estate Market in 1989." University of Minnesota, Economic Report ER90-5, July.

Grillo, K.A. and D.A. Seid. 1987. "State Laws Relating to Preferential Assessment of Farmland." Washington, D.C.: U.S. Department of Agriculture, Economic Research Service, Staff Report No. AGES870326.

Grossman, S.J. and R.J. Shiller. 1981. "The Determinants of the Variability of Stock Market Prices." *American Economic Review* 71: 222-227.

Harris, D.G. 1977. "Inflation-Indexes, Price Supports, and Land Values." *American Journal of Agricultural Economics* 59: 489-495.

Herdt, R.W. and W.W. Cochrane. 1966. "Farmland Prices and Technological Advance." *Journal of Farm Economics* 48: 243-263.

Jones, J. and R.W. Hexem. 1990. "Cash Rents for Farms, Cropland and Pasture, 1960-89." Washington, D.C.: U.S. Department of Agriculture, Economic Research Service, Statistical Bulletin No. 813.

Klinefelter, D.A. 1973. "Factors Affecting Farmland Values in Illinois." *Illinois Agricultural Economics Research* 13: 27-33.

Kuchler, F. and O. Burt. 1990. "Revisions in the Farmland Value Series." *Agricultural Resources: Agricultural Land Values and Markets Situation and Outlook Report.* AR-18. Washington, D.C.: U.S. Department of Agriculture, Economic Research Service, June, pp. 32-35.

Peterson, W. 1986. "Land Quality and Prices." *American Journal of Agricultural Economics* 68: 812-819.

Phipps, T.T. 1984. "Land Prices and Farm-Based Returns." *American Journal of Agricultural Economics* 66:422-429.

Pope, C.A. 1985. "Agricultural Productive and Consumption Use Components of Rural Land Values in Texas." *American Journal of Agricultural Economics* 67: 81-86.

Pope, R.D., R.A. Kramer, R.D. Green, and B.D. Gardner. 1979. "An Evaluation of Econometric Models of U.S. Farmland Prices." *Western Journal of Agricultural Economics* 4: 107-119.

Reinsel, R.D. and E.I. Reinsel. 1979. "The Economics of Asset Values and Current Income in Farming." *American Journal of Agricultural Economics* 61: 1093-1097.

Reynolds, T.E. and J.F. Timmons. 1969. "Factors Affecting Farmland Values in the United States." Iowa State University Agricultural Experiment Station Research Bulletin No. 566.

Robison, L.J., D.A. Lins, and Ravi VenKataraman. 1985. "Cash Rents and Land Values in U.S. Agriculture." *American Journal of Agricultural Economics* 67: 794-805.

Russell, T. and R. Thaler. 1985. "The Relevance of Quasi Rationality in Competitive Markets." *American Economic Review* 75: 1071-1082.

Schian, D.C. and D.A. Seid. 1986. "State Laws Relating to the Ownership of U.S. Land by Aliens and Business Entities, October 31, 1986." Washington, D.C.: U.S. Department of Agriculture, Economic Research Service, Staff Report No. AGES861103.

Scott, J.T., Jr. 1983. "Factors Affecting Land Price Decline." *American Journal of Agricultural Economics* 65: 796-800.

Shalit, H. and A. Schmitz. 1982. "Farmland Accumulation and Prices." *American Journal of Agricultural Economics* 64: 710-719.

Shiller, R.J. 1981. "Do Stock Prices Move Too Much to Be Justified by Subsequent Changes in Dividends?" *American Economic Review* 71: 421-436.

Tegene, A. and F. Kuchler. 1989. "The Contribution of Speculative Bubbles to Farmland Prices." Technical Bulletin No. 1782. Washington, D.C.: Economic Research Service, U.S. Department of Agriculture.

Tweeten, L.G. and J.E. Martin. 1966. "A Methodology for Predicting U.S. Farm Real Estate Price Variation." *Journal of Farm Economics* 48: 378-393.

U.S. Department of Agriculture. 1990 and various back issues. *Agricultural Resources: Agricultural Land Values and Markets Situation and Outlook Report.* AR-18. Washington, D.C.: U.S. Department of Agriculture, Economic Research Service, June.

_____. "Farm Real Estate Market Developments." Washington, D.C., various issues.

_____. 1989. "Economic Indicators of the Farm Sector: State Financial Summary, 1988." ECIFS 8-2, October.

14

Land and Water Costs:
A Case of Whose Costs, When

Charles H. Barnard and Michael Salassi

Introduction

Land and water share a set of conceptual and empirical issues related to commodity costs of production (COP). The supply of both is essentially fixed, especially in the short-run.[1] The market prices of both land and water rights reflect expected values of benefits to be derived from the resource, not just in the current accounting period, but for each period into perpetuity. Further, the demand for both inputs (and hence their price) is sometimes significantly influenced by both nonfarm factors and government intervention. Markets for both land and water are thin, and in the case of water rights, are sometimes nonexistent. Often, the two inputs are traded as one, either because of custom or law. Consequently, even when water rights are exchanged, the market values of the rights are often implicitly hidden in observed rental rates or sale prices of land, complicating the measurement of both.

Despite this common link, we will focus primarily on issues pertinent to land, and secondarily address water issues. Land is the more crucial input from the perspective of COP estimation, representing a significant cost for virtually all crops and for Western beef-raising enterprises. Water, on the other hand, is a proportionally much smaller cost and its measurement is controversial only for a few crops.

The imputation of a cost for the land resource used in the production of agricultural commodities is among the most vexatious problems confronted by COP analysts. Not only can land costs overwhelm the effect of other input costs on the total cost of production, but alternative imputation procedures yield such a wide range of land cost estimates that the whole process is widely perceived as arbitrary (Miller and Skold). For some major crops, the estimated land charge is the largest single input cost, but selection of an alternative imputation procedure can cause the estimate to more than double. For instance, the six land cost estimates published by ERS for the 1974 production of corn varied from $.44 per bushel based on the acquisition price of land to $1.15 based on net share rent (Sharples and Krenz).

No professional consensus exists as to what a land charge should measure or even whether such a charge should be included in COP estimates. Economic theory is not sufficiently definitive, and standard accounting practices are not commonly accepted by economists. Even though economists generally agree that land cost imputations ultimately must be based on the basic economic principle of opportunity cost, that concept is encumbered by inherent measurement problems. As if those problems are not enough, one soon realizes that the conceptual problems are overlaid and confounded by the institutional characteristics of the land market and its associated data collection challenges and, in the case of estimation for Federal policy purposes, by legislative mandate.

Many of the issues that surround the estimation of a land charge for use in COP analysis are unresolvable--the very nature of the land resource belies their resolution. Despite that fact, and despite the complexity of the issues involved, it appears that much of the controversy surrounding land charges stems partly from a failure to recognize the crucial, but sometimes implicit, assumptions that underlie procedures for estimating land costs, and partly from a failure to identify explicitly the implications of those assumptions. Our goal here is to reduce the controversy to a discussion of its lowest common denominators--the underlying assumptions. Through our review of literature, we identified five assumptions that are principal sources of conflict and confusion:

(1) estimation of an ownership cost of land or a use cost of land,
(2) determination of the point in the production cycle that the estimates will represent,
(3) estimation of costs representing individual firms or the commodity industry,
(4) estimation of social or private costs, and
(5) inclusion or exclusion of Government payments in cost budgets.

COP analysts make these assumptions, either explicitly, as a guide to the selection of a conceptual approach, or implicitly, through the selection of a conceptual approach. The correct decision to take in regard to these five assumptions depends very much on whose costs you want to measure and at what point (when) in the production cycle you want to measure them. The decisions taken then determine which of the commonly used procedures are appropriate, determine the data needed, and also affect the interpretation of estimates produced.

Several other issues are not discussed here; instead, we direct the reader to more extensive discussions available in the proceedings of three relevant conferences (Helmers; USDA, 1979; Scott, 1977). First, we do not discuss if, or for which purposes, an imputed charge for land should be included in estimates of total costs of production. This chapter is written from the perspective that if a land charge is to be included in COP estimates, then here are the merits and problems associated with several estimation procedures, and here are the caveats that must surround the interpretation of estimates produced by those procedures. Second, we do not discuss the problems of circularity associated with including a land cost in COP estimates. It is well-known that if market-intervention policies set product prices according to COP estimates that include land costs, an upward spiral of land prices may result.

Methods of Estimation

Land is both an input providing an annual flow of services and a financial asset with many of the characteristics of stocks and bonds. Fundamentally, there are three approaches that can be taken to estimate the cost of using land: (1) measurement of acquisition or actual cost, (2) calculation of residual income, and (3) imputation of supply cost or opportunity cost.

Acquisition or Actual Cost

Basing the land charge strictly on actual expenditures has great intuitive appeal, partly because of its perceived empirical simplicity; the cost of land is then simply an accounting matter. But on closer examination, one quickly discovers that basing land cost on past expenditures is not as clear cut as examining a historical record. Rural land only sells every 20 to 30 years (Wunderlich), and many landowners add improvements to the land after the original purchase. Complete measurement of previous expenditures on land must include the cost of those improvements.

Depreciable improvements such as dikes, ditches, and tile often involve little cash expenditure, usually being installed primarily with operator-owned equipment and the operator's unpaid labor. Even if reasonable estimates of these costs could be established, COP analysts would find it necessary to depreciate the improvements. Overall, such a process is arbitrary and necessarily relies heavily on information not generated by market transactions. Historical expenditure information, some of it reflecting supply and demand conditions that existed twenty or more years ago, may be irrelevant for current decision making, especially if substantial inflation occurred after the expenditure.

Residual Claimant

Another commonly used method for estimating a charge for land services is to assign land as the residual income claimant. The theory of the firm embodies the idea that an entrepreneur combines some factors that he purchases with some that he owns to produce a product. According to marginal productivity theory, it is possible (under certain restrictive conditions) to assign each input a cost such that total cost will exactly equal the income received from sale of the product. If each input is allocated a share of output value in proportion to its value marginal product, total output will be exactly exhausted (Henderson and Quandt, pp. 81-82). Following Ricardo, if all factors other than the fixed factor (land, in our case) are so valued (at their value marginal products), and deducted from total revenue, the remainder (residual) is the marginal return to the fixed factor (land).

If, as is done in practice, we assume that the factor prices actually paid are equal to the value marginal products of the variable inputs, then the costs of purchased inputs are easy to determine from tax or other financial records prepared by the operator. The validity of this assumption, however, requires the very restrictive conditions of optimal employment of all inputs in a perfect market, or a linearly homogeneous production function. It is unlikely that these conditions are met in actual input markets.

Further, marginal productivity theory does not hold when specialized resources such as land are involved. Because land is a specialized factor of production, meaning its supply curve is not perfectly elastic, the factor price of land, and hence its cost, depends on the price of the product. This interdependence between cost of the specialized resource and its demand violates standard assumptions of economic theory, which require input costs to be unambiguously determined from factor prices, untainted by the forces of demand acting through product price. Friedman (p.148) and Pasour (p.245-246) note that when production involves specialized resources, analysts cannot assume that market prices are accurate estimates of input costs.

Even more troublesome problems arise with regard to the residual, which, in practice, accrues to all owned inputs rather than to a single fixed factor. The absence of a market generated price for the owned inputs means that the residual must be divided (allocated) among various fixed (owned) inputs. Ultimately, one must resort to accounting techniques or the concept of opportunity cost to assign a value to the annual contribution of all but one of the fixed factors. Microeconomic price theory, most of which is developed under the assumption of a single fixed factor, does not indicate which of several fixed factors should be assigned the role of residual claimant.

Any of the fixed inputs, such as operator labor, management, and risk-taking and operator-owned land, machinery, and capital could be assigned the residual. One alternative, of course, is to assign land as the residual claimant and determine its cost as the residual when an imputed cost is subtracted from total receipts for each of the other fixed (owned) inputs. Others prefer to estimate an opportunity cost for land and management and leave unpaid operator labor as the residual. Friedman holds that risk-taking and management should be assigned the role of residual income claimant.

Two sets of reasoning suggest that a cost of land determined under the residual method may hold little relationship to its opportunity cost. First, determining the cost of land by assigning it the role of residual claimant can be accurate only to the extent that the conditions underlying marginal productivity theory hold and to the extent that the opportunity cost for the other fixed factors are correctly imputed. Second, by definition, the cost allocated to land under the residual income method depends directly on total receipts--i.e., on actual price and yield. The conceptual implications of this latter circumstance are explored more fully later in the chapter where we discuss the use of net share rents in determining opportunity cost.

Opportunity Cost

If land is not assigned the role of residual claimant, then the only other alternative is to estimate its opportunity cost directly and assign the role of residual claimant to another fixed factor. For example, a traditional method of determining the annual cost of land has been to multiply the current value of the land by the observed interest rate. Such an estimate represents the opportunity cost of funds invested in the land.

Conceptual Issues and Underlying Assumptions. Even though opportunity cost is one of the most basic of economic concepts, its empirical measurement is extremely controversial. According to Pasour, at least some of the controversy stems from confusion created by the usual textbook theory of cost, which is meaningful only in a static equilibrium model and even then, only when resources are unspecialized and can be defined independently of demand. Further ambiguity arises from the perfect knowledge and foresight assumptions, because in reality, costs are not known. In our view, the most notable problem for COP analysis is that standard textbook definitions of opportunity cost leave open the question of exactly whose opportunity costs we are attempting to measure and at what point, or when, in the production cycle we are attempting to measure those opportunity costs. Confusion then reigns when COP analysts fail to define carefully the circumstances to which the opportunity concept is being applied and to acknowledge fully the implication that those assumptions hold for the interpretation and use of COP estimates. To illustrate these points, we carefully examine the textbook concept of opportunity cost and identify a set of crucial questions left to the discretion of COP analysts.

Ferguson very simply defines "the alternative or opportunity cost of producing one unit of commodity X as the amount of commodity Y that must be sacrificed in order to use resources to produce X rather than Y" (p.185). This classic definition is based on a two-good world, and as such defines the social cost of producing X. An individual producer must perform a similar opportunity cost calculus when he purchases resources and uses them to produce and sell a commodity. A full accounting of the producer's economic profit must take into consideration the implicit costs of what could be earned in the best alternative use of his money and what could be earned in the best alternative use of his time. Only in the specialized circumstances of perfect markets are the social costs and private costs of production equal. Common real world phenomena such as the presence of internal and external economies or diseconomies, monopolistic elements in the input markets, or public good attributes are sufficient to cause a discrepancy between social and private cost. Consequently, what constitutes a legitimate cost of production depends on whether you are measuring a social cost of production or a private cost of production.

Pasour discusses further refinements of the opportunity cost concept. In his words, "Cost, in the choice-influencing opportunity cost sense, concerns sacrificed alternatives at the moment of choice. Whether the yield is eventually higher or lower than anticipated has no relevance to opportunity cost. The opportunity cost is not affected unless the decision maker's estimate of the best alternative is affected at the time the decision is made" (p. 246). This implies that opportunity costs must be measured from an *ex ante* perspective. Or, if the analysis of costs of production is conducted after the fact, the goal of that analysis should be an *ex post* evaluation (measurement) of *ex ante* opportunity cost. That is to say, opportunity costs occur in the planning stage and should be measured from that perspective.

Even this refinement leaves open crucial questions regarding the point in the production cycle from which opportunity cost is being measured. Pasour maintains that in empirical work, the decision maker's anticipation of future conditions "tends to be ignored in estimating cost and is replaced by factor prices" (p. 246). When considering the cost of land used in corn production, he argues that its opportunity cost hinges on the expected outlays, prices, and yields of the best alternative crop--regardless of what rent was paid for the land. In our view, the rent paid for use of the land to produce corn may instead be the appropriate measure of the opportunity cost of corn land depending on the range of choices open to the firm. If the firm's choices include the alternative of renting the land to another operator for the production of corn, then the rental rate for corn is an appropriate measure of opportunity costs. If the range does not include this alternative, then foregone opportunities in the production of the alternative crop is the relevant opportunity cost. Essentially, the dilemma is a question of length of run--which resources are fixed and which are variable. If cost estimates are being prepared from the perspective of the industry, then the appropriate opportunity costs must be based on expectations for the alternative crop.

Measuring Opportunity Cost. There are two fundamental ways of measuring the annual opportunity cost of land: (1) at an annual use charge such as that exhibited in the market by the annual rental rate; and (2) according to the amount of interest that is annually foregone due to ownership of the asset.[2] The two methods arise from two related but conceptually different markets for farmland: the rental market where use rights are transferred and the title market where ownership rights are transferred (Scott 1983, p.93). Consequently, the two methods imply fundamentally different interpretations of the cost of land.

Returns to investors in agricultural land can come from two sources: (1) the net income earned from operating the business and (2) appreciation in the value of the land (capital gains).

A land purchase is a market transaction in which full ownership rights are purchased. Rights are obtained not only for one year's agricultural returns, but also for agricultural returns and capital gains for each succeeding year into perpetuity. A cash or share rental agreement, on the other hand, can be viewed as a market transaction in which limited rights are purchased. The rental agreement entitles the farm operator only to the annual agricultural return he can generate from the property.

Capital gains (losses) result when the value of land changes. Land values are a function of all returns expected to be generated by the land, present and future, farm and nonfarm. Factors that alter expected returns include changes in permanent improvements, changes in the potential for nonfarm uses, and changes in the economy-wide price levels. Such changes in land values, however, represent only potential returns (losses) for the farm business because capital gains (losses) are not realized until the asset is sold. Capital gains (losses) create an accounting dilemma for COP analysts because costs are incurred in one accounting period to purchase rights to potential returns that may accrue in future periods.

Even though no professional consensus exists as to how to measure the annual contribution of unrealized capital gains to farm returns, economists must account for their contribution to returns. In any complete accounting system, capital gains must be considered additional income not realized from production but, which nevertheless potentially reduces the economic costs of production (Harrington). Those potential returns must be accounted for or the costs incurred on their behalf will not be offset. Over time the COP budget will reflect an inaccurate picture of the economic conditions facing the commodity industry. Balanced COP budgets, with comparability between costs and returns, are particularly important in periods of high inflation, when annual capital gains can be greater than income from agricultural production. Consequently, a COP analyst must choose between: (1) the ownership-rights approach, representing the cost of holding full ownership to the land, or (2) the use-rights approach, representing the cost of obtaining use of the land for a year/season. In some land markets, particularly those characterized by high rates of inflation or those heavily influenced by demand for nonfarm uses, the annual use cost and the annual ownership cost can be decidedly different. The sets of factors that influence changes in land charges over time may also be different for the use-rights and ownership-rights approaches.

1. Full Ownership Rights: Interest Foregone. The opportunity cost of full ownership must be calculated by multiplying the current value of land by an opportunity interest rate.[3] The interest rate chosen should reflect the rate of return that the capital invested in land would earn in an alternative investment of similar risk, liquidity, and potential for future capital gains or losses. In measuring the annual flow of services from land based on its full opportunity cost, COP analysts have two choices for the opportunity interest rate. They can multiply a current nominal interest rate times the current market value of the land or they can multiply a current real interest rate times the current market value of the land.[4] Both methods can be conceptually equivalent, but in periods of inflation (deflation) care must be exercised to ensure that the concepts are correctly implemented.[5] Even so, differences in the assumptions and data may lead to somewhat different estimates.

Either the real interest rate or the nominal interest rate can be used in estimating the annual opportunity cost of a land investment, as long as comparability is maintained within the entire COP budget between cost components and income components (Harrington, Miller). If the cost of land is calculated by multiplying a current nominal interest rate times the current market value of the land, then analysts must include both production income and capital gains (losses) in the returns portion of the COP budget. The inclusion of capital gains (losses) is necessary to offset the inflation component embodied in market interest rates by the Fisher effect and to account for land opportunity costs created by nonfarm demand for farmland. Or, alternatively stated, an opportunity cost of full land ownership based on nominal interest rates and current market values should be used only if returns include both capital gains and returns from agricultural production.

If capital gains are excluded from the returns portion of the COP budget, then costs associated with full ownership must be excluded. This means that the inflation component must be removed from the market interest rate used to calculate the annualized opportunity cost of land. The inflation premium is an additional cost (above costs necessary to secure agricultural production rights) that is incurred by the landowner in order to obtain the benefits of capital gains. Therefore, when capital gains are excluded from returns, an annual opportunity cost of land should be based on real interest rates and current market values. It is important to note an implication of this procedure, however. Removal of the inflation component, through use of a real interest rate, does not adjust the estimated cost of land for opportunity costs created by nonfarm demand.

2. Use Rights: Share and Cash Rents. Both cash rents and cash equivalents of net share rents are widely used as measures of the annual opportunity cost of land used in agricultural production. The cash rental rate represents a directly-observable, market-generated measure of what a farm operator is willing to pay to obtain the productive services of land for one year. It is widely perceived as a valid measure of true economic rent.

While cash rent seems to measure exactly what the COP analyst wants, it is not without its drawbacks as a measure of the cost of land services, both empirically and conceptually. Analysts must recognize that from a conceptual perspective cash rent does not represent the opportunity cost of full ownership. In particular, a rental agreement does not provide the farm operator with rights to capital gains. While this consideration may not be important universally, nonfarm demand may significantly affect the opportunity cost of landownership for some crops and regions (Robison, *et al.*).

Share rental agreements are often given a cash value and used in COP analysis in conjunction with, or as a substitute for cash rents. The cash equivalent, or net share rent, is usually defined as the landlord's share of the crop receipts minus the landlord's share of expenses. Net share rent is usually calculated by multiplying a harvest-month price times the landlord's share of the harvested crop and then netting out the landlord-paid expenses. Share rental arrangements can be quite complex because landlords often pay real estate taxes and portions of many variable input costs. Thus, detail of share rental arrangements must be known to accurately estimate the net share rent. For example, timing of the rental payment (such as at the beginning of the production cycle versus after harvest) can make as much as 20 percent difference in "true" rent versus observed rent (Scott, 1983).

Even though COP analysts often use cash-equivalent share rentals interchangeably with cash rentals, these two rentals are not conceptually equivalent. Cash rents are negotiated at the beginning of the production cycle and are determined by landlord and tenant expectations of price and yield. Thus cash rents do not depend on the actual outcome of production and marketing. Share rents, on the contrary, are determined by actual yield and actual market price at harvest. Even though negotiations at the beginning of the production cycle predetermine landlord and tenant shares of the commodity and their respective costs for variable inputs, the value of the net share depends on the outcome of the production cycle. This differs conceptually from cash rent, which is a function of expected price and expected yield. From this perspective, the cash-equivalent of the share rent does not measure the opportunity cost of using agricultural land: opportunity cost does not depend on the outcome of the decision-maker's action. Measurement of share rents must be considered an *ex post* evaluation of *ex post* opportunity costs, while measurement of cash rents can be considered an *ex post* measurement of *ex ante* opportunity costs.

Under a cash rental agreement, the operator assumes all of the regular production risks, while providing the landlord with a risk-free return that can be considered certainty equivalent income. Under a share rental agreement, the operator and landlord share the operating risk. Consequently, COP analysts would have to risk-adjust the share rent to estimate its certainty-equivalent income. Certainty-equivalent income for the share rent might be measured as the amount that a landlord would require with certainty (i.e., a cash rent) before becoming indifferent between that amount and a risky outcome with a known expected value (i.e., a share rent). Such a procedure, however, is probably empirically infeasible on any aggregate scale.

Estimating Costs of Water Services

The capitalized value of surface water rights and ground water sources are most often implicitly combined with the capitalized value of land surface rights. The annual cost of water also may be implicitly included in land rental rates. Because water constitutes a legitimate cost of irrigated agricultural production, this mixing does not present a problem from the perspective of estimating total cost of production, as long as the opportunity cost of water is not also included elsewhere in the COP budget. If the costs of water rights are included elsewhere in the COP budget, then costs of water will be double counted to the extent that the value of water rights are capitalized into land values. It is true, however, that the estimated cost of just the land will be biased.

An analogous situation applies with respect to irrigation equipment such as pumps and sprinkler systems. These large, semi-permanent structures are sometimes sold (and sometimes rented) together with land surface rights. Thus, estimates of just the land costs will be biased in all cases and estimates of the total costs of production will be biased upward if the opportunity cost of the irrigation capital is included elsewhere in the COP budget. This can be true regardless of whether the opportunity costs are derived from the foregone interest approach or from the rental data approach.

Special Considerations

Of course, overlying all of the conceptual discussions is the practical. Often, no matter how convincing the conceptual arguments, the decision criterion that determines the method used is data availability. For example, rental data are too sparse for determining the cost of land in livestock enterprises. Consequently, no matter what the merits of the use-rights approach or the demerits of the ownership-rights approach, the analyst simply has no choice. In other cases, data limitations may be less severe but nevertheless necessitate special consideration, and perhaps, adjustment of the data before it can be used in COP analysis.

One special consideration concerns the relationships of mineral rights to land values. Mineral rights are sometimes sold along with land surface rights, even though they can be sold separately. If mineral rights are sold with surface rights, the capitalized value of the mineral rights may be implicitly included in land value data. Unlike the value of water rights, this capitalized value of mineral rights should be excluded from the opportunity cost of land because the surface or agricultural production rights can be purchased separately. Also, there is no offsetting return included in the COP budget because mineral income is clearly not part of the farm sector income.

Other complicating empirical factors relate to specific crops in some regions. Rental agreements often apply to acreage involving more than one crop, necessitating some arbitrary division of the total cash rent among specific enterprises. For some crops, the observed rental rate covers more than the productive value of the land. In the case of peanuts, for instance, domestic marketing quotas are often included in the rent paid. If COP budgets include separate cash expenses for such allotments then double counting may result.

The institutional practices of the Agricultural Stabilization and Conservation Service (ASCS), in distributing government payments, creates yet another problem for the use of cash rental data. By ASCS decree, the tenant in a cash rental agreement receives all direct government payments. Consequently, the cash rent actually exchanged by the tenant and landlord is bound to be negotiated at a higher rate than it would in the absence of government payments or if a portion of government payments was distributed directly to the landlord. This is not a problem as long as government payments are included as a source of income in the returns portion of the COP budget, thus offsetting land costs that reflect returns including government payments. ASCS policy does have significant implications for interpreting ERS traditional COP estimates, however, because ERS attempts to estimate commodity costs of production without direct government payments. This will be discussed more fully in the next section.

ASCS does make direct payments to landlords involved in share leases, distributing government payments directly to the landlord in the same proportion as the crop is allocated between landlord and tenant in the crop share agreement. No adjustment is made for input expenses the landlord may pay. Consequently, the net share received by the landlord (expenses paid by the landlord) is probably negotiated by landlord and tenant to account for ASCS rules regarding distribution of payments. As such, the landlord likely pays more of the expenses and/or receives less of the crop than would be the case in the absence of government payments.

Implications for Use of COP Estimates in Federal Policy

To this point our discussion has led us to the following conclusions. There are basically only two approaches to estimating the cost of land used for the production of agricultural commodities that are both acceptable and practical:

(1) an annual use charge as exhibited by cash rents, and
(2) an annual opportunity cost derived by multiplying the current market value of land by either a nominal or real measure of current interest rates, depending on whether capital gains are included in the returns portion of the COP budget.

Specifically excluded are methods that use the acquisition price of land and those that use net share rents.

Which of the two acceptable methods is appropriate depends on still unanswered questions regarding whose costs are being measured and when in the production cycle the measurement is being taken. In the introduction, we identified five fundamental assumptions which COP analysts should make explicitly before selecting an appropriate procedure for estimating land charges. In the remainder of the chapter, we use the framework provided by those five assumptions to evaluate the appropriateness of current procedures used by ERS to estimate land costs for Federal policy purposes. We begin by briefly reviewing the history of Federal COP analysis as it relates to land cost.

COP History

ERS estimation procedures related to land costs have evolved from four farm bills (McElroy). The Agriculture and Consumer Protection Act of 1973 is the landmark legislation that mandated the Federal COP work. It directed that: "The Secretary of Agriculture....shall conduct a cost of production study ... and establish a current national weighted average cost of production. This study ... shall include ... a return on fixed costs equal to the existing interest rates charged by the Federal Land Bank..."

In response, ERS estimated land costs using several combinations of opportunity interest rate and land value and combinations of cash and share rental rates. For instance, ERS published 6 separate estimates of the national cost of production based on 6 methods of calculating land costs. Two methods involved multiplying a current Federal Land Bank (FLB) interest rate times either the current market value of cropland or the average value of cropland at time of acquisition by the current operator. Two additional methods were based entirely on either cash or net share rents. The remaining two methods were composites of the others using prevailing percentages of cash, share, and owned land. The cost of owned land in the composite methods was based either on current market value of land or acquisition value.

The procedures mandated by the 1973 Act created a major problem (Hoffman and Gustafson). The land costs, calculated as current land value multiplied times current FLB interest rate, conceptually included all 'long run costs (i.e., full opportunity costs) but did not include all appropriate offsetting income in returns. Namely, capital gains were not included in returns, as would have been appropriate given the use of a current market interest rate. As discussed earlier, it is conceptually inappropriate to apply a current interest rate to a current value of land unless capital gains are included as a return.

The Food and Agriculture Act of 1977 addressed the land cost issue only in a peripheral sense. The 1977 Act specified that, after 1978, target prices for each commodity would be adjusted to reflect changes in a two-year moving average of farm costs. Land costs were excluded from the adjustment procedure in an attempt to avoid potential cost-of-production/income-support spirals.

The 1981 Agriculture and Food Act revised the 1973 wording to say, "...include ... a return on fixed costs, ...", essentially nullifying the requirement that land costs be calculated based on current interest rates charged by the FLB. The legislation also established the National Agricultural Cost of Production Standards Review Board (an advisory committee of agricultural producers and economists) to review ERS procedures.

Following the 1981 Act, ERS revised its procedures for estimating the cost of land. For a brief time, ERS land charges were all calculated by multiplying the current value of land times a measure of the real return on farm production assets. The real rate of return was based on a moving average over the previous 20 years (Hoffman and Gustafson). Shortly, thereafter, ERS adopted recommendations of the National Agricultural Cost of Production Standards Review Board to use cash and net share rent for crops; livestock land charges continued being estimated by the current land value/real rate of return method.

The Food Security Act of 1985 did not address COP directly, and the Food, Agriculture, and Trade Act of 1990 is being interpreted as making additions to previous requirements. Consequently, the 1981 Act amending the 1973 Act is interpreted as being still in effect.

Current ERS Procedures

ERS prepares and presents sector enterprise budgets that represent the farm operation and landlord as one business unit. Each enterprise budget contains costs and returns measures separated into three major categories: gross value of production or cash receipts, cash expenses, and economic (or full ownership) costs. All cash interest expenses are included in cash costs, but not in economic costs. USDA cash expenses do not include a cash or share rental expense because all of the land is owned either by the operation or the landlord. Land costs are included in the economic or full ownership section of the accounts.

ERS values land in COP accounts differently depending upon whether the imputation is for a crop or a livestock enterprise. For crops, ERS values land at its net rental value as a proxy for the land's opportunity cost. The land cost for a particular crop is a composite share and cash rent, minus real estate taxes and the value of variable inputs supplied by the landlord. In essence, the two rental rates are weighted together based on the relative amounts of land rented on a cash or share basis. This weighted rate is then applied to all acres in production whether owned or rented. Specifically, the land cost is based on an imputed value of land in the production process and is not calculated as a residual.

The cash rental data used by ERS are actual amounts paid by producers as reported in the Farm Costs and Returns Survey (FCRS). Net share rent is calculated from FCRS data related to the share of production that producers agreed to provide landlords and to the value of inputs landlords provided in the production process. A gross share rent is calculated first, by computing the quantity of output provided to the landlord multiplied by the harvest-month price for the commodity. Net share rent is then calculated by subtracting the value of the expenses paid by the landlord. For some crops in some regions, net share rent includes an implicit payment for the use of water.

For livestock, ERS estimates land charges by multiplying a measure of the real rate of return on production assets (as a proxy for the real rate of interest) times the current land price. The cost of land for livestock enterprises is calculated by multiplying current land value times a 10-year moving average of the rate of return on production assets--a procedure similar to that used for nonland capital inputs. ERS calculates the rate of return for a given year (i.e., for each year in the previous 10-year period) by taking the return to production assets in the agricultural sector, subtracting the imputed value of operator labor, and dividing this figure by the value of production assets. In hog-feeding, cattle-feeding, and dairy operations, land costs are relatively minor. Only the cost of land used for pasture, lots, and building sites are included in the COP estimates. Land costs are much larger, however, for land extensive beef-raising operations, especially in the West.

Federal legislation does not directly address water as an input in commodity costs of production. In ERS budgets, the cost of purchased water is included as a cash expense. Pumping expenses are included among cash expenses for fuel, repairs, and capital replacement, but do not appear as separate line items. The capital value of permanent irrigation developments and water rights are not calculated explicitly, but are presumed to be implicitly included in the economic costs section of the budget under land charges.

Evaluation

ERS procedures must be evaluated not only in light of specific legislative requirements, but also in light of interpreted goals and purposes. Like most laws, COP legislation requires interpretation. ERS, by necessity, has interpreted its mandate, either explicitly or implicitly, in the process of constructing a system for preparing COP estimates.

Sometimes it is difficult to determine the source of a given interpretation --whether it is specifically mandated by the legislation, required by USDA interpretation, or the result of independent decisions by analysts. Consequently, our evaluation of ERS procedures is conducted in light of the limited interpretations of the Congressional legislation that can be found in published sources. While not everyone agrees with ERS interpretations of its mandate, those interpretations at least carry the tacit approval of Congress, the USDA administration, and the National COP Standards Review Board.

The first crucial decision facing COP analysts is whether to include a land charge in cost of production estimates. The ERS position on this question is relatively concrete. Despite all of the conceptual and empirical problems that have been discussed, ERS is nevertheless mandated by Congress to produce COP estimates that include a charge for the land resource. The question is not whether, but how, to estimate this cost. Earlier discussion clearly pointed out that Congress has been active not only in specifying that a land charge will be included, but also in specifying exactly how that charge is calculated.

Other decisions addressed by Federal COP analysts include the following:

(1) To Estimate the Ownership Cost of Land or the Use Cost of Land. ERS land charges for crop enterprises are based on the use-rights approach, while land charges for livestock enterprises are based on the ownership-rights approach. The current procedure for estimating a cost for land used in crop production relies on information about cash and share rentals. Our earlier discussion in regard to ownership-rights and use-rights made two points that bear on the appropriateness of ERS procedures. The first point is that cash and share rents measure only the annual cost of obtaining the use-rights of land. Opportunity costs based on either of these rental rates do not account for capital gains that accrue due to inflation, permanent improvements, or nonfarm demand. Yet clearly, more and more of American agriculture is subject to nonfarm influence. Metro areas contain more than 29 percent of farms, accounting for 30 percent of the value of agricultural products sold (Heimlich and Barnard, p.2). The effect of this phenomenon is much greater in some parts of the country than in others. In the Northeast, for example, estimated agricultural use value accounts for only one-third of the market value of land (Heimlich and Barnard, p.10) For consistency across crops, regions, and years, we conclude that rental data are not an appropriate measure of the opportunity cost of land used in agricultural production. Clearly, this conclusion is subject to debate based on interpretation of the costs ERS is mandated to measure.

A second point is relevant only to the extent that the cost of obtaining use rights is deemed the appropriate cost for ERS to measure. Both cash and share rental rates are inappropriate measures of the opportunity cost of ownership rights in any circumstance. But, because share rents are determined primarily by actual yields and prices, they do not even measure the opportunity cost of obtaining use rights.

Cash and share rental arrangements for land used for livestock production are too uncommon to provide a basis for estimating land costs. Instead, ERS estimates the cost of land used in livestock enterprises by multiplying a real rate of return on production assets times a measure of the current land value. Conceptually, this approach measures the opportunity cost of full ownership.

(2) Point in the Production Cycle at Which Opportunity Cost is Measured. The operational assumption underlying ERS procedures for estimating land charges in crop enterprises is that the alternative use for land for any one landowner is to rent it to someone who will produce the same crop. This assumption has specific implications concerning the point in the production cycle at which opportunity cost is measured.

When opportunity cost is measured after the commitment to plant has been made, the relevant opportunity cost is the return foregone by not planting the best alternative crop. That commitment can be made either by the landowner committing to plant the land to a specific crop, or by the landowner renting with the proviso that the tenant plant a specific crop with no option to sublet. On the other hand, if opportunity cost is measured so early in the production cycle that the landowner could lease the land or a tenant could sublease the land, then the relevant opportunity cost can be derived from the cash rent paid for use of the land in planting the commodity in question.

From a conceptual perspective, determination of the point in the production cycle at which to measure opportunity cost is an arbitrary decision. Practically, it is easier to estimate cash rent for the crop in question than to estimate net returns for the alternative crop, which may vary from region to region and year to year. The fundamental point is that ERS procedures for crop enterprises assume that land opportunity cost is being measured early enough in the production cycle that cash leasing is still an option for the landowner (USDA, 1990).

(3) Are Costs Estimated from the Individual Firm or Commodity Industry Perspective. Although ERS COP estimates are presented as industry budgets rather than operator budgets, this primarily reflects the ERS approach to estimating economic (or full ownership) costs. When presenting economic costs, ERS treats each operation as if it were one business unit owned jointly by the operator and landlord. Consequently, land charges represent all land used for agricultural production, whether owned or rented, and irrespective of debt. But, ERS estimates actually reflect costs from the perspective of the individual operating unit rather than from the perspective of the commodity industry. Estimates designed to reflect the sector or commodity industry would need to account for costs generated by pecuniary externalities, which are not appropriate at the firm level. ERS estimates do not contain such adjustments to firm costs.

(4) To Estimate Social or Private Costs. ERS maintains that its accounts are not really budgets, in the sense of projections of costs, but instead are historical estimates based on actual levels of input use, yields, and prices as measured by farmer surveys. The emphasis on "actual" or "historical" costs, is meant to stress the survey-based nature of the costs, but also serves as a means of distinguishing between social cost and private cost. ERS makes no attempt to adjust its COP estimates to reflect costs that accrue to society but not to individual operating units. In general, such adjustments would require accounting for the effects of technical and pecuniary externalities. But in a broader sense, estimation of social cost would involve the issues and controversy associated with project evaluation, cost/benefit analysis, and macro economic planning.

Water probably provides the most vivid illustration of the importance of this point. Over 25 percent of irrigated acreage in Western States is supplied water by Bureau of Reclamation projects (Moore and McGuckin). The Bureau's water pricing policies generally recover only a portion of

Government expenditures on the water supplied and even less of the social opportunity cost of the water, leading to substantial subsidies to irrigators. Consequently, there is a substantial deviation between social and private opportunity costs for Bureau water.

(5) To Include or Exclude Government Payments. One of the principal characteristics of the COP accounting procedures traditionally used by ERS is the goal of excluding, to the extent possible, the effects of government programs. The purpose of this goal is to provide policymakers with information as to the production costs and returns that stem from the market rather than government programs. ERS is only partially successful in this regard. While the government operates a myriad of agricultural programs that affect farm income and commodity markets, ERS actually attempts only to exclude direct government payments.[6] No attempt is made to exclude the effects of any government programs that act indirectly on commodity markets and prices through mechanisms that do not involve payments directly to operators or landlords. The goal of excluding government payments has important implications for ERS procedures related to estimating a land cost, because government programs have a profound effect on both the rental market and the title market for agricultural land and on returns to agricultural production.

The government operates innumerable programs that affect agricultural land markets by influencing the returns producers expect from agricultural production. Commodity prices are supported through import limitations, commodity stockpiles, marketing quotas, and nonrecourse loans. Programs designed to reduce costs or increase productivity include Soil Conservation Service cost sharing, Farmers Home Administration lending, and Extension Service educational activities. To the extent that benefits from government farm programs increase expected net returns, they also increase land rents. To the extent that producers expect increased returns to accrue for more than one year, the net benefits from government programs become capitalized into land values.

ERS does not attempt to exclude the effects of government programs that do not involve direct payments to operators or landlords for program participation. Consequently, the simultaneous relationship between land returns and land values implies that ERS estimates of land opportunity costs, to some degree, reflect the net effect of government programs. When the net benefits from government programs change, so do rental values and land values. The government also supports farm income with direct payments to farm operators and landlords based on proven yields when they participate in the program. This activity has particularly profound effects on ERS estimates of land cost. ERS omits direct government payments from its estimates of cash receipts, but is only partially successful in removing the effects of direct government payments from estimates of land cost that are based on cash and net share rents.

As a result of ASCS policies for distributing direct government payments, the effects of programs are reflected to substantially different degrees in cash and net share rents. Earlier discussion noted that ASCS pays landlords a portion of direct payments when land is leased on a share basis, but not when land is leased on a cash basis. Assuming that rental rates are negotiated with knowledge of ASCS policies, producers, *ceteris paribus*, should be willing to pay more rent under a cash rental agreement than under an analogous share rental agreement. The additional cash rent should approximate the amount of direct government payments ASCS would distribute to the landlord under the analogous share agreement. Consequently, the landlord's share of direct payments will be reflected in cash rents. For share rental arrangements, ERS values the landlords share of production at (harvest-month) market price, which does not reflect the value of direct government payments. This approach is consistent with USDA's goal of eliminating the effects of direct government programs where possible.

The implication of ERS procedures, however, is that the value of all direct government payments are reflected in cash rental rates but not in net share rates. Consequently, the composite rental rate used by USDA, which is composed of both net share and cash rates, only partially excludes the effects of direct government payments. This also implies that ERS estimates of land costs inappropriately vary across regions and crops depending on tenure practices.

Rice, more than any other commodity, demonstrates how the cost of land and water can interact to influence rental rates. Unlike other crops, rice is grown under flooded field conditions throughout the growing season. As a result of this interdependence between land and water, rice land must have irrigation water available. The source can be surface water in the form of a canal or river running adjacent to the land or underground water in the form of a well located on the land. Much of the rice produced in the U.S. is grown under share rental arrangements in which reimbursement for the use of water figures prominently. Operators (tenants) often pay landlords equivalent shares for the use of water and land. For example, under a common share rental agreement in the Gulf Coast area, the operator might pay 20 percent of rice production for use of the land and another 20 percent of production for use of the water. The landlord might reimburse the operator for a portion of various specified production expenses.

Furthermore, direct government payments (in the form of deficiency and marketing loan payments) comprise a sizeable portion of gross returns from rice production. Given that producer participation rates in rice support programs regularly exceed 90 percent, this implies a substantial differential between the market value of rice and a total return including direct payments. Consequently, a rice land charge based predominantly on share rental arrangements, which excludes the effect of direct government payments, is substantially less than a rice land charge calculated to include the effects of direct government payments. An ERS comparison of 1988 rice production costs with and without direct government payments showed that inclusion of government payments increased net land charges by up to 234 percent, depending on rice production region (Salassi, *et al.*).

In closing, we note that estimation of land charges can never be a tidy conceptual or empirical process. Nevertheless, careful consideration of several basic assumptions can help identify the implications of using one procedure rather than another. Estimation for Federal COP analysis, of course, is constrained by legislative and executive requirements. But, further discussion of actual and perceived requirements would be useful.

Notes

1. The supply of water may be fixed in the sense that access to water may be legislatively controlled by a pseudo-input called a water right.

2. Pasour maintains that opportunity cost is a purely subjective concept, unmeasurable by economists. Recognition of this idea is especially important for planning decisions at the firm level.

3. The acquisition price of land times the interest rate prevailing at the time of acquisition has occasionally been used to reflect costs actually incurred by the purchaser. But, as discussed earlier, estimates derived from such a procedure bear little resemblance to the land's current opportunity cost, and should be dismissed from serious consideration.

4. Nominal refers to the observed interest rate actually charged on loans, while real refers to the level of the same interest rate that would prevail in the absence of inflation.

5. Several methods of calculating a real interest rate are in common use. Readers should note, however, that there is an entire set of controversies and related literature concerning the meaning and measurement of real interest rate. While we recognize these issues exist, further discussion is beyond the scope of this chapter.

6. Income and expenses related to crop insurance and disaster payments are also excluded.

References

Friedman, Milton. 1976. *Price Theory.* Chicago: Aldine Publishing Co.

Ferguson, C.E. 1969. *Microeconomic Theory.* Homewood: Richard D. Irwin, Inc.

Harrington, D.H. 1983. "Costs and Returns: Economic and Accounting Concepts." *Agr. Econ. Res.* 35: 1-8.

Heimlich, R.E., and C.H. Barnard. 1990. "Agricultural Adaptation to Urbanization: Farm Types in Northeast Metropolitan Areas". Selected paper, Northeast Agr. and Resour. Econ. Assoc.

Helmers, G.A., ed. 1980. "Developing and Using Farm and Ranch Cost of Production and Return Data: An Appraisal." Great Plains Agricultural Council Publication 95. University of Nebraska.

Henderson, J.M., and R.E. Quandt. 1958. *Microeconomic Theory.* 2nd Edition. New York: McGraw-Hill.

Hoffman, G., and C. Gustafson. 1983. "A New Approach to Estimating Agricultural Costs of Production." *Agr. Econ. Res.* 35: 9-14.

McElroy, R.G. 1987. "Major Statistical Series of the U.S. Department of Agriculture: Costs of Production, Vol. 12." USDA, ERS. Agr. Handbook No. 671.

Miller, T.A., and M.D. Skold. 1980. "Uses and Users of Costs and Returns Data: A Needs Analysis" in Helmers, G., ed. *Developing and Using Farm and Ranch Cost of Production and Return Data: An Appraisal.* Great Plains Agricultural Council Publication 95. University of Nebraska.

Miller, T.A. 1983. "Costs, Returns and Land Values On Colorado Wheat Farms 1950-1980: How Historical Relationships Have Changed." Colorado Agr. Exper. Sta. Tech. Bull. 150. Colorado State University.

Moore, M.R. and C.A. McGuckin. 1988. "Program Crop Production and Federal Irrigation Water." *Agricultural Resources: Cropland, Water, and Conservation Situation and Outlook Report.* U.S. Department of Agriculture, Econ. Res. Serv. AR-12.

Pasour, E.C., Jr. 1980. "Cost of Production: A Defensible Basis for Agricultural Price Supports?" *Amer. J. Agr. Econ.* 62 :244-248.

Ricardo, D. 1911. *The Principles of Political Economy and Taxation.* New York: E.P. Dutton.

Robison, L.J., D.A. Lins, and R. VenKataraman. 1985. "Cash Rents and Land Values in U.S. Agriculture". *Amer. J. Agr. Econ.* 67 :794-805.

Salassi, Michael, Mary Ahearn, Mir Ali, and Robert Dismukes. 1990. "Effects of Government Programs on Rice Production Costs and Returns, 1988." USDA, ERS. Agr. Info. Bull. No. 597.

Scott, J.T., Jr. 1977. "Returns to Land in the Corn Belt, Government Crop Price Guarantees Based on Cost of Production, and Land Values." Paper presented at a symposium held at the annual meeting of the AAEA. Dept. of Agr. Econ. AE-4448, University of Illinois, August.

Scott, J.T., Jr. 1983. "How to Determine the Cost of Land by Observing Rents" in *Rents and Rental Practices in U.S. Agriculture: Proceedings of a Workshop on Agricultural Rents.* Chicago: The Farm Foundation.

Sharples, J.A., and R. Krenz. 1977. "Cost of Production: A Replacement for Parity?" in *Agr. Food Policy Rev.* USDA ERS. AFPR-1, pp. 62-68.

U.S. Dept. of Agriculture, Economic Statistics and Cooperative Service. 1979. "Estimating Agricultural Costs of Production--Workshop Proceedings." ESCS-56.

U.S. Dept. of Agriculture. 1990. "Cost of Production: Major Field Crops, 1988." USDA, ERS. ECIFS 8-4.

Wunderlich, G. 1990. "Trends in Ownership Transfers of Rural Land." USDA, ERS. AIB No. 601.

Estimating Costs of Land and Water Services: Discussion

Allen M. Featherstone

The chapters by Robison and Koenig and Barnard and Salassi approach the topic of the estimation of land and water services from two points of view and both end up at virtually the same point. Both chapters provide insight into appropriate measures of land cost while neither discuss water services cost estimation in depth.

Before discussing the chapters more in depth, I would like to commend USDA's Economic Research Service for making their estimates widely available and for providing documentation on how the estimates are produced. Anytime, an estimate is published there is always a risk that it will be taken out of context or used incorrectly. Providing the documentation of how the results are constructed allows those users who are interested in trying to use the data correctly to do so. Recently I have observed an agribusiness, who uses the USDA numbers extensively, check on how the numbers are calculated before they use and present their analysis to others.

The chapter by Robison and Koenig makes the point that use value of land for agricultural purposes may not be equal to market value. They argue that nonfarm demand for land for purposes such as industrial or residential development will lead to a divergence between market value and value in agricultural production. They also suggest that government programs may also be viewed as driving a wedge between use value and market value. Barnard and Salassi also make roughly the same point although coming at it from a slightly different angle. They argue that either cash or share rent is an appropriate measure of land cost.

Both chapters accept the premise that the cost of obtaining operating rights for land is different than obtaining ownership rights, (i.e., use value is not equal to market value). The 1987 Census of Agriculture reports that 54.6% of the land in farms is owned by the operator (U. S. Dept. of

Commerce). Thus for 45.4% of all land in farms the appropriate measure of the cost of obtaining land use from a farmer's perspective is the cash or share rental rate and that the return to land is not a residual return. In fact, the land input could likely be viewed as a variable input.

It is important for farmers, policymakers, and even agricultural economists to begin to fully understand this point. Land use need not be tied to land ownership. Land investment decisions and land ownership decisions can be treated as independent decisions. To facilitate this understanding, it may be appropriate to begin putting together cost and return budgets for specific crops based upon the use value of land. A separate budget could then be put together analyzing the returns for land ownership. If a producer was interested in land ownership, they could choose the appropriate crop budget and use the land ownership budget to look at the full cost and returns for that individual.

Robison and Koenig in their chapter estimate the percentage of agricultural land's market value attributed to agricultural rents and government payments. Value is expressed as a sum of use value in production agriculture and nonagricultural rents and government payments. Theoretically, there are no major problems with analyzing the value of land in this manner. However, their empirical results must be interpreted carefully. This arises from the use of the agricultural cash rent to measure the use value of land. This concern can best be illustrated from arguments made in the Barnard and Salassi chapter. Barnard and Salassi illustrate that cash rent is affected upwards by government programs.

This is further illustrated in a paper by Featherstone and Baker who have found that a one dollar increase in returns for an Indiana farm (whether through market forces or government payments) will increase cash rent by 8.1 cents the next year. If this increase is viewed as a permanent increase, as might be the case with government payments, cash rent will be 60.3 cents higher for each dollar increase in returns. Thus, both theoretical and empirical evidence suggests that cash rents are influenced by government programs. Thus, the percent of returns attributed to use value in the Robison and Koenig chapter are incorrect because they have government programs captured in both their "R" value and their "Q" value.

Although not directly related to the cost and returns budgets, Robison and Koenig suggest concern about USDA's estimates of agricultural farmland value. I would like to echo their concerns about the reliability of these numbers. The manner in which these estimates are derived provides many opportunities for response errors. Relying on farmers who at most have little contact with actual transactions of land to provide an estimate of value is worrisome. The differences between the Minnesota numbers and USDA's numbers for farmland value illustrates a potential problem. Although the land values collected by the University of Minnesota are likely to be more reliable, there still is a chance for response errors. Asking an individual to give an estimate of the value of land as of a specific date allows for many outside influences to enter into the process. It would be useful if these numbers could be matched with actual sales from time to time to anchor these data series.

Barnard and Salassi provide a useful discussion of legislation which deals with the inclusion of land in the cost and return budgets. They also provide a fair amount of detail with regard to how the returns are actually calculated. I am concerned with the procedure that USDA uses to obtain cost of land because share rents excluding government payments are averaged with cash rents which were pointed out above to include a return to government payments. Land cost estimates for two identical operators could differ substantially based upon the terms of rental agreement. The cost of land for the operator using all share rent will be lower than the cost of land for the operator using all cash rent because share rent does not include a return to government payments while cash rent does. This problem could be reduced dramatically if USDA included direct government payments when calculating cost and return budgets.

In conclusion, both chapters make a contribution to literature dealing with the estimation of land and water services. A consensus exists between the chapters which suggests that agricultural use value may not be equal to market value and that some form of rental rate seems to be an appropriate measure to use in cost and return budgets for the value of land.

References

Featherstone, Allen M. and Timothy G. Baker. 1988. "Effects of Reduced Price and Income Supports on Farmland Rent and Value." *North Central Journal of Agricultural Economics* 10: 177-89.

U.S. Dept. of Commerce. 1990. "Agricultural Economics and Land Ownership Survey (1988)." Bureau of the Census. 1987 Census of Agriculture. Volume 3, Part 2.

Estimating Costs of Land and Water Services: Discussion

Michael Duffy

The Barnard and Salassi chapter deals strictly with the land price issue from the cost of production standpoint. They do a very nice job outlining the underlying theoretical issues. They also identify five assumptions or areas where the conflict and confusion arise. I will present them here because I feel they bear repeating.

The first issue is estimation on an annual ownership cost or an annual land use cost. The second assumption is what point in the production cycle the estimate represented. Third, are cost estimates for an individual farm or the commodity industry. The fourth point is whether these estimates are for private or social costs. And, the fifth point is whether or not to include government payments. Although this fifth point is stated in the chapter as "government payments in cost budgets." I would argue that the payments are more a concern on the receipt side of the ledger.

Barnard and Salassi use their five assumptions or points to arrive at the crux of the matter. That is, whose costs do you want to measure and when in the production cycle. I would also add why to their list, but they may have subsumed that question under whose.

They proceed by listing and discussing alternative ways to view land in COP estimates. They arrive at two approaches: using an annual use charge such as cash rents or annual opportunity cost using current market value. The end of this chapter looks at how the ERS estimates address the five assumptions. The ERS estimates are derived where "the land cost is based on an imputed value of land in the production process and is not calculated as a residual."

The Robison and Koenig chapter takes a different approach. First, they discuss many of the problems with the current data sets available for land value estimates. The second point they make is that land values are composed of more than just the agricultural value segment. I think we would all be well-served to keep in mind the quality of our data as we use them in statistical routines.

The model they present and test essentially is saying there is more than one component to land values. They estimate the models looking at government payments and other nonagricultural rents.

I wish that this chapter could have been more thoroughly developed. I think they are correct in the approach they are taking. However, I feel using state estimates will mask some of the nonagricultural demand for land. An example would be development demand which is more likely to be found at the county or crop reporting district level.

There are some observations I have on the land market that were not mentioned in these chapters. Farmers are the primary purchasers of land in the Midwest and many other regions. Robison and Koenig mention that 30 percent of the purchasers are nonfarmers as a rationale for questioning farmer opinion data on land values. I would turn this statistic around (as a note, in Iowa, farmer purchases are a much higher percentage) and say if 70 percent of the land is purchased by farmers then can we use standard investment theory to analyze land values?

Farmers do not purchase land to resell it. Rather they purchase it to own it. There is a certain utility gained from owning land. This seems to be a bigger factor than has been noted. In a sense this could be part of the nonagriculture component discussed by Robison and Koenig.

Given that farmers are the primary purchasers and that the purchase decision is different from investors makes me question the appropriateness of including capital gain in land value estimates. Barnard and Salassi discussed including real interest and capital gains. It is my contention that nominal interest rates and current market value are the most appropriate measure of opportunity cost of ownership for the farmer.

Water rights were given only passing mention by Barnard and Salassi. This is an extremely important consideration in some parts of the country. Similarly, mineral extraction rights can also be an important determinant of the land value. Water rights and related issues need to be recognized. One possible suggestion would be incorporation into the model developed by Robison and Koenig.

There was a general consensus among the discussants that in cost of production work the appropriate charge for land would be a charge reflecting the annual agricultural use. A charge for cash rent or equivalent was felt to be the best alternative. No consensus emerged on valuing land in the long run nor on the water rights issue.

PART SIX

Estimating Costs of Non-Human Capital Services

15

Allocation of Capital Costs in Enterprise Budgets

Oscar R. Burt

Introduction

The problem of intertemporally allocating the functional services of an asset with economic life greater than a single accounting period has always been an intriguing puzzle, and economists have typically concluded that allocation is inevitably arbitrary and driven by specific purposes underlying the allocation (see for example, Lutz and Lutz, p. 7). The leading economic theory of depreciation has been based on discounted value of quasi-rents associated with services provided by the asset; but for a single asset, as contrasted to an infinite chain of like assets, depreciation charges are not *accounting admissible* as defined by Atkinson and Scott because they sum to more than the difference between investment cost and salvage value, i.e., total depreciation charges include goodwill of the firm itself (Edwards). When an infinite planning horizon is used with a sequence of identical assets as replacements, the implicit depreciation charges based on present value are accounting admissible; we call this model the asset valuation method. The other economic theory is based on Hick's definition of income, which in this context says that the firm should be equally well off during each period it owns the asset in-so-far as depreciation allocations affect income (Hicks).

The author tried to reconcile these two theories, as well as what appeared to be other separate theories in the literature (Burt, 1972). A main conclusion was that using the Hicksian income based theory requires that the firm be in such a position that age of the asset per se does not affect the firm's gross returns in a significant way. This restriction seems substantive until one considers the problem of defining quasi-rents over many assets in a consistent way when they are all part of an interrelated production process, which is necessary if the asset valuation method is used. Fortunately, the income measurement approach is clearly the one for cost of production studies which seek to represent a national or regional industry, or a representative firm according to some definition.

Wright introduced the concept of *opportunity value* which "is measured by the least costly of the alternatives avoided through owning the services." From the point of view of an individual firm, this could be commercial rental of the services, but even for the firm, opportunity value would predominantly be cost of the same services from a replacement asset. For purposes of calculating industry costs of production, the relevant alternative would always be services from a replacement. The opportunity value theory of Wright's falls under the income based theory when put in operational form as the *unit cost theory* which was originally introduced by J. S. Taylor in 1923 and generalized by Harold Hotelling in 1925.

In this chapter discussion of the problem will be limited to farm assets such as equipment and machinery for which there exist market prices for new assets and the farmers are essentially price takers in these markets. In this situation and for the task at hand, opportunity value for the services of a used asset is the cost of the same services from the most alike new replacement asset, although we might quibble about just how comparable the services are at different points in time because of technical change and manufacturers' specific packaging of consumptive and productive characteristics. This is so because the used asset market is thin and heavily influenced by local conditions, or nonexistent in some cases, thus making it necessary to estimate used asset values implicitly from new prices and a theory of depreciation.

Nonstochastic Economic Life

Unit Cost Theory of Depreciation

The unit cost measure defined here was called "unit cost plus" by Taylor to distinguish it from a simple measure that ignored interest costs, but following later writers, we will call Taylor's unit cost plus simply unit cost. Hotelling used the term "theoretical selling price" for what we call unit cost. As those familiar with his research would expect, Hotelling's brilliant analysis was presented in a very general mathematical form using continuous time and variable interest rates through time. He also allowed for a component of costs such as taxes and insurance which were proportional to book value; the latter implicitly makes the model stochastic although it meant solving a functional equation in his formulation. We use a discrete time model and begin with the simpler nonstochastic case to communicate the central ideas before introducing the complications of random economic life for the asset. All monetary values are implicitly defined with respect to the purchasing power of money at a single point in time, i.e., adjusted for inflation when measured over time.

The following notation is introduced:
t = age of the asset in years
T = planned replacement age (exogenous in most of the discussion)
I = acquisition cost of a new asset
C_t = annual operating and maintenance outlays (including labor)
Q_t = physical measure of output, i.e., amount of quantifiable services provided by an asset of age t (assumed positive for simplicity)
r = discount rate ("real" opportunity cost of capital)
β = $1/(1+r)$, the discount factor
S_T = net salvage value of the asset at planned replacement age, T
D_t = depreciation charge in year t
V_t = book value of the asset at age t
u = nonnegative unit cost for the service Q_t (Hotelling's theoretical selling price)

The integer age variable t refers to the continuous time interval (t-1,t], and all transactions, including the accounting charge for depreciation, are defined to take place at the end of the year. For example, Q_t is output on the time interval (t-1,t] and evaluated for discounting purposes as if it all occurred at the moment t. For some applications, Q_t would be a quality adjusted measure of the services provided, or actual physical output produced with the machine. The essential idea is to use a standardized measure of output such that a constant unit cost over age is unambiguous.[1]

Think of $uQ_t - C_t$ as an implicit rent on the asset, then unit cost is determined by making it as small as possible under the constraint that the present value of rents plus salvage value is not greater than the initial investment cost, i.e., unit cost is determined by[2]

(1)
$$I = \sum_{j=1}^{T} \beta^j (uQ_j - C_j) + \beta^T S_T,$$

which when solved for u gives

(2)
$$u = [(I - \beta^T S_T) + \sum_{j=1}^{T} \beta^j C_j] / [\sum_{j=1}^{T} \beta^j Q_j].$$

Hotelling also treated replacement age T as part of the minimization problem, which if we were to do the same, would require a search over the integer T to minimize u in (2). However, we prefer to leave the criterion of optimal replacement unspecified to avoid controversy which is tangential to the purpose at hand. Unit cost is quite intuitive economically; it is the present value of all costs minus present value of salvage, all divided by a weighted sum of output over the life of the asset where the weight at age t is the present value weight β^t (discounted value of total output measured in physical units, if you please).

After incurring the initial investment cost, present value of the remaining services in the asset would be

(3)
$$V_t = \sum_{j=t+1}^{T} \beta^{j-t}(uQ_j - C_j) + \beta^{T-t}S_T$$

for t = 0, 1,...T. Note that V_0 defined by (3) is simply the right hand side of (1), and therefore, equal to I. Annual depreciation charges are given by

(4)
$$D_t = V_{t-1} - V_t,$$

and it is easily verified that these charges are accounting admissible because they sum to I-S_T, the difference between initial cost and salvage value of the asset. The summation of D_t in (4) from t=1 to t=T yields canceling terms in V_t, t = 1, 2,...T-1, which leaves V_0 - V_T, but V_0 = I and V_T = S_T.

The problem is simplified considerably when output Q_t is constant over age of the asset because uQ_t reduces to a constant over t, say uQ. = K. In this case (1) can be solved for a constant annual charge,

(5)
$$K = [(I - \beta^T S_T) + \sum_{j=1}^{T} \beta^j C_j] / [\sum_{j=1}^{T} \beta^j],$$

and then unit cost is simply

(6)
$$u = K/Q_.,$$

while the appropriate measure of depreciation is still given by (4) with uQ_j in (3) replaced by K from (5). The contrast of (2) with (5) and (6) makes it clear that in the general case of (2), one cannot simply calculate unit cost by a proportional adjustment to annual cost of owning the asset.

In (5) K is amortized present value of costs of owning and operating the asset, but in (2) unit cost of service from the asset depends on the sum of the product of the discount factor and the amount of services from the asset each year throughout the future life of the asset, not just the discount factor alone. Consequently, the cost of a specific number of units of service from an asset cannot be calculated from amortized present value of costs associated with ownership of the asset unless the quantity of services is constant during each period in the life of the asset; the distribution of services from the asset over its life is an intrinsic part of the weighting required to calculate unit cost of the services. It should be intuitively clear from (2) that a relatively large number of services provided early in the life of the asset relative to late in its life, will tend to reduce unit costs and vice versa.

Farm Enterprise Costs of Production

If the task were to do "synthesized" enterprise budgeting,[3] assuming current technology with respect to machinery and equipment, the calculation of unit cost from the formula in (2) would be sufficient; this cost would be multiplied times the number of units of service per year required from a machine in the particular enterprise under study. Adding these annual costs over the entire complement of machinery used in the enterprise would give the sum of annual capital costs and direct repair and maintenance costs for the equipment. But the task faced by ERS is considerably more complicated, and these results provide a benchmark for comparison to current procedures. To that end we go through some algebraic manipulations that partition out components in cost which can be compared with corresponding ERS measures.[4]

Annual costs of owning and operating the asset are

$$(7) \qquad A_t = (rV_{t-1} + D_t) + C_t,$$

that is, the sum of interest on the investment and the decrease in value of the asset, plus operating and maintenance costs. In a cost of production study, we would be interested in estimating the equivalent annual value of these annual costs, i.e., the weighted average over the T years using weights, β^t, $t = 1, 2,...T$, where the reciprocal of the sum of these weights is the familiar formula,

$$1/\sum_{j=1}^{T} \beta^j = (1-\beta)/\beta(1-\beta^T) = r/[1-(1+r)^{-T}],$$

and this would be multiplied times the present value of costs to get equivalent annual value.

But if u denotes cost per unit of service provided by the asset during its life, then total cost of the services provided during year t must be equal to uQ_t. Therefore, present value of annual costs is the identity

(8)
$$\sum_t \beta^t u Q_t = \sum_t \beta^t (r V_{t-1} + D_t + C_t).$$

Using (1), the present value of operating costs is

(9)
$$\sum_t \beta^t C_t = \sum_t \beta^t u Q^t - (I - \beta^T S_T),$$

which when substituted into (8) and canceling the summation in $\beta^t u Q_t$ on both sides of the equation leaves

(10)
$$\sum_t \beta^t (r V_{t-1} + D_t) = I - \beta^T S_T.$$

We conclude that the present value of capital costs for an asset is simply the right hand side of (10), and equivalent annual value is[5]

(11)
$$[r/(1 - (1+r)^{-T})](I - \beta^T S_T).$$

The one remaining component of annual costs is amortized present value of operating and maintenance costs which is

(12)
$$[r/(1 - (1+r)^{-T})]/\sum_t C_t (1+r)^{-t}.$$

It would appear from reading USDA's Costs of Production publication (USDA, pp. 10-11) that ERS is using a substitute for (11) which is an approximation for

(13)
$$r[\sum_{t=1}^{T} f_t V_{t-1}] + (I - S_T)/T,$$

where f_t is the relative frequency of asset age t and these weights sum to unity.[6] Note that (11) can also be written as

(14)
$$[r/(1-(1+r)^{-T})]\sum_{t=1}^{T}(rV_{t-1}+D_t)(1+r)^{-t}.$$

Conceptually, ERS is using a relative frequency and simple average in place of weighted averages based on the discount factors,

$$(1+r)^{-t}, t = 1, 2,...T.$$

This is so for the second term in (13) because the depreciation charges D_t have a sum equal to $I - S_T$. Making the assumption of a steady-state age distribution is not sufficient for using (13) instead of (14) because firms individually must always incur implicit interest costs. Each firm with an asset of a given age must reckon with the opportunity cost of capital regardless of asset age; this is made clear by (14) where the term involving rV_{t-1} is this annual opportunity cost.[7]

Therefore under static technology, the annual cost measures developed here from unit cost theory are representative of the entire industry regardless of age distribution at any point in time. The steady-state merely assures us that the survey data provide an accurate average of asset value as defined in (13). The correct measure of annual capital cost is (14) which is expressed most simply by (11), and the cost per unit of service from the asset is u as defined in (2) which contains both operating and maintenance costs together with capital cost per se.

But as explained in the last paragraph preceding this section, annual costs of owning the asset are not adequate to measure costs of production unless services produced by the asset, Q_t, are constant over t. It might be tempting to make some kind of *pro rata* allocation of annual costs to different enterprises according to the amount of services from the asset used in that enterprise, but this will not give a proper weighting in relation to the variation in services provided over the life of the asset. In the general case, we need the measure of cost to be on a per unit of service instead of per time period basis. ERS recognizes this to some extent by dividing $I - S_T$ by total units of service embodied in the asset to get "replacement cost" (USDA, p. 11). Inspection of (2) shows that the component of unit cost associated with capital cost of the asset (depreciation and opportunity cost) is

(15)
$$[I - \beta^T S_T]/[\sum_{j=1}^{T} \beta^j Q_j],$$

which is different than (11) which can be written as

$$[I - \beta^T S_T]/[\sum_{j=1}^{T} \beta^j].$$

Therefore, ERS should use (15) instead of the sum of what they define as replacement cost and opportunity cost; the latter being some approximation to the first term in (13) adjusted to reflect the relative amount of services used in a specific enterprise.

ERS handles operating and maintenance costs much like "replacement cost" by summing these costs over the life of the asset and dividing the result by total number of hours of service embodied in the asset. The per hour cost is then multiplied times the hours the asset is used in a particular enterprise. Direct labor costs of operating the asset are calculated much the same way by making these costs proportional to the hours the asset is used. This procedure is in contrast to using the component of unit cost given by the second term in (2), namely,

(16)
$$[\sum_{j=1}^{T} \beta^j C_j]/[\sum_{j=1}^{T} \beta^j Q_j].$$

Note that the numerator calculates present value of the costs, and the denominator makes the adjustment to a per unit of service measure taking into account both the discount factor and the variable rate at which services are provided over the life of the asset. The set of costs $\{C_t\}$ should include all variable costs associated with operating the asset, including those which are constant over age. Costs which do not change with age of the asset could be handled separately if Q_t were constant over t, but in general, it is obvious from (16) that taking C_t equal to a constant will not yield a cancellation of terms in the numerator and denominator and avoid the necessity of recognizing the effect of variation over Q_t in the weighting used. The only exception where there is a canceling of terms is if the cost component is proportional to Q_t, but this situation would appear to be rare.

As an example, suppose that we want to estimate the value of services from a harvester used in the production of grain on one acre. For a typical grain harvester used in the region of study, the purchase price and salvage value at the average age at which the harvester is taken out of service (scrapped) are estimated, and an estimate is made of the number of units of service that the harvester would provide at different ages, i.e., Q_t, t=1,2,...T. The latter would largely depend on the amount of downtime during the harvest season in relation to age under an assumed average intensity of utilization for the machine, but it could also depend on how heavily the machine was "loaded", e.g., the speed at which the harvester moves for given field conditions. If the latter is an important consideration, hours of service would not be an appropriate unit of measure for Q_t, unless it were adjusted downward for aging of the harvester and the hours of service were standardized to that of a new machine. Lastly, a schedule of operating and maintenance outlays by age of the harvester, i.e., C_t, would need to be estimated. From this information, u can be calculated from (2) after specifying a suitable real interest rate which determines β. The last step is to multiply u by the number of units of service required to harvest an acre of grain, and that product is the estimated capital, maintenance, and operating costs associated with the harvester for an acre of grain. Direct labor costs would be included in these operating and maintenance costs, C_t.

In view of the fact that ERS is charged with the responsibility for estimating annual costs of production year after year, it would be feasible to use a weighted moving average of vintage capital measured by unit cost as defined in (2). The weights would be estimates of relative frequencies of the various vintages in the mix of machinery on farms, thus accounting for technological change which tends to reduce unit costs of production. At least this approach could be used on major items such as tractors and an associated complement of machinery, harvesters, etc. After the process reached "maturity", only the current year's unit cost would need to be estimated each year and the earliest year used last time deleted, thus maintaining a rolling average of unit cost properly weighted by the relative numbers in the population of machines.

Generalizations of Unit Cost

We will now generalize these results to accommodate taxes and insurance costs which tend to be proportional to book value of the asset if a consistent economic theory of depreciation is used; clearly, arbitrary depreciation charges frequently used for income tax purposes or just simplicity will yield book values which are spurious as far as measuring replacement cost or opportunity value of an asset. Property taxes are usually purported to be proportional to market value of the asset, and a logically consistent economic theory of depreciation should yield book values which approximate market prices for used assets in a competitive environment, at least under static conditions. Strictly speaking, an allowance should be made for average salvage value of a destroyed asset when considering insurance, but this will be left to the section on random economic life of assets and the results here can be thought of as a first approximation.

Hotelling demonstrated with his continuous time model of unit cost depreciation that costs proportional to book value of an asset can be introduced into the model by simply adding a component to the discount rate which reflects the sum of insurance and tax rates for each age of asset. Either tax or insurance rates could be different across ages of the asset, such as special taxes on new purchases or insurance rates which reflect a greater peril for theft of a newer asset. Hotelling obtained his results by solving a functional equation which happened to be linear and simple enough to solve analytically. The discrete time model presented here could be used to deduce analogous results, where the continuous time functional equation would be replaced by a system of linear equations.

We define δ_{jt}, with $j > t$, as the product of three discount factors comparable to β used above for the real interest rate, and also the product of that composite discount factor over periods $t+1$, $t+2$,...j. Letting x_t and i_t denote the tax and insurance rates at age t, then

(17)
$$\delta_{jt} = \prod_{k=t+1}^{j} [1/(1+r)(1+x_k)(1+i_k)], j>t, t=0,1,...,T-1.$$

The basic equation defining unit cost, and the equation for unit cost explicitly, are obtained by replacing β^j by δ_{j0} in (1) and (2). Implicit value of the asset V_t in (3) is obtained by replacing β^{j-t} by δ_{jt} and we define δ_{jT} equal to unity to make the formula for $V_T = S_T$ appropriate. The general tendency of these added costs is to increase unit cost, and also increase depreciation charges for younger relative to older assets, if maintenance costs are monotone increasing while output is monotone decreasing as is typical for farm machinery and equipment. Under this same assumption, optimal planned replacement age will be increased.

Another generalization is systematic obsolescence which would imply decreasing unit cost over time with respect to the latest technology being used. A trend for this decrease could be estimated from price and cost data for the same type of asset purchased new over a sequence of several years. This estimated trend function for unit cost would be taken as given except for the intercept, which would play the same role as the unknown fixed unit cost u. Let $\alpha - \phi(t)$ be the trend for unit cost and a is treated as an unknown to be calculated by an equation like (1), i.e.,

$$I = \sum_{t=1}^{T} \beta^t[(\alpha - \phi(t))Q_t - C_t] + \beta^T S_T$$

Then this equation is solved for a to get a value to be used in calculation of V_t and the depreciation function D_t, where u is replace by $\alpha - \phi(t)$ in the equations. It is seen that the declining trend in unit cost has the same effect as changing the distribution of the output path Q_t in the direction of a more rapidly falling rate of output with respect to age. This effect will tend to accelerate depreciation of the asset and reduce the optimal replacement age, T.

Random Economic Life

One purpose of the stochastic model is to handle the following type of situation. The firm has an economic replacement strategy which depends on several random state variables which summarize the physical condition of the asset at any age, and given a set of values for the random state variables, an optimal sequential decision rule exists which indicates whether or not to replace the asset or wait at least one more stage of the decision process. State variables other than age could be such things as hours of use, whether or not critical components such as the engine were functioning well, hours of use since last overhaul, etc. Application of any such decision rule for replacement generates a probability distribution for survival of the asset to various ages, which implies a set of survival probabilities by age. It is assumed in the model for depreciation theory that this decision rule includes a finite maximum age of replacement such that there exists a finite age, T, at which the asset would be replaced if it survives under the decision rule up to that maximum age. The model is easily adjusted to accommodate the case where T→∞.

The same model is equally useful for a situation where sample data can be used to estimate the necessary probabilities without any knowledge of actual replacement decision rules being applied or the stochastic processes generating the data. However, I am not optimistic about the wide availability of such data, although one might expect some large firms or industry groups to have some. These data would require tracking individual assets to find out how long they were in service, and thus provide relative frequencies on replacement age.

The stochastic model used here for depreciation theory is essentially the same as the discrete time model in my *Optimal Replacement Under Risk* (1965) with the exception that returns and costs are assumed to occur at the end of the year instead of the beginning. Some additional notation is required:

F_t = salvage value when the asset randomly "fails" and is prematurely replaced at age t.

γ = the average fraction of services Q_t received and costs C_t incurred in the interval (t-1,t] when the asset randomly fails, where $0<\gamma<1$.

p_t = conditional probability that an asset survives over the interval (t-1,t] without random failure, given that it has reached age t-1

Π_t = the unconditional probability of survival through period t, which is the product $p_1 p_2 ... p_t$.

The variables C_t, Q_t, and S_t represent measures under the condition that the asset survives through age t. The present value measure V_t is now a conditional expected value, given survival through age t, and depreciation D_t is the difference between the two conditional expected present value measures, V_{t-1}- V_t. The conditional probability of failure at age t, given survival to age t-1, is clearly 1-p_t by the definition of p_t given above, and the unconditional probability of failure at age t is $\Pi_{t-1}(1-p_t)$.

If u were the selling price for output, conditional expected net returns received at the end of period t, given survival through age t-1, would be

(18)
$$R_t = p_t(uQ_t - C_t) + (1-p_t)[F_t + \gamma(uQ_t - C_t)]$$

and expected value of net returns over the entire life of the asset, starting immediately after a new purchase, would be

(19)
$$V_0 = \sum_{t=1}^{T} \beta^t \Pi_{t-1}[p_t(uQ_t - C_t) + (1-p_t)[F_t + \gamma(uQ_t - C_t)]] + \beta^T \Pi_T S_T.$$

The equation defining unit cost, which is the counterpart of (1) in the nonstochastic theory, is obtained by replacing the notation V_0 with I in (19); and then solving that equation for u yields

(20)
$$u = [I + \sum_t \beta^t \Pi_{t-1}[p_t C_t + (1-p_t)(\gamma C_t - F_t)] - \beta^T \Pi_T S_T]$$
$$/[\sum_t \beta^t \Pi_{t-1}[p_t + (1-p_t)\gamma]Q_t].$$

The expected present value function starting with an asset of age t is

$$V_t = \sum_{j=t+1}^{T} \beta^{j-t} p_{t+1} \cdots p_j (uQ_j - C_j) + \beta^{T-t} p_{t+1} \cdots p_T S_T$$

(21)

$$+ \sum_{j=t}^{T-1} \beta^{j-t+1} p_{t+1} \cdots p_j (1 - p_{j+1}) [F_{j+1} + \gamma (uQ_{j+1} - C_{j+1})]).$$

Note that when $j=t$ in the last summation, the product of probabilities for survival must be taken equal to one, and when $t=T$ the product of survival probabilities on the salvage value term for S_T is taken as one.

As in the deterministic case, the depreciation charges, $D_t = V_{t-1} - V_t$ must sum to $I - S_T$ because $V_0 = I$ by the construction of u, and $V_T = S_T$ because planned replacement takes place immediately after the asset has survived to period T, leaving only the term S_T in (21).

An extra set of terms appears in (21) which is not in (3); it is a set of term for expected present value of salvage associated with random failure of the asset and is the last summation. This summation of terms is absent in (3) whether or not the discount rate is adjusted for insurance and taxes. However, such a term would exist in the case of insurance if a more complete specification were made with Hotelling's model in which an allowance were made for insurance on the *net* book value of the asset after deducting salvage value when a loss occurred, i.e., the firm would not waste money insuring the salvage value. Careful comparison of the coefficient

$$\beta^{j-t} p_{t+1} \cdots p_j$$

in the first summation in (21) with the formula for δ_{jt} in (17) will make it clear that they are the same if x_t is zero and

$$p_t = 1/(1+i_t),$$

thus making the deterministic model with insurance consistent with the sequential stochastic model.

Notes

Helpful comments of Stephen Sosnick on an earlier draft of this chapter are gratefully acknowledged.

1. The problem of obsolescence is introduced later.
2. If $Q_t=0$, $t=1,2...k$, as would be the case for an orchard, then the present value of costs for $t=1,2,...k$ are transposed to the left hand side of (1).
3. Sometimes called the engineering approach to cost of production estimation.
4. Although ERS uses a real rate of interest in their calculations of opportunity cost of capital, the rate is an estimate of the real rate of return to production assets

in the farm sector over the previous 10 years. Apparently, the idea is to reflect short-run opportunity costs, which would be somewhere else in the farm sector, but it would seem that the overall purpose of the ERS cost of production estimates is not that myopic. If we want a short-run measure, we can use gross margins, and a long-run measure should reflect opportunity cost of capital for the entire economy.

5. Watts and Helmers showed that (10) is an identity for any asset value function, V_t, as long as $V_o = I$ and $V_t = S_t$, but we note that implied income of the firm will be constant over life of the asset (the Hicksian criterion) if and only if V_t is defined as (3) which determines D_t through (4).

6. Klonsky's national survey of cost of production methods used by land grant universities does not provide a very clear picture of how capital costs are calculated, except that 90 percent of them use straight line depreciation and the methods used to calculate opportunity cost of capital is highly variable (Chapter 10, this volume). Watts and Helmers show some comparisons of using the correct measure of costs in (11) with the common method of straight line depreciation jointly with opportunity cost of capital estimated by the implied value of the machine at median age times the real interest rate.

7. For an authoritative source on this argument in the context of the classic problem of optimal forest rotations, see Samuelson, pp. 473-4.

References

Atkinson, A. A. and W. R. Scott. 1979. "The Depreciation Theory of Cost Allocation: A Survey and Synthesis." Unpublished Working Paper.

Burt, Oscar R. 1965. "Optimal Replacement Under Risk." *Journal of Farm Economics* 47: 324-346.

_____. 1972. "A Unified Theory of Depreciation." *Journal of Accounting Research* 10: 28-57.

Edwards, Edgar O. 1961. "Depreciation and the Maintenance of Real Capital" in J.L. Meij, ed. *Depreciation and Replacement Policy.* Chicago: Quadrangle Books.

Hicks, J. R. 1946. *Value and Capital.* 2nd ed. Oxford: Oxford University Press.

Hotelling, Harold. 1925. "A General Mathematical Theory of Depreciation." *Journal of the American Statistical Association* 20: 340-53.

Lutz, F. and V. Lutz. 1951. *The Theory of Investment of the Firm.* Princeton: Princeton University Press.

Samuelson, Paul A. 1976. "Economics of Forestry in an Evolving Society." *Economic Inquiry* 14: 466-91.

Taylor, J. S. 1923. "A Statistical Theory of Depreciation Based on Unit Cost." *Journal of the American Statistical Association* 18: 1010-23.

U.S. Dept. of Agriculture. 1990. "Economic Indicators of the Farm Sector: Costs of Production, Major Field Crops, 1988." ECIFS 8-4. Washington, D. C.

Watts, Myles J. and Glenn A. Helmers. 1981. "Machinery Costs and Inflation." *Western Journal of Agricultural Economics* 6: 129-145.

Wright, F. K. 1964. "Towards a General Theory of Depreciation." *Journal of Accounting Research* 2: 80-90.

16

Estimating Costs of Durable and Operating Capital Services

Cole R. Gustafson, Peter J. Barry, and Mir B. Ali

Introduction

Interest costs associated with farmers' ownership of nonland capital assets (both durable and operating inputs) must be allocated indirectly because these investments are valued infrequently by a market transaction. The process is complicated because the value of nonland capital assets changes over time due to depreciation, length of seasonal use, inflation, and other economic factors. Agricultural economists use either a real, a nominal, or an assigned interest rate to determine nonland interest costs. Wide disparity exists in state university methods where interest rates applied to equity capital vary from 5 to 10 percent (Casler). It is an important choice because interest on nonland capital exceeds 5 percent of total economic costs for many crop and livestock enterprises (USDA, 1990).

This chapter critiques alternative methods of quantifying opportunity interest costs associated with ownership of nonland farm assets. It focuses on methods adopted by USDA, primarily because of excellent documentation and because the agency has experimented with both nominal and real rates of interest in an attempt to accurately portray the cost structure of U.S. agriculture. Results of the study will show the rate choice depends on the data's end use as well as information availability. Following sections review USDA's past methods, discuss theoretical considerations of capital budgeting, and describe two empirical methods of quantifying nonland capital returns.

Historical Evolution of USDA
Nonland Capital Accounting

The Agricultural and Consumer Protection Act of 1973 directed USDA to conduct cost of production studies of major crops and dairy, and specified how to calculate fixed costs. In particular, Section 808 of the Act stated that "a return on fixed costs equal to the existing interest rates charged by the Federal Land Bank (FLB)" should be used. This provision created major conceptual and estimation problems for USDA when high rates of inflation prevailed during the latter 1970s.

Nominal FLB interest rates contain an inflation premium. When current asset values were multiplied by these rates to determine opportunity costs, the resulting charge included both a cost for real interest and inflation. However, enterprise cost and return budgets did not include capital gains stemming from nominal appreciation of asset values, leading to an understatement of enterprise profitability (Hoffman and Gustafson, Hottel and Gardner, and Miller). In essence, long-run costs, including full opportunity costs, were being compared with current returns from production.

The Agriculture and Food Act of 1981 provided USDA greater flexibility in estimating production costs. In response, USDA developed a new budget format that separated current costs from long-run costs, thus making cashflow analyses independent of full economic costs analysis (Hoffman and Gustafson). At present, USDA implements these procedures by applying a nominal 6-month U.S. Treasury bill interest rate to input expenses (e.g., seed, feed, fertilizer), a long-run real rate of return on farm production assets to other nonland capital (i.e., machinery and equipment), and a composite share and cash rental rate to the value of farmland (USDA, 1990).

These revised procedures, though improved, still contain three conceptual flaws.[1] The first inconsistency relates to the application of nominal interest rates to operating inputs and real rates of return to other nonland capital. Although rates logically would be expected to vary across assets with differing risk characteristics and longevity, the calculation of each return should handle inflation consistently within the same enterprise budget (i.e., either include an inflation premium in both calculations or exclude it from each).

Second, the use of both an opportunity cost and a rate based on residual income returns to production assets confuses the purpose of the budgeting exercise. Is the goal to compare agricultural returns with nonfarm firms (in which case an opportunity rate available outside the sector is relevant) or to compare the financial well being of firms that possess differing structural characteristics within the sector (which necessitates using a rate that reflects alternative opportunities within the sector)?

The third issue relates to the appropriate method of calculating residual income returns to production assets. At present, USDA uses returns to all production assets as a proxy for nonland returns. Is this appropriate, given the dominance of real estate in farm asset portfolios? Results of this study show a considerable disparity between rates of return to farmland and returns to other nonland production assets. Returns to nonland assets are understated when assets included in the rate calculation are dominated by farm real estate.

Theoretical Considerations

Capital budgeting is a useful framework for determining returns to nonland capital. The net present value method is one of the more popular methods of capital budgeting (Barry, Hopkin and Baker; Miller; Walrath; Watts and Helmers). In its simplest form, the model can be represented as:

(1)
$$NPV = -C_0 + \sum_{i=0}^{n} \frac{NC_t}{(1+i_n)_t}$$

where NPV is the net present value of the investment, C_0 is the initial cash outlay, NC_t are the nominal net cash flows at the end of period t generated by the investment, n is the number of periods in which cash flows related to the investment occur, and i_n is the risk-adjusted nominal discount rate. The farmer's decision criterion is to accept investments that yield a positive NPV. To more fully understand the components of this model, following parts of this section discuss nominal and real discount rates, nominal and real asset values, and implications for valuation of nonland returns.

Nominal and Real Discount Rates

A nominal discount rate (r) reflects an investor's time preference for money and can be decomposed into a risk-free real interest rate (i_r), an inflation premium (p), and a risk premium (u) associated with the uncertainty of receiving C_t as follows:

(2)
$$i_n = (1+i_r)(1+p)(1+u) - 1$$

At any given point of time, nominal and real interest rates differ by an inflation premium (assuming risk premiums are negligible, u=0). Table 16.1

TABLE 16.1   Nominal and Real Long Term Interest Rates 1959-1989, Adjusted Using Implicit GNP Price Deflator [a]

Year	Nominal	GNP Deflator	Real
		- Percent -	
1959	4.07	2.36	1.67
1960	4.01	1.64	2.33
1961	3.90	0.97	2.90
1962	3.95	2.24	1.67
1963	4.00	1.57	2.40
1964	4.15	1.54	2.57
1965	4.21	2.74	1.44
1966	4.66	3.55	1.07
1967	4.85	2.57	2.22
1968	5.25	5.01	0.22
1969	6.10	5.57	0.50
1970	6.59	5.53	1.01
1971	5.74	5.71	0.02
1972	5.63	4.73	0.86
1973	6.30	6.45	-0.14
1974	6.99	9.09	-1.93
1975	6.98	9.81	-2.58
1976	6.78	6.41	0.35
1977	7.06	6.66	0.38
1978	7.89	7.28	0.57
1979	8.74	8.86	-0.11
1980	10.81	9.03	1.63
1981	12.87	9.68	2.90
1982	12.23	6.38	5.50
1983	10.84	3.90	6.68
1984	11.99	3.66	8.04
1985	10.75	2.97	7.55
1986	8.14	2.61	5.38
1987	8.64	3.16	5.31
1988	8.98	3.32	5.48
1989	8.58	4.12	4.28

[a] Yields based on closing bid prices of five dealers.   Averages (to maturity or call) for all bonds neither due or callable in less than 10 years.
Source: Statistical Abstract of the United States.

lists nominal and real U.S. treasury bond interest rates for the past 30 years. When capital budgeting, nominal discount rates are used if future cash flows C_t include inflation and are expressed in nominal terms, whereas real discount rates must be used when C_t is specified in real terms and adjusted for inflation (Barry, Hopkin and Baker; Miller; Watts and Helmers).

Once the method of handling inflation is determined, decisionmakers must specify i_r. Individual farmers have the choice of basing discount rates on either the firm's weighted average cost of capital (debt and equity combined) or on their cost of equity capital. The choice depends on whether the firm's goal is to maximize returns to assets or returns to equity capital. A weighted average marginal cost of capital is the appropriate rate if the firm's goal is to maximize returns to assets. Firms maximizing equity returns should base their rate on the firm's marginal cost of equity capital. Barry, Hopkin and Baker (1988) illustrate the advantages of each approach and note that capital budgeting based on returns to equity accounts for changes in financial structure, leverage and costs of debt capital that occur with investment. When interest paid is not deducted before calculating residual returns, as in USDA's economic cost and return budgets, the proper capital budgeting objective is to maximize asset returns.[2]

Quantifying discount rates for groups of farmers is more complex. The decision is difficult because operator goals, sources of capital, financial structure, and risk premiums (determined jointly by levels of risk and attitudes toward risk) differ among farmers. Farm management record-keeping associations choose a discount rate based on perceived current returns to agriculture (direct estimation) or on comparative investment returns available outside of the agricultural sector (opportunity cost). Figure 16.1 compares nominal U.S. treasury bond rates with residual income returns to production assets in agriculture. The comparison highlights the recent divergence between these rates.

The choice of approach depends on the objective of the budgeting exercise. Use of an opportunity cost rate assumes the budgeting objective is to compare agricultural returns with rates available in other sectors in the economy. This process is useful when the movement of resources in and out of the agricultural sector is being investigated. On the other hand, discount rates based on alternatives within agriculture facilitates comparisons of peer firms, many of whom have limited employment opportunities and/or noneconomic reasons for remaining in agriculture. At present, USDA mixes these concepts when it applies an opportunity cost rate to one class of assets and a directly estimated rate to another, within the same budget.

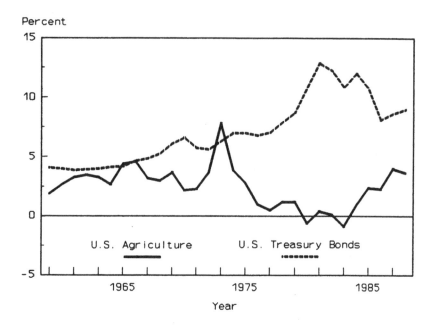

FIGURE 16.1   Residual Income Returns in Agriculture vs. U.S. Treasury Bond Yields. *Source:* USDA, 1989 and Statistical Abstract.

Nominal and Real Asset Values

Unlike interest rates, asset values do not contain an inflation premium. The real value and the nominal value of an asset are identical at any given time. Assume a hypothetical asset exists with a price today of $x (to simplify the analysis, also assume the asset does not depreciate). This price is both a real value and a nominal value. Over time, the asset's value changes in accordance with variations in future cash flows, interest rates, inflation and risk. At some future date when the asset is repriced again, as depicted in equation 1, the resulting figure will also be both a real and a nominal value. Intertemporal changes in asset values are defined as capital gains/losses (again abstracting from depreciation).

Most recently, the variability of capital gains/losses in agriculture has exceeded that of residual income earnings (Figure 16.2). These changes in value either may be nominal (due to inflation) or real (arising from supply/demand shifts, discount rate changes, or differing risk environments). Nevertheless, capital gains must be properly accounted for in financial statements (Aukes; Barry, 1988; Gustafson, Barry and Sonka).

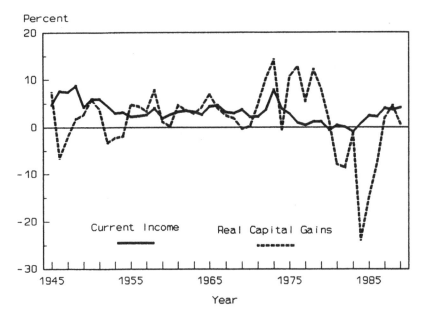

FIGURE 16.2 Composition of Equity Returns. *Source:* USDA, 1989.

Record-keeping associations rarely include capital gain/loss statistics due to data limitations and infrequent sales of agricultural assets.[3] However, this omission leads to an understatement of profitability. Many record-keepingassociations try to compensate for this bias by selecting a real discount rate to value returns to capital (Hottel and Gardner). The bias, though improved, still exists because of real growth in the payment series. To illustrate, note that nominal future net cash flows increase due to rising real net cash flow increases (RC_t), inflation (p), and real growth (g):

(3) $$NC_t = RC_t \, [(1+p)(1+g)]_t$$

Now, substitute equations 2 and 3 into equation 1 to show:

(4) $$NPV = -C_0 + \sum_{t=1}^{n} \frac{RC_t \, [(1+p)(1+g)]_t}{[(1+i_r)(1+p)(1+u)]_t}$$

By canceling the inflation term $(1+p)$ in both the numerator and the denominator of equation 4, price level changes are removed from the analysis. However, real growth in the payment series still exists in the numerator, which in turn will cause real capital gains when the asset is valued at succeeding points of time. Although negligible for most forms of nonland capital, real capital gains/losses do accrue to owners of breeding livestock and farm machinery.[4] Failure to show this source of return in enterprise budgets leads to biased estimates of firm profitability since costs of obtaining these gains (primarily real interest on debt and equity capital) are listed as expenses.

To summarize, either real or nominal discount rates can be used to value nonland capital returns. If the former is used, real growth in returns must be included to avoid bias. If the latter is used, nominal gains accruing from inflation also should be added.

Consistency Across Asset Classes

A final consideration is the uniformity of discount rates across differing classes of assets with unique risk attributes. Some record-keeping associations apply a constant rate of return to all assets while others vary rates among asset classes according to the liquidity, longevity, and/or productivity characteristics of various farm assets. The following discussion illustrates the underlying assumptions of each approach.

Farmers are assumed to form an array of investment alternatives and select those possessing non-negative net present values. Presumably, this stratification occurs across all classes of assets. At the margin then, the net present value for the last investment (NP_1) can be equated with that of its predecessor (NPV_2), since both are at or near zero.[5] Scale differences between the two can be removed by multiplying the larger by a size parameter S ($0<S<1$):

$$(5) \qquad\qquad NPV_1 = NPV_2 S$$

Substituting the components of equation 4 for each asset and canceling inflation terms yields:

$$(6) \qquad -C_{0,1} + \sum_{t=1}^{n} \frac{RC_{t,1}(1+g_1)^t}{[(1+i_{r,1})(1+u_1)]^t} = -C_{0,2}S + \sum_{t=1}^{n} \frac{RC_{t,2}(1+g_2)^t S}{(1+i_{r,2})(1+u_2)^t}$$

where $C_{0,i}$, $RC_{t,i}$, $i_{r,i}$, g_i and u_i are unique to asset i. If each asset's respective risk premium fully reflects variations in cash flows and rates of growth between the two assets, the scaled initial costs of each asset can be assumed to be equal and subtracted from each side of the equation to obtain:

(7)
$$\sum_{t=1}^{n} \frac{RC_{t,1}(1+g_1)^t}{[(1+i_{r,1})(1+u_1)]^t} = \sum_{t=1}^{n} \frac{RC_{t,2}(1+g_2)^t s}{[(1+i_{r,2})(1+u_2)]^t}$$

Rearranging these terms shows:

(8)
$$\sum_{t=1}^{n} \frac{(1+i_{r,2})^t}{(1+i_{r,1})^t} = \sum_{t=1}^{n} \frac{RC_{t,2}(1+g_2)S^t(1+u_1)^t}{RC_{t,1}(1+g_1)^t 1+u_2)^t}$$

which implies that discount rates for each asset can be equated only (i.e., $i_{r,1} = i_{r,2}$) if:

(9)
$$\sum_{t=1}^{n} \frac{(1+u_1)^t}{(1+u_2)^t} = \sum_{t=1}^{n} \frac{RC_{t,1}(1+g_1)^t}{RC_{t,2}(1+g_2)^t S}$$

Thus, researchers have two choices, depending on whether or not chosen discount rates embody a risk premium. First, unique discount rates can be applied to each category of assets reflecting the risk attributes of each asset class. In essence, returns to capital and risk are commingled. Alternatively, returns to capital and returns to risk can be separated. With this approach, a uniform discount rate (that excludes a risk premium) is applied to all nonland capital assets. Another line item in the budget values returns to risk independently, often stated as a residual on the last line of an enterprise budget. Conceptual difficulties arise either when uniform discount rates are applied to capital without an explicit line item for risk or when both a return to risk and varying discount rates are used.

Empirical Estimation of
Nonland Capital Discount Rates

Farm management record-keeping associations typically choose a discount rate based on perceived residual income returns to agricultural assets (direct estimation). The limitation of this approach is that real estate assets dominate this portfolio. In 1988, real estate assets comprised 73 percent of total agricultural assets (USDA, 1989). Rates of return to real estate assets are less than relative returns to other agricultural assets because farmland ownership offers numerous intangible and social benefits, does not readily depreciate, can be an effective hedge against inflation, and is a promising candidate for risk reduction in well-diversified investment portfolios (Barry,

1980; Barry and Robison; Boehlje and Eidman). Consequently, current rates of return to nonland capital are biased downward when real estate is included in the portfolio of assets.

To avoid this problem, nonland returns to capital must be estimated independently of returns to real estate assets. Although stocks of nonland capital are reported by USDA and numerous farm record-keeping associations, returns to nonland capital cannot be calculated directly because these agencies do not segregate residual returns by asset class. Alternatively, nonland returns to capital can be allocated indirectly among nonreal estate and real estate assets using regression. An aggregate and a micro approach are illustrated below.

Aggregate Approach

Residual income returns to farm production assets (RIR), the value of nonreal estate assets (NREA), and the value of real estate assets (REA) from 1940-88 for U.S. agriculture, deflated by the implicit GNP price deflator, are shown in Table 16.2. Regressing RIR on NREA and REA yields the following equation:

$$(10) \qquad RIR = 0.183NREA - 0.023REA, \; R^2 = .17$$
$$(0.022) (0.008)$$

where the standard error of each parameter estimate is shown in parentheses. Both parameter estimates are statistically significant at $p = .01$.

The results of this regression imply that residual income returns to nonland capital averaged 18.3 percent over the preceding five decades whereas residual income returns to real estate were actually negative at -2.3 percent. Hence, USDA currently understates returns to nonland capital when a 2.81 percent rate reflecting residual income returns to all production assets is applied. Of course, returns to nonland assets are overstated if nonland capital assets are depreciated too rapidly (Gustafson, Barry and Sonka).

Micro Approach

A single aggregate estimate does not reflect the variation in nonland capital returns that is likely to exist across differing farm sizes, enterprises, and geographic regions. One method of determining these values is to (1) identify farms in the strata of interest that do not contain real estate (e.g., farms that rent all of their land), (2) calculate total capital returns for these farms (which equals nonland capital returns by definition), and (3) use the

TABLE 16.2 Real Residual Income and Asset Values, Billions of 1982 Dollars

Year	Residual Income	Real Estate Values	Nonreal Estate Values
1940	11.5	258.5	149.2
1945	22.3	328.0	187.9
1950	25.5	315.5	191.6
1955	12.9	341.9	159.9
1960	16.5	398.4	167.0
1961	20.2	412.5	170.8
1962	21.9	418.5	173.4
1963	21.3	435.8	170.4
1964	19.5	456.2	166.6
1965	28.1	475.7	177.5
1966	30.9	486.0	182.3
1967	24.5	502.2	183.3
1968	23.6	451.5	182.8
1969	27.1	490.7	183.7
1970	19.3	482.1	181.7
1971	19.8	490.1	188.5
1972	29.2	522.8	209.0
1973	59.0	602.8	243.0
1974	36.5	621.5	207.2
1975	29.2	646.9	211.8
1976	18.4	723.5	210.3
1977	16.2	756.8	211.0
1978	23.3	833.5	242.2
1979	26.8	898.3	256.5
1980	9.5	913.0	248.8
1981	20.3	834.8	225.5
1982	18.7	748.8	212.2
1983	7.5	711.0	197.9
1984	23.7	591.0	193.7
1985	24.6	498.6	170.9
1986	26.6	443.1	159.1
1987	32.1	438.7	158.9
1988	28.9	447.2	165.8

Source: USDA, 1989.

value as a proxy for nonland capital returns on remaining farms that do contain real estate assets. Obviously, farms that solely rent land are scarce and their representativeness is questionable. In addition, the opportunity cost method would still be required to allocate returns to unpaid resources (i.e., a farmer's labor and management).

TABLE 16.3 Regression Variable Definitions

<u>Dependent Variable</u>

Returns to owned inputs, risk, and management per wheat acre (YAC) =

Sum of:

1. Return to operating capital per wheat acre
2. Return to nonland capital per wheat acre
3. Return to land (net land rent) per wheat acre
4. Return to unpaid labor per wheat acre
5. Return to risk and management per wheat acre

<u>Independent Variables</u>

1. Operating capital-whole farm (OPRCAP)

2. Nonland capital-whole farm (NONLAND)

3. Acres operated on farm (ACOPR)

4. Unpaid labor hours (operator, other, and family members)-whole farm (LABOR)

5. Dummy variables:

Irrigation (IRRIG)	= 1 for irrigated acres on farm = 0 for no irrigated acres (dryland) on farm
Type of wheat (WWHT)	= 1 for winter wheat (proportion of winter wheat acres to total wheat acres >.50) = 0 for spring wheat
Farming practice (FALLOW)	= 1 for fallow acres on farm = 0 for no fallow acres on farm
Farm type (CASHGR)	= 1 for cash grains (based on gross farm income) = 0 for livestock and other crops

YAC = F(OPRCAP, NONLAND, LABOR, ACOPR, IRRIG, WWHT, FALLOW, CASHGR)

Similar to the aggregate approach, regression also can be employed to obtain a more robust value at the micro level. For example, consider data from USDA's annual Farm Cost and Returns Survey, which is a list and area frame probability-based survey designed to elicit information on farm financial characteristics, expenses, receipts, resource bases, production practices and operator characteristics. In 1986, 25,000 contacts were made in the United States. Data from this survey enable firm-level calculations of farm costs and returns (Glaze and Ali).

To illustrate the regression procedure, a database containing whole farm financial and wheat enterprise budget information for the 254 farms surveyed in the Northern Plains region was compiled (Table 16.3). The dependent variable, per acre returns to owned inputs (YAC), is regressed on whole farm operating capital (OPRCAP), nonland capital (NONLAND), acres operated (ACOPR), unpaid labor hours (LABOR), and four dummy variables reflecting irrigation (IRRIG), type of wheat produced (WWHT), farming practices (FALLOW) and farm type (CASHGR).

In essence, regression is used to allocate residual income returns among levels of all unpriced land, labor, and capital resources. Variations in residual returns among farms are linked with corresponding differences in unpaid resources. The regression coefficient is interpreted as a rate of return or income to the respective resource--as unpaid resource levels increase, residual returns would be expected to increase by an amount equal to the product of the regression coefficient and increased resource level.

Results of the regression are depicted in Table 16.4. The low R^2 reflects the cross-sectional diversity in wheat production practices, earnings, and expenses in the region. In addition, the estimates of the significance of the coefficients should be viewed as very rough since adjustments were not made for the complex sample design of the survey. Like the aggregate analysis above, returns to nonland capital are positive while returns to land are negative. The magnitude of the coefficient must be interpreted with care because the independent variable is a whole-farm value while the dependent variable is a per acre value. Therefore, the value is not a rate of return but a level of residual return per acre of wheat. Multiplying the coefficient by

TABLE 16.4 Preliminary Results Based on 1986 Wheat, Northern Plains

Variable	DF	Parameter Estimate	Standard Error	"t" Statistic	Prob > |t|
INTERCEP	1	4.025129	6.55588891	0.614	0.5398
OPRCAP	1	-0.000011239	0.00002277	-0.494	0.6221
NONLAND	1	0.000032329	0.00001301	2.485	0.0136
LABOR	1	0.000549	0.00141132	0.389	0.6977
ACOPR	1	-0.000590	0.00070833	-0.833	0.4059
IRRIG	1	29.003375	8.96303618	3.236	0.0014
WWHT	1	13.047263	5.40343672	2.415	0.0165
FALLOW	1	-7.309881	4.46411304	-1.637	0.1028
CASHGR	1	4.858355	4.63873693	1.047	0.2960

$R^2 = .09$

Source: FCRS, USDA.

the average firm's nonland capital level of $220,000 yields a per acre return to nonland capital of $7.11 for wheat. This value is substantially greater than the return of $6.01 calculated by present procedures. The overall return to all nonland capital at the firm level can be determined only by calculating similar regressions for other enterprises on each farm of interest.

Summary

This chapter identifies three limitations in current USDA methods of calculating residual economic returns to nonland capital: (1) inconsistency in rates applied to different classes of assets, (2) a failure to differentiate between direct return and opportunity cost concepts, and (3) a rate calculation based primarily on farm real estate. Many state and university farm record keeping systems follow similar approaches.

As an alternative, this study developed a regression procedure that indirectly allocates nonland returns among nonreal estate and real estate assets. An aggregate and a micro approach were illustrated. Results of the analysis show that current methods overstate residual income returns to real estate and underestimate returns to nonland capital.

Notes

1. Methods of valuing assets also affect the specification of nonland returns. However, this topic is beyond the scope of this chapter. Burt touches on the topic in another chapter of this book when alternative depreciation methods are discussed.

2. This objective is assumed for the remainder of the chapter. If interest paid is listed as a production expense, the concepts presented in following portions of the manuscript are equally applicable to specification of residual returns to equity capital.

3. A fundamental distinction between asset valuation and the role of record-keeping associations may be that asset valuation depends upon anticipated values for the various variable, while the record-keeping associations are measuring past performance and, thus, use actual, measurable results.

4. Capital gains/losses accrue when accounting depreciation rates differ from actual economic rates of depreciation.

5. Assuming each asset provides cash flows over an identical number of future time periods (n). If time periods differ, the NPV of each asset must be annualized (Miller; Watts and Helmers).

References

Aukes, R. 1987. "Double Counting Agricultural Income." *Can. J. of Agr. Econ.* 35: 463-479.

Aukes, R. 1988. "Double Counting Agricultural Income: A Reply." *Can. J. of Agr. Econ.* 36: 357-58.

Barry, P. J. 1980. "Capital Asset Pricing and Farm Real Estate." *Amer. J. Agr. Econ.* 62: 549-53.

Barry, P. J. 1988. "Double Counting Agricultural Income: A Comment." *Can. J. of Agr. Econ.* 36: 353-56.

Barry, P. J. and L. J. Robison. 1986. "Economic and Accounting Rates of Return for Farm Land." *Land Economics* 62: 388-401.

Barry, P. J., J. A. Hopkin, and C. B. Baker. 1988. *Financial Management in Agriculture.* 4th edition. Danville, IL: Interstate Printers and Publishers.

Boehlje, M. D., and V. R. Eidman. 1984. *Farm Management.* New York: John Wiley and Sons, New York.

Casler, G. L. 1991. "Use of State Farm Record Data For Studying Determinants of Farm Size" in Arne Hallam, ed. *Determinants of Size and Structure in American Agriculture.* Ames: Iowa State University Press.

Glaze, D. and M. Ali. 1988. "Distribution of Costs of Production For Wheat Farms." *Agricultural Income and Finance: Situation and Outlook Report* AFO-31. Econ. Res. Ser., U.S. Dept. of Agr., Washington, D.C.

Gustafson, C. R., P. J. Barry, and S. T. Sonka. 1990. "Utilizing Expectations to Measure Economic Depreciation and Capital Gains of Farm Machinery." *Agribusiness: An International Journal* 6: 489-503.

Hoffman, G. and C. Gustafson. 1983. "A New Approach to Estimating Agricultural Costs of Production." *Agr. Econ. Res.* 35: 9-14.

Hottel, J. B. and B. L. Gardner. 1983. "The Rate of Return to Investment in Agriculture and Measuring Net Farm Income." *Amer. J. Agr. Econ.* 65: 553-57.

Miller, Thomas A. 1981. "Farm Enterprise Budgeting Under Inflation." Unpublished paper presented at ESS-NED-USDA COP Task Force Seminar, Washington D.C., June 2-4.

U.S. Dept. of Agriculture. 1990. "Economic Indicators of the Farm Sector: Costs of Production--Major Field Crops, 1988." ECIFS 8-4, Econ. Res. Ser., Washington, D.C., April.

_____. 1989. "Economic Indicators of the Farm Sector: National Financial Summary, 1988." ECIFS 8-1, Econ. Res. Ser., Washington, D.C., Sept.

Walrath, A. J. 1973. "The Incompatibility of the Average Investment Method for Calculating Interest Costs With the Principle of Alternative Opportunities." *S. J. Agr. Econ.* 5: 181-85.

Watts, M.J. and G.A. Helmers. 1979. "Inflation and Machinery Cost Budgeting. *S. J. Agr. Econ.* 11: 83-88.

Estimating Costs of Non-Human Capital Services: Discussion

Utpal Vasavada

Introduction

The chapters by Burt and Gustafson *et al.* consider different aspects of the problem of estimating costs of nonland capital services in agriculture. The chapter by Burt focuses on measuring depreciation costs while the Gustafson *et al.* chapter looks at interest cost measurement accruing to ownership of nonland capital services. Both chapters go the distance since they evaluate methods utilized and offer concrete suggestions on how these methods can be improved upon. Except for a few general remarks, I will confine my attention to the Gustafson *et al.* chapter.

Measuring Depreciation and Interest Costs: General Remarks

Before moving to a discussion of the chapters, it is perhaps appropriate to make a few general observations. First, since significant gaps exist in both the conceptual and empirical literature, the importance of this research area can hardly be exaggerated. On the conceptual front, as Burt suggests, cost of production of agricultural crops cannot be properly defined without getting into the controversial area of depreciation theory. In addition, there are important empirical questions, namely what data are appropriate, what techniques are available to make the measurements, and what are acceptable compromises between the demand for conceptual rigor and resource availability constraints imposed on an analyst.

Second, while both chapters look at the capital measurement issue, there are striking differences in their focus and content. The Gustafson *et al.* chapter is more tightly focused on empirical questions emerging from their critique of existing USDA measurement practices while the chapter by Burt begins with a conceptual treatment of depreciation costs and proceeds to explore the coherence between the derived conceptual measure and the measure actually used by USDA. It is rather unfortunate that USDA estimation practices have become a lightning rod since much careful thought has gone into developing these estimates. Furthermore, in many instances, the derived estimates present the best cost effective compromise between analytical rigor and resource constraints.

Finally, consider the relationship between these two chapters and others in this volume. An observation made by Karen Klonsky merits mention, namely, that the bulk of university based estimates of nonhuman capital costs use the straight line method to estimate depreciation costs. There seems to be greater consensus over estimation of variable cash expenses than on capital cost measurement. The chapter by Mitch Morehart, *et al.* clearly explains how the USDA estimates both depreciation and interest costs. The chapter by Vernon Eidman and an ensuing discussion by Eldon Ball consider the multioutput enterprise issue. This issue is important given the well-known difficulties associated with allocating capital service costs among enterprises and the significant body of econometric evidence favoring the joint production hypothesis.

Measuring Interest Costs in Agriculture

The Gustafson *et al.* chapter contains a good discussion of previously used methods to compute interest costs, some problems associated with their use, methods currently in use by USDA to estimate interest costs, remaining flaws in methods currently employed by USDA, and some suggestions on how to correct these flaws. It is, therefore, evident that it is comprehensive in scope. While recognizing that any paper cannot do everything, the scope of this chapter could have been broadened to include methods used by different universities in their extension programs, a discussion of potential areas of conflict and consensus between methods used by USDA and universities, and a discussion of "costing" interest charges for multioutput enterprises.

The chapter identifies at least four potential areas where existing methods may benefit from some reworking. First, it is incorrect to match up nominal interest rates with real cash flows and real interest rates with nominal cash flows. Second, there is an apparent contradiction when both opportunity cost and residual returns are mixed up. Third, it is inappropriate to use a uniform rate or a variable rate plus a risk premium. In the former case, the effect of risk is excluded and, in the latter case, there is double counting. Fourth, the returns to all assets is a poor proxy for the rate of return to nonland capital.

A few remarks are in order on possible solutions to the fourth problem identified by Gustafson *et al.* Consider first the aggregate approach. It would help if there were more discussion about a possible conceptual justification for the estimated equation. Otherwise, one is left wondering whether the equation is an ad hoc regression. Another possible limitation of this model is its linearity. The model implies that the rate of return to all production assets can be increased without limit by investing in nonreal estate assets. Similarly, disinvestment in real estate assets will increase this rate of return. However, as Gustafson *et al.* observe, the data point to a preponderance of real estate assets. How does one reconcile these observations? Is the linear model structure justified?

A second solution proposed by Gustafson *et al.* is the micro approach. Here too, some discussion of the conceptual foundations of this equation would be appropriate. The right hand side contains variables that are normally choice variables in a microeconomic framework. Some effort to test exogeneity of these variables would help to clarify this issue. Furthermore, over two thirds of the variables in the model are insignificant at the 5% level of significance. This does not constitute strong evidence in favor of the estimated model and casts a spell of doubt on inferences that can be drawn from the model.

I'll make a final comment on the appropriate interest rate to use in capital budgeting studies. There are many studies in the econometric tradition that illustrate how the discount rate can be estimated from the data. A paper by Burt and another by Vasavada and Somwaru on adjusting discount rates for risk come to mind. It appears that there is little communication between the capital budgeting and econometric traditions on this point.

Conclusions

Both chapters provide a good feel for the issues and problems emerging from nonhuman capital service measurement. However, there is little doubt that this area of research merits further attention and many issues remain unresolved. What is disturbing is that different methods are employed by universities and the USDA. Some effort to communicate the merits of alternative methods, as well as the consistency of conceptual with empirical concepts, will serve to illuminate how researchers might proceed to estimate these important components of cost of production. The chapters by Burt and Gustafson *et al.* have contributed to such a clarification but we would be deluding ourselves if we believed that these problems have been resolved. A promising direction to explore in this area of research is the allocation of interest and depreciation costs among enterprises in a multioutput framework. In the absence of statistical and other circumstantial evidence negating the joint production hypothesis, the use of single enterprise budgets must be treated with some degree of suspicion.

Notes

The author gratefully acknowledges the support of the Department of Rural Economics, Laval University, Quebec, Canada in preparing this paper.

References

Burt, Oscar. 1986. "Econometric Modelling of the Capitalization Formula for Farmland Prices, 1963-82." *American Journal of Agricultural Economics* 68: 1-9.

Vasavada, U and A. Somwaru. 1990. "Capital Accumulation and Asset Pricing in U.S. Agriculture." USDA Technical Bulletin No. 1771. Washington D.C., February.

Estimating Costs of Non-Human Capital Services: Discussion

Gregory M. Perry

I will begin by reviewing both chapters, highlighting their arguments. Burt used a classic approach to asset valuation, based on the NPV of returns generated by the asset over its expected life. Depreciation is the change in asset value over a particular time period (Burt used a year) and, with opportunity and operating costs, constitutes the total cost incurred by a farmer when owning and using a capital asset. He then compared this approach with that used by the USDA. Burt seemed to have no major argument with the USDA method, but did stress the importance of identifying the proper discount rate (in real terms) when estimating opportunity costs.

Gustafson, Barry, and Ali used the capital theory approach in their analysis, which also forms the basis for ERS depreciation estimates. The authors addressed three issues they believe are important in estimating the discount rate for capital theory models. They are: (a) use of nominal rates for variable costs and real rates for nonland capital, (b) use of opportunity cost concept vs. that of residual income returns, and (c) including other assets (such as land) in calculating residual income returns.

In discussing both chapters, I want to focus my comments on depreciation costs, rather than the opportunity cost issue. Before moving to depreciation issues, however, I would like to make a point or two about opportunity costs. Both Burt and Gustafson *et al.* argue for use of a real discount rate in calculating opportunity cost. From a theoretical standpoint, their arguments are correct. Both depreciation and opportunity costs should be calculated in real dollars. Implementing the theory correctly, however, is certainly open to question.

In the real world both interest rates and depreciation costs are reflected in nominal dollars. Estimating the rate of inflation occurring in both costs (particularly in interest rates) is no easy task. As Gustafson *et al.* noted, the real opportunity cost should reflect returns to durable assets only, rather than including assets such as land. I'm not sure a long-run estimate of residual returns is the answer here. Use of nominal opportunity cost is acceptable if the value of last period's equipment is expressed in t dollars, then reduced by one minus the inflation rate. That is,

(1) $$rV_{t-1}+(V_{t-1}-V_t)=(r+f)V_{t-1}+(1-f)V_{t-1}-V_t$$

In my view anything that can allow us to use nominal opportunity costs is going to aid in understanding what this value is and how it might be calculated.

Of greater concern to me is the approach used in estimating depreciation. The capital budgeting approach is the *modus operandi* of ERS and was also supported by both Burt and Gustafson *et al.* I believe capital budgeting has several flaws that make it an unacceptable measure of depreciation in USDA's cost of production estimates. The reasons for such a bold statement are the subject of the remainder of this discussion.

By definition depreciation is a market-oriented concept. As Burt defined it, depreciation is the change in asset value over a specified time period. The capital budgeting approach can be calculated as

(2) $$D_t = (V_O - V_N(1+r)^{-N}) * \left(\sum_{i=1}^{N} (1+r)^{-i} \right)^{-1}$$

where D_t is depreciation, V_O is new asset price, r is the discount rate, N is expected equipment life (in years), and V_N is salvage value in year N. The formula used by ERS (and reported by Burt) is an approximation of (1):

(3) $$D_t = r\left[\sum_{i-1}^{T} \gamma_i V_{k-1}\right] + (V_{0T}-V_{Nt})/N$$

where γ_i is the proportion of assets of vintage i in the sample data set, T is maximum life of all assets in the sample and V_{k-1} is the value of vintage i equipment in year t-1. In essence, the ERS approach is capital budgeting with separate depreciation and opportunity cost estimates. Key data needs for estimating depreciation in (3) are (a) price of new equipment, (b) estimate of equipment life and (c) estimate of salvage value. In addition, a linear depreciation pattern is assumed. As Watts and Helmers point out, the depreciation pattern will have an impact on annualized depreciation and opportunity cost. Equation (3) correctly handles opportunity cost estimates. The assumption of a linear depreciation pattern, however, does not hold up in the real world.

Expected life is an unknown value. ERS commonly uses American Society of Agricultural Engineers (ASAE) data in these estimates. These values are apparently based on potential life, rather than actual life experienced by equipment in use. For example, ASAE has estimated tractor life at 10,000 hours.[1] I have a data set containing auction sales for over 1600 tractors sold between 1984-1990. These data were selected from a larger set in excess of 6,000 because they contained information on equipment make, year manufactured, total hours of use, and other factors. Of the 1600+ tractors, none had more than 10,000 hours of use and less than 5% had over 6,000 hours of use. About 60% of these tractors sold in farmer retirement auctions, a substantial number of which are estate sales. In addition, about 10% sold in bankruptcy auctions. Consequently, about 70% of these tractor sales should represent a good cross section of use. In fact, they may even tend to represent older than average equipment, because farmers nearing retirement are likely to wear out what equipment they have rather than replace with something newer. If tractors do indeed last 10,000 hours, it seems likely that at least a few would show up in this data set. I believe the 10,000 hour estimate is much too high; if it is, ERS is underestimating annual depreciation, perhaps by a substantial amount.

Salvage value is estimated based on expected life and annual use. Consequently, if expected life estimates are wrong, salvage value estimates are also wrong. Other factors might also cause a misestimate of salvage value. High fuel prices could cause inefficient equipment to be retired much earlier than was anticipated at purchase. High farm prices and a shortage of equipment could keep equipment in use longer than expected. Of course, these may be temporary phenomenon, but they still impact on the makeup of the aggregate capital stock. And, of course, using a salvage value here seems somewhat hypocritical. If market data can't be used to estimate changes in market value, how can this use be justified in estimating salvage value?

Of greater concern to me is the attempt to use the capital budgeting concept to estimate average depreciation across a number of farms. The approach outlined by both authors and by others (such as Watts and Helmers) is really concerned with estimating opportunity and depreciation costs for a single durable asset. Yet clearly ERS is mandated to estimate average depreciation costs across a number of farms. Burt argues that "the annual cost measures developed in his chapter from unit cost theory are representative of the entire industry regardless of age distribution at any point in time" (p. 9). I disagree. I believe that the assumptions required to use the capital budgeting approach to estimate average depreciation do not hold in the real world.

Actual average depreciation for a sample of farmers is really the average of changes in asset market value across the sample over a specified time period. If we grouped their equipment by vintage (or year of manufacture), depreciation may be expressed as

(4)
$$D_t = \sum_{i=t-1}^{t-T} \gamma_i (V_{it-1} - V_{it})$$

Under what conditions does the depreciation portion of equation (3) equate to (4)?

Assume $V_{it} = V_{i-1\,t-1}$ meaning, for example, that a 1988 tractor has the same value in 1989 as a 1989 tractor in 1990. For this assumption to be valid (a) technological change must occur at a constant rate over time, (b) there can be no inflation, (3) the effects of the macroeconomy on supply and demand for tractors must be zero (i.e., no capital gains or losses), and (d) there must be no year to year change in average usage or care given to equipment. Then (4) becomes

(5)
$$D_t = \sum_{i=t-1}^{t-T-1} V_{it}(\gamma_{i-1} - \gamma_i) + \gamma_{t-1} V_{i-1\,t-1} - \gamma_{t-T} V_{t-Tt}$$

Note that if $\gamma_i = 1/T$ and T becomes the maximum life assumed for the sample data set (N), then (5) reduces to the depreciation portion of (3). This last assumption (e) means that the proportion of each equipment vintage represented in the sample data set must be the same regardless of the vintage. To illustrate, if the maximum tractor age encountered in the cost and returns survey is 20 years old, 5% of all tractors must be one year old, 5% must be two years old, and so on. These conditions are sufficient to equate the two approaches. They may also be necessary conditions. Gustafson et al. argued that (c) does not hold. Assumptions (d) and (e) are even more absurd. The point is, the depreciation estimates bear no resemblance to the real world.

So with all these problems, why does ERS use a capital budgeting approach. Why isn't market information used to calculate depreciation? One argument made is that a market valuation approach can't be used because no reliable market data is available. As Burt put it "the used asset market is thin and heavily influenced by local conditions, or nonexistent in some cases." I disagree. Strong markets exist all over the country for many types of farm equipment. Equipment that have nonfarm uses (such as pickup trucks, larger trucks, crawlers, etc.) almost always have strong used markets. Several markets also exist for equipment used strictly in agriculture. Equipment dealers are obviously one market. Classified ads in farm newspapers and other advertising media are a second market.[2] A third market are auctions, of which only a part are reflected in our data set mentioned above. In addition, at least two groups (The National Farm and Power Equipment Dealers Association and Hot Line Inc.) publish "blue books" on used equipment prices, thereby providing important market information to buyers and sellers. I argue that there is sufficient market information available to value used farm equipment.

Models are also available to provide estimates of market value for used equipment values. A recently completed Ph.D. dissertation at Oregon State University focused on estimation of remaining value models for combines, swathers, balers, disks, plows, planters, and four different sizes of tractors (Cross). The results are quite satisfactory, both in goodness-of-fit measures and model structure. I believe use of these functions by ERS will result in an improvement in estimating equipment values and depreciation costs. ERS is already using a weighted average of equipment values to estimate opportunity costs for farm equipment. It would not take a great deal more effort to calculate market depreciation costs. The result would be a more accurate estimate of average costs of production.

Another argument is that capital asset costs are really long-run in nature and, therefore, should be amortized or evened out over their expected life. Market value changes reflect short run depreciation costs, but in the long run total changes in market value will be the same regardless of the approach. So why have short-run fluctuations in asset market conditions mixed in with a long-run cost?

One problem with the long-run approach is that it is improperly implemented. Repairs, maintenance, fuel, and other costs are directly related to the age and condition of equipment. When the asset is new, repair costs are low, the equipment is efficient in fuel use, etc. As equipment ages, however, these costs increase. Consistency demands that, if equipment depreciation and opportunity costs are to be treated as long-run costs, all expenses associated with that equipment should also be treated as long-run costs. To do otherwise results in an underestimate of total equipment costs when the equipment is new and higher costs when old age sets in. Why do we amortize all these costs together in an equipment replacement model, yet treat them differently in a cost of production budget? Aren't both approaches doing essentially the same thing?

The more important issue to me, however, is the underlying philosophy about durable asset costs and how they should be represented. Should they reflect actual costs in the year or be a long-run amortized cost? Which approach better reflects current costs of production for a farmer? Given we are dealing with an average cost across farms, I believe the choice is clear. Durable asset costs should be treated as a current cost, just as all other nonland costs are treated in the FEDS budgets. To do otherwise is to "mix apples with oranges".

Let me make this argument from a practical standpoint. Agricultural durable asset investment is cyclic in nature. In the good years farmers buy equipment because they need it and can afford the purchase. In the bad years they make do with the equipment they have. Depreciation costs are heaviest in the first few years after purchase. In a good year the cost and returns budgets should show a healthy return above variable costs and retaining returns to the farmer. By making heavy investments in equipment farmers tilt the aggregate equipment mix toward the newer end, thereby increasing average depreciation costs. In bad years, the aggregate complement will age and average cost will go down, thereby increasing returns to the operator. In short, capital costs tend to be negatively correlated with returns above variable costs, resulting in a more stable return to the operator. By taking a long-run approach to equipment expense, farm incomes look much more variable than they really are. An approach like that in (4) would better reflect the equipment costs perceived by farmers.

The focus on equipment has caused many of us to overlook another important depreciation cost for farms. Farm buildings in 1988 had an estimated value of $75 billion dollars, or nearly identical to that of farm equipment. Yet virtually no research has been done to determine how these buildings change in value over time. Clearly their characteristics imply a much different depreciation pattern than exists for equipment. Often buildings are lumped with land for valuation purposes, but we shouldn't expect that changes in building value will track with land value changes. Solid research is needed to analyze how buildings depreciate.

In summary, the chapters by Burt and Gustafson *et al.* raise some important issues and should be given careful consideration. I believe capital budgeting is an inappropriate approach to estimating durable asset costs unless no market-oriented approach is available. In the case of farm equipment, used asset markets can provide a good indication of depreciation costs and should be integrated into ERS depreciation estimates. Equipment costs can and should be treated the same as most other costs of production.

Notes

1. Expected life is then apparently calculated as total hours divided by annual hours of use.

2. As an example, the 2/1/91 issue of the *Capital Press*, a weekly farm newspaper serving Oregon, Washington, and parts of Idaho and California, lists individual advertisements for close to 100 tractors, with several hundred more listed under dealer advertisements.

References

Cross, Tim. 1991. "Estimation of Agricultural Machinery and Equipment Depreciation Using Flexible Functional Forms." Unpublished Ph.D. dissertation, Oregon State University.

Hot Line, Inc. *Farm Equipment Guide's Quick Reference Guide.* Ft. Dodge, IA.
National Farm and Power Equipment Dealers Association. *Official Guide, Tractors and Farm Equipment.* St. Louis, MO. Published biannually.
Watts, Myles J. and Glenn A. Helmers. 1981. "Machinery Costs and Inflation." *Western Journal of Agricultural Economics* 6:129-145.

PART SEVEN

Estimating Costs of Human Capital Services

17

Human Capital Issues in Measuring Costs of Production

Daniel A. Sumner

Introduction

The key to success in developing cost of production estimates is in limiting the magnitude and importance of errors. Measurements of cost *will* be inaccurate but attention to the specific uses of the estimates can help reduce the practical importance of inaccuracies for decision-makers. The thrust of this chapter then is explicitly pragmatic. It attempts to deal with the question: What can be done to more usefully generate cost and returns information, and, in particular, information about the costs of the human capital input into production?

Directions from Congress provide the basic mandate for the USDA cost of production studies. However, we know from experience that cost of production indicators, once produced, will be broadly used for a variety of questions and issues. The variety of potential uses creates additional problems for cost of production analysis in general and perhaps especially in the case of farm labor. The problems are both conceptual and empirical. Conceptually, a key issue is how to value the hours of labor input by farm operators and others who are not paid a direct cash wage. A related issue is which hours of farm work to count for farm operators. In particular, is time spent gathering information, making business decisions, and related management and marketing activities to be considered part of the cost of production? On the empirical side, the issues include the difficulties in collecting accurate measures of the quantity of both unpaid and hired labor, given the seasonality of labor demand and often the lack of quantity records. In addition, costs per unit of hired labor are difficult to measure because of piece rate systems of pay, fringe benefits, and payments in kind, and because employers may not have records of wage rates.

In the United States, estimation and analysis of farm costs and returns and related farm budgeting is prepared for all major commodities by all major agricultural research universities and by the USDA. Similar efforts are conducted in other countries as well.[1] Particular attention is paid to the USDA estimates in this chapter because their efforts are extensive, elaborate, and well documented.

How to Value the Farm Work of Operators and Other Unpaid Workers?

Under what conditions and for what purposes would we want to value operator labor at the hired wage rate? Consider the case for which we want only non-management functions to count as a labor cost. In this case, it may be reasonable, as an approximation, to consider operator and hired labor as perfect substitutes. Then, if hired labor is available at a given prevailing wage, operator labor will be used to the point where the value of operator's time is equal to that wage. In this setting, the farm is facing an upward sloping supply function of the farmer's own time. As operator time available for non-farm activities is reduced in order to increase its use on the farm, the value of time at non-farming activities will tend to rise.

Figure 17.1 illustrates a case of a farm operator who works on the farm, but not off the farm. The time allocation choice the farmer faces is between time spent at farm work (shown on the horizontal axis) and time at non-work activities. The marginal value of operator farm work is shown as the downward sloping function labeled VMP. At point "c", where the marginal value of the operator farm work meets the wage of the hired labor (w_h), farm operator time becomes a perfect substitute for hired labor. To the right of point "c", the farm demand for operator time is horizontal at wage w_h.

The amount of time the farm operator will devote to farm work is equal to the hours at which the relevant VMP equals the marginal value of non-work activities (MVN). Two alternative MVN functions are shown in the Figure. For MVN_a, the chosen hours of farm work is indicated by "a" on the horizontal axis and the marginal value of farm work is w_h. In this case, the farmer has spent enough time at farm work that he or she is working at tasks for which the farmer's efforts are substitutable for hired labor on an hour for hour basis. For situation of MVN_b, the farmer devotes "b" hours to farm work activities. In this case, hired labor is not perfectly substitutable for farmer time and w_h is a clear underestimate of the value of the farmer's hours of work on the farm.

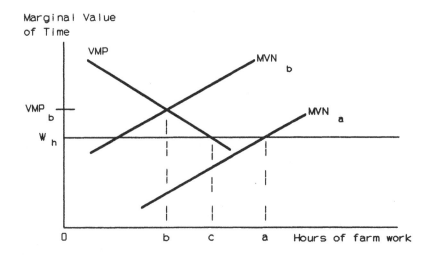

FIGURE 17.1 Farm operator time allocation choice.

In Figure 17.1, two cases are illustrated by different MVN curves. Obviously, VMP curves and hired farm wage rates also vary across farms. When determining how much time to devote to farming, whether w_h is the relevant value of time, depends on where the intersection shown alternatively by point "a" and "b" lies relative to the intersection of VMP and MVN.

The infra-marginal area above w_h and below the VMP curve would appear in the cost and returns accounting as residual returns attributable to management and risk. If w_h is used to value farm operator's work, then in case b, an additional area shown by the rectangle from w_h to VMP_b for hours of work, b would also be attributed to management and risk.

When off-farm work is performed, an explicit alternative wage rate to evaluate the opportunity cost of unpaid labor is available. In order to adequately reflect the choice between farm operator and other unpaid labor (l_u) and hired farm labor (l_h), one must compare the opportunity wage of unpaid labor (w_u) to the wage of hired labor (w_h). When unpaid labor and hired labor are both used on the farm, the equilibrium condition is

(1)
$$\frac{dQ/dl_u}{w_u} = \frac{dQ/dl_h}{w_h}$$

Also, of course, each type of labor will be used such that

(2)
$$P_Q \frac{dQ}{dl_u} = w_u, \quad P_Q \frac{dQ}{dl_h} = w_h.$$

In (1) and (2), Q measures output quantity, and P_Q is the price of output.

In general, unless hired and unpaid labor are perfect substitutes in farm production, the hired farm wage is not the wage appropriate for allocation of unpaid labor time. The allocation of unpaid labor time affects farm size enterprise choice and the choice to farm at all. Under or over valuing unpaid labor time can give misleading impressions of the relative costs of different farm operations.

Cost of production measurements that value operator and other unpaid labor at the hired farm wage w_h treat hired labor and unpaid labor on the farm as perfect substitutes. Both equations (1) and (2) showed that when $dQ/dl_u = dQ/dl_h$, it also implies that $w_u = w_h$ in farm decision making.

In the case of field activities and physical labor performed by unpaid labor, we may want to associate with the farm or enterprise what cash outlays would have been if the work were performed by hired labor rather than unpaid operators or family labor. In that case we would indeed use the wage rate paid to hired labor for the same tasks to value the time of the unpaid labor. If the tasks are comparable, and the time spent to accomplish the same amount of work is the same, then this approach will not bias cost estimates. However, for example, if unpaid labor uses fewer hours than hired labor to accomplish the same tasks, farms using unpaid labor would have a measured cost advantage in the accounting scheme used by the USDA and others. These lower measured costs would be an under estimate from the perspective of farm decision making when the unpaid worker has an alternative wage that also reflects his or her additional productivity.

Which Hired Labor Wage for Unpaid Labor Time?

The NASS *Farm Labor* report provides separate wages for field workers, livestock workers, and supervisory workers as well as workers paid on an hours basis, piece-rate and other methods. For some cases, the supervisory wage would clearly be the appropriate one to value the labor hours of farm operator labor. In general, if the hired wage approach is used, one would want to have some detailed information on the tasks performed by the farm operator and other unpaid workers to associate an appropriate wage. The Farm Cost and Returns Survey on occasion collects some wage information, including the wage rates of hired managers. This may provide further data for wage interpretation of unpaid workers.

Interpreting Evidence
that the Hired Farm Wage is Appropriate?

It is often observed that farmers do farm tasks that are also performed by hired labor. This can be interpreted as evidence that a hired wage is the appropriate value of time for farm operators. However, several problems of this conclusion are apparent. First, the farm operators productivity per hour may be larger (or smaller) than hired workers, even at the same task. Second, what looks like the same task to an outsider may, in fact, be different. For example, the farmer milking cows may be also examining for disease and checking productivity or other problems. Third, hired labor may not be available at some fixed wage for certain tasks at the times and under the conditions that the farm operator might be employed. This may be particularly true for small jobs that cannot be anticipated well in advance. In such cases, the fixed costs of acquiring hired labor (especially in terms for farm operator time) make employment too costly.

A Dairy Example Using USDA Estimates

The following example may help make our consideration of valuing unpaid labor more concrete. Table 17.1 lists abbreviated costs and returns statements per hundredweight of milk produced for the upper Midwest and Pacific regions. These two major milk-producing regions of the United States differ in a number of respects. In particular, the Pacific region uses more hired labor and the upper Midwest uses unpaid labor. Typical herd size also is much larger in the Pacific region.

Many analysts are uncomfortable with the imputations used in creating the "economic costs" and therefore use only cash expenses in cost comparisons. But if only variable cash expenses are compared, the upper Midwest costs are low relatively to the Pacific region. And, since farms chose whether to use owned resources and farm produced inputs or to specialize and buy more inputs from market sources, cash expenses can be quite misleading as a proxy for full costs.

In the list of costs shown in Table 17.1, unpaid labor is valued at the hired labor rate of about over $5.00 per hour in the upper Midwest. If the off-farm salary of non-farm managers were used to value the time of the farm operators this wage would likely double or triple. This would dramatically increase the full (imputed) costs of dairy farming and cause the residual returns in the upper Midwest to change from positive to negative. Further, since Pacific dairy farms use about twice the hired labor but only one-forth the unpaid labor as upper Midwest farms, the relative costs and returns of the two regions would change as well.

TABLE 17.1 Costs and Returns to Milk Production, 1989

	Upper Midwest	Pacific
	- $ per cwt. -	
Cash receipts	14.81	13.68
Hired labor	0.68	1.40
Total variable cash expenses	8.44	10.13
Unpaid labor	0.63	0.18
Total economic costs	13.90	12.44
Residual	0.91	1.24

Source: USDA, 1991.

For the decision of whether to enter into or remain in dairy farming, the potential earnings at other occupations is the relevant value of human capital. The full opportunity cost of labor is relevant if we are considering the question of net entry of operations into dairying in the two regions. But, this is just one specialized use of farm cost and returns estimates and it is probably one that has more than the usual potential for pitfalls. Current dairy farm operators gauging the potential non-farm opportunity cost of their time may have particular difficulty. Further, there is likely a positive connection between investments of human and financial capital in farming, so leaving dairy farming as an occupation would usually mean withdrawing and reinvesting equity as well. Shifting one's occupation and financial portfolio is not a costless transaction and these fixed costs are relevant as well. The point here is that if one were serious about analyzing the movement of human capital into or out of dairy farming it takes much more analysis than just looking at cost and returns tables with the unpaid labor valued at the non-farm wage (Sumner and Leiby).

It should also be pointed out that one could add unpaid labor costs to the residual to get total returns to the operator and other unpaid labor and management. The total economic costs include imputations for the use of all invested operator capital (valued at the average rate of return).

Now let us compare residual returns between the Midwest and the Pacific regions when these include unpaid labor costs. In the Midwest, a payment of $1.54 per cwt. compares favorably to a return of $1.42 per cwt. in the Pacific. But now compare these returns on a full time equivalent basis for unpaid labor. Consider a farm in each region that employs one full-time unpaid person at 2,000 hours. In the upper Midwest, at 8.33 cwt./hour (based on 0.12hr/cwt.) we get yields production of 16,667 cwt. and $25,667 for the farm. (At the upper Midwest average of 138 cwt. per cow this translates into a herd of 121 cows, well above the regional average.) In the Pacific region, at 33.33 cwt./hour we get 66,667 cwt. and $94,667. (At the

Pacific region average of 166 cwt./cow this translates into a herd of 402 cows, also above the regional average.)

The large implied herd size per full time operator may reveal that unpaid labor per cwt. may be under reported in these estimates. However, it also seems likely that Pacific region dairy farmers did not earn nearly four times their upper Midwest counterparts. The imputed wage in the Pacific region may well be higher but not by a factor of four.

So what is learned from this exercise? Probably the first point is that comparing costs even for the same type of farm is full of pitfalls. It is certainly not clear that we would use the comparisons of these two dairy farms for all purposes. A second point may be that it may make more sense to put all returns to unpaid labor into the residual for farm returns rather than attempt to more carefully impute the cost of using operator and family labor. It is clearly misleading to value the time of an experienced farmer and his sixteen year old son using the same wage. But, it is also hard to see what decisions are improved by using some proxy for off-farm value of time for each. It clearly does not seem that policy or management decisions would be enhanced by this imputation.

The Undercounting of Human Capital Inputs

Now turn to the potential problem of measuring the quantity of labor or human capital used on farms. The FCRS accounts for hired labor expenses, including total payments for contract labor and payments for non-wage costs of employing labor. However, several cost items that are based on hired labor services are included in other cost categories. The FCRS gathers information on custom work (defined as services that combine labor and machine use), general services (such as legal, accounting and management services), equipment repairs (which includes parts and labor) and veterinary services. In each case, a substantial part of these costs are payments for human capital services but none of these costs are allocated to labor.

In assessing the variable costs per unit of output or the net returns to farm operations, the category in which expenses are placed does not matter. However, among the uses of FCRS data is developing indicators of the labor or human capital share of farm costs. It is also the case that, for each of the four categories listed above, hired labor on the farm or farm operator labor can be a close substitute for the services that are not included in the labor total. In that case, in comparing accounts for different locations, farm size, or time periods what may appear as differences in labor input or labor cost, may instead more appropriately be attributed to differences in the use of labor imbedded in the farm services category. Further, when purchased services, such as management or equipment repairs are substituted for the time of the farm operator, this usually means that the costs are moved from the imputed economic cost category to variable cash costs. The implication is to get a complete measure of the labor and human capital input and contribution to farm production, it may be useful to attempt to separate the labor services component from costs listed in other areas on the budget and costs and returns surveys.

Conclusions

The foregoing comments were prepared with an aim to better understand what cost of production measures are telling us. However, it is appropriate to conclude by raising a more basic concern. In the tradition of Friedman and Stigler (and as stressed by Pasour), it must be noted that costs of production for a farm firm are fundamentally unique to that farm. It can be misleading indeed to expect relative costs as measured by our surveys and estimation methods to indicate the decisions that will be taken by producers or how industries will evolve. This is especially true when cost information is averaged by size category, region or other aggregate. The problems are principally that outside observers cannot know enough about the alternative uses of the farm's resources to correctly gauge how the operator might respond to economic conditions. These alternative uses depend very specifically on detailed characteristics of resources that are generally not standardized. Even more important, however, is how the farmer values these alternatives and determines the relevant opportunity costs. This issue is particularly important in assessing unpaid operator and farm family human capital used on farms. In the short run, the relative preferences of particular farmers for certain activities will influence how much farm work and which tasks will be performed by unpaid labor. In the longer run, the choice to enter or leave farming is determined by preferences across occupations, as well as on potential earnings.

Notes

The views expressed are those of the author and not necessarily of the U.S. Department of Agriculture or of North Carolina State University.

1. Demand for information about costs and returns is not unique to the United States. For example, the Australian Bureau of Agricultural and Resource Economics (ABARE) and the New Zealand Ministry of Agriculture and Fisheries both conduct extensive farm level surveys which are then used to assess costs and returns. In addition, private sector organizations develop estimates of costs and returns for particular industries. The Australian and New Zealand accounts are presented on an average per farm basis--neither the Australian nor New Zealand accounts are designed to be used as average unit cost estimates, but rather to assess the economic health of the industry. In New Zealand, the survey and analysis underlying costs and returns

estimates seem to be on a relatively *ad hoc* basis conducted separately for each industry. The New Zealand estimates deal only with cash accounts, and the data from the dairy, meat, and wool boards (not shown) are also presented on a cash basis. This approach avoids many complications, but also provides no assessment of returns to fixed capital and to operator labor.

In Australia, ABARE conducts a nationwide survey based on personal interviews of producers, follow-up data collection from accountants, and others with whom the producers conduct business. The ABARE survey asks producers to supply wages paid per worker and the number of full-time equivalent weeks worked for all permanent workers including the operator, partners, and family, even if not paid a wage. No weeks or hours of work information is collected for non-permanent hired workers. In valuing the time of the farm operator and unpaid family members, ABARE uses Australia's "Federal Pastoral Industry Award rates." These rates amount to minimum salary level allowed for full-time hired farm workers, and are, in fact, the prevailing wages for such workers.

The Farm Costs and Returns Survey (FCRS) of the USDA gathers information on the total wages paid by the farm broken down by operator, other household members, and all others. For farm operators and other unpaid workers, information is collected on the average hours worked per week in each month of the year. For versions of the FCRS used to collect data on costs of particular enterprises, respondents are asked the percentage of hours used to produce that commodity and, of those hours, the percent devoted for machine operation, hand labor, and other jobs.

The USDA and ABARE approaches to costs and returns accounting are clearly similar. With respect to human capital, both agencies collect expenditures on hired labor without collecting hours of work; both collect information on the quantity of unpaid labor used on the farm and imputes the value of this time based on a hired labor wage. The approach of using the hired wage rate to impute unpaid labor costs is conceptually the same between the two countries. It is clearly at odds with an opportunity cost concept of looking at the off-farm wage rate or some other non-farm alternative wage that might be earned by farm operators or family members. However, both agencies collect data on farm operator characteristics and off-farm earnings of farmer household members.

Demand for information about costs and returns is not unique to the United States. For example, the Australian Bureau of Agricultural and Resource Economics (ABARE) and the New Zealand Ministry of Agriculture and Fisheries both conduct extensive farm level surveys which are then used to assess costs and returns. In addition, private sector organizations develop estimates of costs and returns for particular industries. The Australian and New Zealand accounts are presented on an average per farm basis--neither the Australian nor New Zealand accounts are designed to be used as average unit cost estimates, but rather to assess the economic health of the industry. In New Zealand, the survey and analysis underlying costs and returns estimates seem to be on a relatively *ad hoc* basis conducted separately for each industry. The New Zealand estimates deal only with cash accounts, and the data from the dairy, meat, and wool boards (not shown) are also presented on a cash basis. This approach avoids many complications, but also provides no assessment of returns to fixed capital and to operator labor.

In Australia, ABARE conducts a nationwide survey based on personal interviews of producers, follow-up data collection from accountants, and others with whom the producers conduct business. The ABARE survey asks producers to supply wages paid per worker and the number of full-time equivalent weeks worked for all permanent workers including the operator, partners, and family, even if not paid a wage. No weeks or hours of work information is collected for non-permanent hired workers. In valuing the time of the farm operator and unpaid family members, ABARE uses Australia's "Federal Pastoral Industry Award rates." These rates amount to minimum salary level allowed for full-time hired farm workers, and are, in fact, the prevailing wages for such workers.

The Farm Costs and Returns Survey (FCRS) of the USDA gathers information on the total wages paid by the farm broken down by operator, other household members, and all others. For farm operators and other unpaid workers, information is collected on the average hours worked per week in each month of the year. For versions of the FCRS used to collect data on costs of particular enterprises, respondents are asked the percentage of hours used to produce that commodity and, of those hours, the percent devoted for machine operation, hand labor, and other jobs.

The USDA and ABARE approaches to costs and returns accounting are clearly similar. With respect to human capital, both agencies collect expenditures on hired labor without collecting hours of work; both collect information on the quantity of unpaid labor used on the farm and imputes the value of this time based on a hired labor wage. The approach of using the hired wage rate to impute unpaid labor costs is conceptually the same between the two countries. It is clearly at odds with an opportunity cost concept of looking at the off-farm wage rate or some other non-farm alternative wage that might be earned by farm operators or family members. However, both agencies collect data on farm operator characteristics and off-farm earnings of farmer household members.

Statistical comparisons of costs and returns for the U.S., Australia, and New Zealand are available in the author's paper presented at the *Conference on Economic Accounting for Commodity Costs and Returns*, Kansas City, Missouri, February 19-21, 1991.

References

Friedman, Milton. 1976. *Price Theory.* Chicago: Aldine Publishing Co. pp. 45-151.

Pasour, Jr., E.C. 1980. "Cost of Production: A Defensible Basis for Agricultural Price Supports?" *American Journal of Agricultural Economics* 62: 244-248.

Stigler, George J. 1968. *The Organization of Industry.* New York: Richard D. Irwin, Inc. pp. 71-94.

Sumner, Daniel A. and James D. Leiby. 1987. "An Economic Analysis of the Effects of Human Capital on Size and Growth Among Dairy Farms." *American Journal of Agricultural Economics* 69, No. 2: 465-470.

U.S. Dept. of Agriculture. 1991. "Economic Indicators of the Farm Sector: Costs of Production--Livestock and Dairy, 1989." Economic Research Service, Commodity Economics Division. ECIFS 9-1.

_____. "Farm Costs and Returns Survey." National Agricultural Statistics Service, and the Economic Research Service, Washington, DC, various years.

_____. 1991. "Farm Labor." National Agricultural Statistics Service. Washington, D.C. February 12.

18

Costs and Returns: A Perspective on Estimating Costs of Human Capital Services and More

Wallace E. Huffman

Introduction

Labor is one of the important inputs in production. Labor services in agriculture are provided by (1) farm operators/managers who perform very important allocative, production/marketing decision-making, and supervisory functions (Schultz, 1972; Huffman, 1985, 1991b) and frequently engage directly in production, (2) by members of the operator's family or relatives who largely work without direct compensation, and (3) by hired workers. Hired workers are of two types: regular hired workers who live on or close to the farm and hold full-time year-round jobs, and seasonal hired workers who work only part of the year (Martin, 1985). Seasonal hired workers are workers who live within daily commuting distance of the farm, domestic migratory workers who travel from farm to farm taking up temporary residence nearby as opportunities for employment change geographically, and aliens who cross national boundaries to work. Some aliens enter the United States under special national programs and others enter illegally.

Systematic attempts by the USDA to estimate employment on farms date back at least to the late 1930s and regular attempts to estimate wage rates of hired farm labor date back to 1909. The USDA used estimates of employment on farms up to the 1950s, then it switched to a personhours measure. Benchmarks of labor used per acre and per head were prepared for 1910, 1919, and 1929 based upon extensive field surveys and for 1939, 1944 and 1950 based upon secondary data (USDA, 1939, 1943, 1944, 1954, 1955). This type of labor-requirements methodology was applied by the

USDA from the late 1950s until the mid-1980s, when farmers were asked to indicate the hours of labor used on farms (Hauver). Although the first investigations of wage rates for hired farm workers date back to 1866, regular annual inquiries do not start until 1909, and after 1923 they were largely made quarterly (U.S. Dept. of Commerce, 1975).

Estimates of hours of labor input and costs of labor have historically had primarily two uses in the USDA: to provide components for estimates of costs of agricultural production and for one of the inputs in an index of agricultural production efficiency (e.g., USDA, 1990; Zepp and Simmons; Loomis and Barton; Ball; Hauver). The objective of this chapter is to present concepts and practical details about how to improve estimates of the quantity and cost of human capital services in agricultural production. Although measurement of hours of hired labor and the wage rate for this labor will receive some attention, the primary focus is on the hours and cost/value of farm operators/managers and family labor.

The Economic Model

An agricultural household model is one that combines the agricultural producer, consumer, and labor supply decisions of households into a single conceptual framework. This model is quite versatile (Huffman, 1991a; Singh, Squire, and Strauss). It can accommodate effects of uncertainty about prices, technology, pests, weather, and employment. Also, it can accommodate joint production both in the sense of producing commodities for consumption and sale and multiple-output agricultural technology.

The model has special cases that represent the range from highly specialized large-scale commercial agriculture to small-scale or peasant agriculture. In highly specialized large-scale commercial agriculture, all inputs are purchased and all outputs are sold in the market so that household consumption and farm business decisions are separable. Farm costs and returns issues then become specialized to farm business decision making. In small-scale or peasant agriculture, households produce a significant share of the commodities they consume (and sell none or a small share), and they supply most of the inputs that are used in household and farm production. For these households, consumption and farm business decisions are unlikely to be separable. Also, characteristics of a rural life-style may be an important part of the goods that these households consume.

In the model that is to be developed here, a few stylized facts will be incorporated. They serve to focus the modeling on key issues. They can be modified, but that would be at some additional complexities of model specification. At this stage, these complications seem likely to be a major distraction. First, the decision unit is a single family husband-wife farm household that owns and operates a self-employed farm business. Second, households are assumed to make production and consumption plans at the beginning of the year and to carry out these plans. Third, farm and household production decisions are separable; and farm production is multiproduct. Fourth, farm production contains a time lag, and prices for farm output are uncertain at the time that plans are made. Input prices are known with certainty at planning time. Fifth, human time of farm operators, of their spouses, and of nonfamily hired farm labor are heterogenous (Huffman, 1985, 1991b; Schultz, 1972). Sixth, farm (and nonfarm) household members have the opportunity for off-farm work at nonfarm jobs, but employment in the nonfarm sector is uncertain because of unemployment caused by unanticipated business-cycle shocks. Seventh, farm households have risk neutral preferences toward uncertainty.[1]

Some might quarrel with the assumption of a single family being the owner-operator of a farm business. The major complication with having more than one family is that we would need to make some strong assumptions about aggregating consumption preferences across the households. Although much research has been done on attitudes toward risk of farmers, the results are not definitive, and risk neutrality is one reasonable assumption that is consistent with some of the empirical studies.

The economic decisions of these farm households are summarized in the following equations:

(1)
$$U = U(T_h^M, T_h^F, Y, \xi^M, \xi^F, \Delta, \tau),$$

(2)
$$\bar{T} = T_f^j + T_m^j + T_h^j, T_m^j \geq 0, T_f^F \geq 0, j = M, F,$$

(3)
$$(1-\mu^M)W^M T_m^M + (1-\mu^F)W^F T_m^F + V + P_1^* Q_1 + P_2^* Q_2 - W_1 X_1$$
$$- W_2 X_2 - C_F = P_y Y,$$

(4)
$$Q_1 = F(Q_2, T_f^M, T_f^F, X_1, X_2; \xi^M, \xi^F, \Delta)$$

Farm households derive utility from current consumption goods which are leisure of the husband and wife (T_h^j) and from goods purchased in the market (Y). The transformation of these goods into utility is conditioned by the husband's and wife's human capital (ξ^j), informational technology (Δ), and tastes or other household characteristics (τ), e.g., number of children in the household and commuting distance to service centers. These last three variable types are assumed to differ across households, but they are not considered to be part of the decisions that households are currently making.

Farm households have resource and technology constraints. A household receives a human time endowment at the start of the planning period of $\hat{T}$ for each household member. Here we limit the analysis to a husband and wife, but clearly children and their time allocation could be an important factor. The time of the husband and wife are heterogenous for a variety of reasons, including skill differences (Becker; Huffman, 1985, 1991b). The time of each adult can be allocated to any of three activities: own-farm work (T_f^j), off-farm work at nonfarm jobs (T_m^j), and leisure (T_h^j).[2] The optimal hours of off-farm work for a husband or wife and optimal hours of farm work of the wife might be zero. Hence, nonnegativity constraints are imposed:

$$T_m^M \geq 0, T_m^F \geq 0, \ T_f^F \geq 0.$$

The cash income at planning time of a farm household is uncertain because of (1) uncertain farm output prices and (1) uncertain off-farm work employment prospects. For now, farm production is assumed to be certain, but this can be modified with some more work. With risk neutral preferences and nonstochastic production, the farm output prices used in the planning process are expected prices at planning time for harvest dates, and the wage for off-farm work is the expected wage, i.e., the wage given employment adjusted for the expected probability of employment or $(1-\mu^j)$ W^j. Thus, expected household wage income is $(1-\mu^M) \ W^M T_m^M + (1-\mu^F) W^F T_m^F$.

Individuals are assumed to face a perfectly elastic demand for their labor services at off-farm jobs. The wage rate for an individual of a given sex is, however, assumed to be affected by the amount of his/her human capital (ξ^j) and local labor market and cost of living conditions (ω). The assumption that an individual's wage does not vary with his/her hours of work follows a long tradition in labor economics. The relationship is summarized as:

(5) $$W^j = W^j(\xi^j, \omega), j = M, F.$$

This wage relationship is also the one that individuals who perform hired farm labor face when they consider nonfarm work.

Household nonfarm asset income is denoted V. Farm households have expected net farm income of $P_1^{\cdot}Q_1 + P_2^{\cdot\cdot}Q_2 - W_1 X_1 - W_2 X_2 - C_F$ where $P_1^{\cdot}Q_1 + P_2^{\cdot\cdot}Q_2$ is expected gross receipts at planning time for farm outputs that are produced this period and $W_1 X_1 + W_2 X_2 + C_F$ are variable costs associated with carrying out the production plan.[3]

The technology of farm production is multiple output and multiple input. Q_1 is the numeraire output, e.g., bushels of feed grains, and Q_2 is a vector of other outputs, e.g., other crops/pasture outputs, livestock outputs. T_f^j are the inputs of the farm husband's and wife's time to farm work, but optimal T_f^F might be zero. The input X_1 is the quantity of nonfamily hired farm labor, which might be zero. The input X_2 is a vector of nonlabor farm inputs and might include land services. The efficiency of the farm production process is affected by the human capital of the husband and wife (ξ^j) and informational technology (Δ).[4]

The representation of agricultural technology as multi-output multi-input seems imperative. From a planning perspective, a multicrop crop-rotation is required on almost all U.S. cropland in order to maintain soil fertility, control pest problems, and control soil erosion. More generally, on some of the farmland, meadow or grass must be included in the long-term land use rotation in order to control soil loss within tolerable limits. Ruminant livestock are required in order to make efficient use of this forage. More specifically, in the Corn Belt, beef cattle and hogs graze meadows and corn stalks. In the Great Plains, calves and yearlings graze winter wheat during the fall and winter months.

Furthermore, some inputs contribute to more than one farm output. For example, time spent by the farm operator in deciding how much of each output to produce or how to allocate available farmland to alternative uses cannot be allocated by anything other than by an arbitrary fashion to Q_1 or Q_2. Even nitrogen fertilizer applied to affect primarily the corn yield this year will partially carry over to affect next years crop yield. Information technologies of telephones, radios, computers, farm information services, accounting services, farm magazines, etc., cannot be allocated to any single output or activity.

Thus, these are reasons why for almost all of U.S. agriculture, the technology of production is best described as being joint. Separability even by major commodity groups does not seem to make much economic sense. Furthermore, costs-returns methodologies that rely on a strong separability assumptions should be viewed with much skepticism and applied only where long-term credibility is not an important virtue.

The production function (4) is substituted into the cash-income constraint (3) to obtain a combined expected cash income-technology constraint of:

$$(6) \quad (1-\mu^M)W^M T_m^M + (1-\mu^F)W^F T_m^F + V + P_1^* F(Q_2, T_f^M, T_f^F, X_1, X_2;$$
$$\xi^M, \xi^F, \Delta) + P_2^* Q_2 - W_1 X_1 - W_2 X_2 - C_F - P_y Y = 0$$

The conditions for an optimal agricultural household plan is to maximize equation (1) subject to resource constraints imposed by (2) and (6). Assuming interior solutions for all choices, except possibly for T_m^M, T_m^F, T_f^F, and X_1, the first-order conditions for a constrained maximum are:

(7)
$$\lambda[P_1^* F_{Q_2} + P_2^*] = 0$$

(8)
$$\lambda[P_1^* F_{X_2} + P_2^{*'} F_{Q_2} \frac{\partial Q_2}{\partial X_2} - W_2] = 0$$

(9)
$$\lambda[P_1^* F_{T_f^j} + P_2^{*'} F_{Q_2} \frac{\partial Q_2}{\partial T_f^j}] - \gamma^j = 0, j = M, F,$$

(10)
$$\lambda[P_1^* F_{X_1} + P_2^{*'} F_{Q_2} \frac{\partial Q_2}{\partial X_1} - W_1] \geq 0$$

(11)
$$\lambda(1-\mu^j)W^j - \gamma^j \geq 0, j = M, F;$$
$$T_m^j > 0, T_m^j[\lambda(1-\mu^j)W^j - \gamma^j] = 0$$

(12)
$$\lambda W_1 - \delta_1 \geq 0; X_1 > 0, X_1(\lambda W_1 - \delta_1) = 0$$

(13)
$$U_{T_h^j} - \gamma^j \geq 0, j = M, F,$$

(14)
$$U_y - \lambda P_y = 0$$

(15)
$$\bar{T} - T_f^j - T_m^j - T_h^j = 0, j = M, F$$

and the budget constraint (6), where γ^js, λ, and δ_1 are lagrange multipliers for marginal utility of husband's and wife's time, marginal utility of income, and shadow value of hired farm labor, respectively, and F_i and U_k are partial derivatives of F and U, respectively.

Equations (9)-(13) give the optimal conditions for human time allocation in this model. From (10) and (12), if the equality holds, a farm business hires farm labor up to the point where the marginal value product of hired labor is equal to the hired farm wage rate. If the shadow value of hired labor (δ_1) at $X_1 = 0$ is less than W_1, then the optimal quantity of hired farm labor is zero.

Equations (9), (11), and (13) give the conditions for optimal allocation of husband's and wife's time. For a husband, equation (9) and (13) are assumed to be equalities, i.e., he has positive optimal hours of work at own-farm and leisure, and thus at optimal allocations the marginal value should be the same in these two uses. For him, equation (11) provides the condition for optimal off-farm work. If the expected off-farm wage is less than the marginal value of his time in consumption/farm production (($1-\mu^M)W^M < \gamma^M/\lambda$), then $T_m^M = 0$, i.e., his optimal hours of off-farm work is zero. However, if $(1-\mu^M)W_M = \gamma^M/\lambda$ then at optimal allocations the marginal value is equal in all three uses. The expected off-farm wage rate now sets the marginal value of his time in a way that is exogenous to a household's consumption and farm production decisions, i.e., farm production (input and output) decisions are separable from consumption and labor supply decisions.

For a wife, optimal hours of her leisure time is assumed to be positive and the marginal value of this time is γ^F/λ. If at $T_f^F = 0$ the marginal value of wife's time in farm work is less than its marginal value in leisure or off-farm work and if $(1-\mu^F)W^F = \gamma^F/\lambda$, then she also works off-farm for a wage. In such a case, the marginal value of her time at off-farm work exceeds its value for potential farm work. A wife engages in all three time-using activities if $(1-\mu^F)W^F = \gamma^F/\lambda$ at an interior solution. It is also possible that even when she allocates all of her time to leisure, the marginal value of her time in consumption still exceeds the expected wage rate for off-farm work. In this case, all of her time should be allocated to consumption.

Conditions (7) and (8) provide marginal conditions for optimal Q_2 and X_2. Optimal Q_2 occurs where the marginal revenue lost from producing fewer units of Q_1 equals the marginal gain in revenue from the additional unit of Q_2. Optimal X_2, the nonlabor inputs, occurs where the marginal value product of X_2 equals its price.

If the equality conditions hold for conditions (7)-(12), then the farm business decisions can be separated from the household consumption decisions. The conditions for optimal outputs and inputs are then:

(16) $$Z_j^* = \psi_{z_j}[P_1^*, P_2^*, W_1, W_2, (1-\mu^M)W^M, (1-\mu^F)W^F, \xi^M, \xi^F, \Delta],$$

where $Z_1 = Q_2, X_1, X_2, T_f^M, T_f^F, Q_1$

In particular, the price (marginal cost) of husband's and wife's time allocated to farm work is the opportunity cost of this time which is equal to the expected off-farm wage. For each of them, the opportunity cost is applicable to all hours of farm work irrespective of whether it is used in the production of Q_1 or Q_2 or used for management, supervisory, or production activities. The expected wage rate for each individual will depend on his/her human capital (ξ^j) and local labor market conditions (Δ).

In the United States, the amount of human capital embodied in the husband and wife operating a farm and in hired farm labor can on average be ordered as follows: the husband, the wife, and the hired nonfamily labor. Thus, in a given region (state), the expected wage rates of these individuals are ordered in the same way with farm operator's time having the largest marginal value and hired farm labor having the lowest. Hence, valuing the farm operator's time at the wage for hired farm workers, as the USDA procedures do, (USDA, 1980; Hauver) will substantially underestimate "farm labor costs."[5] In addition, if the time of the operator that is allocated to management activities, including marketing, is totally ignored (e.g., as in the old USDA procedures), there is a further underestimate of farm labor costs.

When the husband and/or wife in an agricultural household do not allocate any time to off-farm work (and family and hired labor are not perfect substitutes), farm business and consumption decisions are not separable. What this means economically is that all of the conditions (6), (9)-(15) must be solved simultaneously. For example, assume that all optimal quantities are at interior solutions, except $T_m^j = 0$, then the conditions for optimal production and consumption decisions are:

(17) $$Z_j^* = \phi_{z_j}[P_1^*, P_2^*, W_1, W_2, P_y, V, \xi^M, \xi^F, \Delta, \tau]$$

where $Z_j = Q_2, X_1, X_2, T_f^M, T_f^F, T_h^M, T_h^F, Y, Q_1$

We see that the optimal quantity of farm outputs and farm inputs depends on the price of household consumption goods (P_y), household nonfarm asset income (V), and on the household taste parameters (τ). Furthermore, the marginal value of a farm husband's time (in consumption and farm work) is larger than its opportunity cost at off-farm work, i.e., $(1-\mu^M)W^M < \gamma^M/\lambda$, so that in these cases where the operator allocates all of his (her) time to consumption and farm work, the wage rate for hired (nonfamily) farm labor will almost certainly be a gross under estimate of the marginal cost of his time for economically sound cost of production estimates. In this case, his (her) expected wage rate for off-farm work is also an under estimate of the cost of his (her) time for farm work (and leisure).

Empirical Implementation

Given the conceptual framework presented for agricultural households in the previous section, we need to think about how to implement a sound empirical costs-returns methodology. An econometric approach has several advantages over alternative, e.g., back of the envelope, black-box computer, or strictly accounting, methods. First, the data can be asked in a fairly scientific way to reveal coefficients that directly or indirectly relate to the responsiveness of decisions to the economic environment. Second, these parameters and other relationships in the parameters can be bounded by confidence intervals. One reason this is of practical importance is that government program costs vary by a large magnitude when responsiveness or costs are miscalculated by a very small amount. Third, the econometric structure that ties together farm household decisions can be better identified. When identification is better, we hope that our forecasts are more accurate.

Data

Although the data needs are sizeable for my methods, they do not seem to be very different from what is currently being collected by the USDA in its Farm Costs and Returns Survey (FCRS). However, the size of the sample might need to be increased significantly.

The following data are needed:

1. The total quantities of each major farm input (including irrigation water) and output group and consumption group. For labor/human time, this requires the total annual hours for the categories of farm operator, wife, other family, and nonfamily hired workers. Hours of hired labor could be disaggregated into seasonal and regular hired labor. The annual hours of off-farm work of the farm operator, wife, and other family members are also needed.

2. Prices are needed for each of the major categories of purchased farm inputs and for consumption goods. Futures market prices or local prices of farm outputs are also needed. Wage rates for off-farm work are needed for farm operators, wives, and other family members that participate in this activity.

3. Some data on other farm household characteristics are needed. They include the age and education of the farm husband and wife, household nonfarm asset income, number of children at home by major age categories (e.g., less than six years of age, 6 to 18 years of age), state location, race of head of household.

4. Other information includes data on local labor market conditions (e.g., unemployment rates, employment growth rates), informational

technologies that aid decision making, and normal and pre-planning weather conditions.

Econometric Procedures

Econometric models must be specified for each major agricultural household decision structure and then fitted to data on a large enough sample to be able to obtain meaningful results. See Huffman, 1988; Huffman and Evenson, 1989; Shumway; Lopez; Capalbo and Antle; and Singh, Squire, and Strauss for some examples. Note that the subgrouping of observations in a random sample of agricultural households by homogenous decision structures can result in potential problems. A homogeneous decision structure is one where exactly the same set of decision variables are at nonextreme values (largely non-zero). When these structures differ, the set of variables that drive the optimal choices are different; for example, consider the differences in the general specification for Q_1, Q_2, T_t^M, T_t^F, X_1, and X_2 in equations (16) and (17). (See Huffman, 1988.) Furthermore, the coefficients of the variables that these structures have in common can be expected to differ.

Since these different structures are largely the result of rational economic decision making; they are clearly nonrandom. Thus, the subgrouping implies that agricultural households are nonrandomly assigned to these structures or subsamples (Maddala). This means when choice functions are fitted directly to the subsample that seems best to fit the conditions, the estimated parameters cannot in general be expected to be good estimates of the response coefficients for all farms that could be in a structural class. There is potential sample selectivity bias (Huffman, 1988). The choice functions for each decision structure need to be examined for possible selection bias, and when it is present, the econometric specification needs to be modified to account for it (e.g., see Huffman, 1988; Huffman and Evenson, 1989; Huffman and Lange; Tokle and Huffman). This requires a qualitative choice analysis of structural assignment using multivariate probability analysis.

For farm households where off-farm work occurs, the probability of this event must be examined. Furthermore, gender-specific labor demand functions that depend on an individual's schooling, age, and race, and on local labor market and cost of living conditions (see Tokle and Huffman) must be specified and fitted. Possible sample selectivity is an issue here and must be examined and incorporated into the demand equations. These off-farm wage equations will provide predicted wage rates that can serve as the cost of operator, wife, and possibly other family labor in farm production for at least half of the U.S. farm households. For the others, they will provide a lower-bound estimate of the marginal value of time that, at least for farm operators, would be significantly larger than the wage rate for farm hired labor.

Labor's (Human Agent's) Share

Kuznets (1966, pp. 181-83; 1955) showed that in western developed countries labor's share of national income had over the long term increased from 55 to 75 percent.[6] By 1970, 75 percent of U.S. national income consisted of employee compensation, i.e., wages, salaries, and benefits. Schultz (1977) argues that the human agent's factor cost share of U.S. national income in 1970 was 80 percent when the value for management services performed by self-employed individuals are included. He also argued that the long term rise of labor's share in the United States is largely the result of a long term rise of wage rates relative to the price of other inputs. This rise is closely associated with an increase in the average amount of human capital invested in human agents (i.e., schooling, experience, health). Between 1910 and 1970, the real wage rate of production workers in U.S. manufacturing, one good indicator of real wage trends, rose by a factor of four. The real wage rate peaked in 1985 and fell slightly after that time (Table 18.1).

The post-1970 period is one, however, where the supply of labor was growing rapidly (relative to the recent past) in the United States due a sharp rise in the labor force participation rate of women (Table 18.2) and because the World War II baby-boom cohort was entering the labor force, and it was large relative to older cohorts (See Smith and Ward, Goldin, Welch). These two factors caused average labor-market experience levels of workers to decline for a while. Thus, a temporary slowing down of the rise of real wage rates is not surprising. The long term trend of rising real wage rates will emerge again.

Using USDA data, Tweeten reports labor's share for U.S. agriculture as being 53 percent in 1910, 40 percent in 1940, 38 percent in 1950, 19 percent in 1970, and 13 percent in 1986. Also, Lianos emphasizes that labor's share decreased by slightly more than 50 percent between 1949 and 1968. My hypothesis is that labor's share of the cost of U.S. agricultural production is considerably higher and has been falling less rapidly (or not at all) than the USDA's data show. This is a hypothesis that was also stated in 1972 by Schultz.[7]

My reasons are that the USDA has significantly underestimated the labor input in U.S. agriculture and the value of human time, especially of the farm operator/manager, over time. Schultz' earlier (Schultz, 1972) concerns were based primarily on an undervaluation of the cost of human time. We will examine several concrete pieces of evidence in the following paragraphs.

TABLE 18.1 Historical Data on Average Wage Rates for Production Workers in Manufacturing, 1890-1989

Year	Average weekly hours paid	Average hourly wage ($/hr)	Real hourly wage (1914=100) [a]
1890	54.0	.16	79
1900	53.2	.18	96
1910	51.0	.21	100
1920	47.4	.55	126
1925	44.5	.54	141
1930	42.1	.55	151
1935	36.6	.54	179
1940	38.1	.66	214
1945	43.5	1.02	259
1950	40.5	1.44	273
1955	40.7	1.86	318
1960	39.7	2.26	348
1965	41.2	2.61	378
1970	39.8	3.36	396
1975	39.4	4.81	411
1980	39.7	7.27	425
1985	40.5	9.54	433
1989	41.0	10.47	408

[a] Deflated by the Consumer Price Index, 1890-1960, and the Implicit Price Deflator For Personal Consumption Expenditure, thereafter (Source: Economic Report of the President, 1991).
Source: Ress, A. and U.S. Department of Labor (1983) and (1989).

First, consider the labor cost shares in the Evenson-Ballou data set (Huffman and Evenson 1990, Ch. 10). Major USDA series were part of the raw material for constructing this data set. However, they went considerably beyond standard USDA procedures for measuring inputs as they attempted to follow the recommendations of the AAEA Task Force on Measuring Agricultural Productivity (USDA, 1980).

What the Evenson-Ballou data show are that labor's cost share for U.S. agriculture was 37.3 percent during 1950-60, and it decreased gradually through 1982 (Table 18.3). It was 30.8 percent for 1961-70, 25.5 percent for 1971-77, and 20.7 percent for 1978-82. However, 20.7 percent is considerably larger than the 11 percent that I hear stated as the USDA's estimate of labor's share of the cost of agricultural production currently.[8] Labor's share of the cost of agricultural production seems unlikely to have declined much during the 1980s.

TABLE 18.2 Historical Data on U.S. Labor Force Participation, 1890-1989 [a]

Year minimum	All males	All females [b]	All civilians
1890	87.4	18.9	49.0
1900	87.3	20.6	49.3
1910	86.3	NA	NA
1920	86.5	23.7	50.2
1930	84.1	24.8	49.3
1940	79.0	25.8	55.9
1950	79.0	29.5	59.2
1960	77.4	35.1	59.4
1970	80.6	41.6	60.4
1975	77.9	46.3	61.2
1980	77.4	51.5	63.8
1985	76.3	54.5	64.8
1989	76.4	57.4	66.5

[a] All persons $\geq$ 14 years old 1890-1960; $\geq$ 16 years old thereafter. Includes only civilians.
[b] *Source:* Goldin (1990)
NA = Not available.

Hours of Farm Work (Agent's Time)

In their productivity analysis, the USDA substantially underestimates both the hours and the wage or cost of farm work. Roughly from 1960-1985, the USDA based labor use for productivity analysis on input-output coefficients from engineering-type studies (actually labor requirement per unit of output or per acre, see Hauver). They had weak ties to reality. Starting in 1985, the USDA asked farmers about the hours of labor performed on their farm by themselves, unpaid family members, and hired labor. This was an improvement, and Hauver shows that this change in methodology resulted in an increase by 93 percent for ERS's estimate of total agricultural labor use in 1984 (6.4 bil. hours versus 3.3 bil. hours).

Farmers may not know very well the hours worked by hired labor on their farm when this labor is obtained through farm labor contractors (FLC). Farm labor contractors have become a more important source of seasonal farm labor after the mid-1970s. The share of the FLC wages in the total farm labor wage bill has increased from 9.9 percent in 1974 to 11.6 percent in 1982 (Vaupel and Martin, Martin and Holt). However, 65 percent of the FLC wages were paid in the three states of Florida, California, and Texas. Farm labor contractors are intermediaries that handle all of the details of locating labor, contracting with growers, transporting workers to fields, supervising the labor, and paying the workers. Thus, growers/farmers are

TABLE 18.3 Input Shares: U.S. Agriculture (42 states) [a], 1950-1982

Input Shares	1950-60	1961-70	1971-77	1978-82
Total[b]				
Fertilizer	.030	.038	.058	.055
Seeds	.015	.015	.025	.024
Land	.151	.182	.174	.177
Miscellaneous	.067	.095	.133	.175
Labor	.373	.308	.255	.207.
Capital services	.095	.098	.097	.111
All feeds	.267	.265	.258	.252
Crop Production				
Fertilizer	.069	.087	.118	.111
Seeds	.035	.035	.051	.049
Land	.175	.221	.213	.213
Miscellaneous	.083	.113	.154	.200
Labor	.453	.357	.292	.235
Capital services	.183	.187	.172	.193
Livestock Production				
Feed purchases	.272	.320	.352	.335
Feed fed on farm	.175	.117	.120	.124
Hay	.028	.032	.038	.041
Land	.133	.152	.136	.141
Miscellaneous	.055	.082	.112	.150
Labor	.311	.270	.219	.179
Capital services	.026	.029	.023	.030

[a] The New England states are excluded.
[b] Crop output/livestock output shares: .438/.562, .435/.565, .494/.506, .497/.503.
Source: Huffman and Evenson 1990, Chapter 10.

not as close to the labor utilization when they use FLCs as when they hire, supervise, and pay the workers themselves. The growers/farmers do, however, know what they pay to farm labor contractors (the expenditure), which includes the cost of the services provided by the farm labor contractor.[9]

Farmers that employ piece-rate workers have an incentive to systematically underestimate the worker's hours. The reason is that if the workers do not pick (produce) enough output so that their average wage rate exceeds the legal minimum wage, the grower/farmer must make up the difference in earning so that the average wage at least meets the legal minimum wage.[10] This problem is most serious for U.S. farmers/growers that are located close to the Mexican border where there is a perfectly elastic supply of agricultural labor at the U.S. legal minimum wage (which is about 10 times higher than the Mexican wage). This problem seems to be less severe when farmers/growers are located farther from the Mexican border because wage rates are more likely to be above the legal minimum.

Survey procedures frequently miss farm labor performed by unpaid family members. If unpaid family members do not perform more than 15 hours of farm work during the survey week, their labor is not included in the USDA (quarterly surveys), Bureau of Census (CPS), and the Bureau of Labor Statistics (BLS) surveys. This seems to me to be a mistake. It under-represents the contribution of women to agriculture. Furthermore, much of the work these individuals perform is seasonal, and they can give quite an accurate estimate of the magnitude when the question is properly phrased.[11] The Farm Costs and Returns Survey, which is the data source for individual commodity costs and return estimates by USDA, doesn't seem to have this problem because it asks about all hours worked by unpaid family members.

The Price/Cost of Time

The USDA substantially underestimates the value or cost of activities of the human agents performed on farms (farm labor broadly defined). The most serious aspect of this error comes from using the wage rate for hired farm workers as the cost/price of all farm labor performed on farms. This is true for both USDA's productivity analysis and commodity costs and returns analysis. The farm operator/manager (and wives of farmers, too) have on average a much larger stock of human capital than hired farm workers (Schultz 1972; Huffman 1985; Huffman 1991b).

I present two examples to suggest the magnitude of this problem. They contrast USDA procedures with my suggestion of using opportunity cost at off-farm work, suggested in the previous section. First, consider Iowa, North Carolina, and Oklahoma farms reported in Huffman (1976c). This study was based on the 1964 Census of Agriculture, which was the only Agricultural Census to collected data on earnings from off-farm work of farm household members. In 1964, the average daily wage rate for hired farm labor in Iowa, N. Carolina, and Oklahoma was $10.90, $6.60, and $9.10, respectively. The average daily off-farm wage rate of farmers was in contrast, $16.71, $13.67, and $15.96, respectively. Although some positive selectivity bias may be present in these averages, the off-farm wage rates of farmers are 153 percent, 207 percent, and 175 percent, respectively, of the average daily wage rate of hired farm labor. For wives, the average daily wage rate for off-farm work was $13.84, $11.48, and $12.84 for these three states. These off-farm wage rates of wives are also considerably larger than the average wage rates for farm hired labor in 1964.

Second, consider the data for Southwestern Wisconsin farms in 1986 (Saupe). The average hourly wage rate for off-farm work of male farm operators was $10.09 and for their wives $6.71. In contrast, the average hourly wage rate for hired farm labor in the Lake States (WS, MN, MI) was $4.12 in 1986. These off-farm wage rates are 145 percent and 62 percent larger than the average wage rate of hired farm labor. The data also show that the opportunity cost for farm work by male farmers and their wives increases as their years of formal schooling increases. The average hourly wage for male farmers who have completed less than 12 years of schooling was $6.95, which is significantly higher than the average wage rate for hired farm labor, but it was $8.29 for those who have completed high school (but not more), and $13.06 for those who had completed more than 12 years of schooling. For wives of farmers, their average off-farm wage rate increases by a similar proportion in going from <12, to 12, and then to >12 years of completed schooling.

Wages reported as being paid by farmers to family members and corporate officers are suspect. My expectation is that these payments are an underestimate of the opportunity cost of the individual's time. The reason being that these individuals can expect at some point in time to share farm profits or appreciation in the value of the farm business, including land, either when it is sold or when they inherit it. For those who are youth and plan to continue with the family business, they are engaging in an apprenticeship where they work and at the same time learn about key characteristics of the land and business. This is creating human capital that will be valuable to them in the future (see Huffman 1991b), and they can be expected to accept (and to be paid) lower than opportunity wages now for their work.

Even greater problems are associated with any kind of salary reported as being paid to managers-workers-owners/officers of family corporations. These individuals, for tax purposes, are most likely receiving a "salary" that roughly covers minimum current living expenses, but they are also entitled to their share of corporate profits. Furthermore, profits may be retained and carried forward and distributed in some future period or not at all. Thus, the allocation of these payments has a large arbitrary component, and reported wages or payments cannot be taken as a meaningful measure of labor cost.

Although I do not have data that will represent the full effect of differences between USDA procedures and my procedures for arriving at costs of production and labor cost shares, the results reported in Huffman (1976b) will shed some light on the issue. Although these data have the disadvantage of being "history," they have some major advantages. First, a large amount of effort went into deriving measures of inputs that were conceptually sound (see Huffman, 1976b). Second, the labor intensity of agricultural production, however measured, differs markedly across the three states. Two different methods for costing hours of farm labor by farmers and their wives were used. In the first method, the USDA's procedure of using the wage rate for hired farm labor as the cost of all operator and farm family labor is applied. In the second, my procedure of valuing operator and farm family labor at the opportunity cost represented by the off-farm wage rate was applied.

A comparison of the labor cost shares of these two procedures shows (i) that labor's share is much larger when the opportunity cost method is applied: 31.2 percent vs. 24.2 percent for Iowa, 45.7 percent vs. 32.9 percent for N. Carolina, and 39.6 percent vs. 29.6 percent for Oklahoma.[12] (ii) The relative importance of farmers' time in farm labor is increased: 71.8 percent vs. 66.5 percent for Iowa, 67.6 percent vs. 55.9 percent for N. Carolina, and 67.7 percent vs. 60.0 percent for Oklahoma. Furthermore, the cost shares of the nonlabor inputs (e.g., land and building services, machinery services) are reduced when the opportunity cost method of valuing farm family labor is applied.

More on the Cost of Labor

Labor markets are inter-related geographically through temporary and long-term migration and farm/off-farm work (Huffman, 1977, 1984, 1991a; Barkley; Emerson) and through occupational choice and skill enhancement that converts labor of one form into another with different skills (Orazem and Mattila). Much of the work with livestock and crop harvesting in the United States utilizes relatively low-skilled labor. In the case of harvesting workers, dexterity and endurance are important; and experienced workers are undoubtedly faster than green workers. However, there is not much reward to formal schooling in these activities, but migratory workers seem likely, as unlikely as it may seem, to receive some compensation for their transient lifestyle and relatively short work-season (Huffman, 1984; Rosen).

The farm operators/managers have had long term rising schooling completion levels in the United States, and farm management experience is valuable. Furthermore, farm operators are linked to the nonfarm labor market by dual jobholding and by exiting from agriculture for nonfarm jobs (Huffman, 1977, 1991a; Barkley; Huffman and Lange; Tokle and Huffman). We saw in Table 18.1 the large long term rise of real wage rates for production workers in manufacturing, and this occupation is generally accessible to individuals who are farmers. Econometric analysis of wage rates or labor demand facing male farmers by Sumner and Huffman and Lange have shown that human capital characteristics (schooling, experience) are very important factors for explaining their nonfarm wage rates.

When the hired farm work force had a large Black component, the average schooling level was low, and after the Blacks exited, Mexican immigrant agricultural works came in large numbers. During the 1980s, the hired farm work force has become "Mexicanized" in much of the United States, especially in the Southwest, West, and Florida (Coltrane; Martin, 1986, 1987; Martin and Holt; Polopolus, Moon, and Chunkasut). These workers are from Mexico, and they have on average 3-5 years of completed schooling in Mexico. Furthermore, these workers migrate to other regions of the United States for work and have replaced local high school students and married women who during the 1960s and 1970s supplied much of the peak seasonal labor needs in field crops (Martin and Holt). Over the longer term, hispanics having low schooling levels have replaced American Blacks who had low schooling levels.

The average wage rates of U.S. hired farm workers have followed a very different path over the past 40 years than that of wage rates of manufacturing workers. During this time, the average wage rate for hired farm workers (without board) have been either slightly below or slightly above the national legal minimum wage (Table 18.4). In particular, it was not until after 1965 that the average wage rate for hired farm labor was regularly above the legal minimum. Two major reasons seem to exist.

First, when the federal minimum wage was enacted in 1938, agricultural labor was not covered. Raising the minimum wage for the covered sector caused temporary unemployment of low-wage or low-skilled labor. Many of these workers found employment in U.S. agriculture which lowered the market clearing wage from what it would otherwise have been. A large share of these affected workers were American Blacks. In 1968, hired agricultural labor was first covered by the federal minimum wage, so later increases of the federal minimum wage tended to raise, rather than lower, the wage of hired agricultural labor. This seems, also, to coincide with the final wave of Black exits from U.S. agriculture.

Second, since at least the early 1970s, there has been a perfectly elastic supply of low-wage Spanish-speaking foreign agricultural labor available in the United States. The real wage rate for hired farm workers in United States has been 5 to 10 times higher than in Mexico since at least the 1960s (Torok and Huffman). Although the Bracero Program ended in 1964, and this eliminated the only legal option for large numbers of Mexican workers to come to the United States until recently, large numbers of illegal aliens from Mexico have come to the United States to work. Since the enactment of the 1986 Immigration Reform and Control Act, a relatively large number of alien Mexicans can now work legally again in the United States as season agricultural workers (SAWs).

In California and Texas, the field workers and supervisors are Mexicans or Mexican-American, the workers have low schooling levels, and all supervision is in Spanish. With the aid of Spanish speaking hired farm-labor

TABLE 18.4 Historical Data on Hired Farm Labor and Legal Mimimum Wage, 1890-1989

Year minimum	Rate without board [a] ($/da)	($/hr)	Real wage [b] (1914 = 100)	Federal legal minimum wage ($/hr)
1890	.95		72	
1900	1.15		94	
1910	1.35		100	
1920	3.30		114	
1925	2.35		93	
1930	2.15		89	
1935	1.35		68	
1940	1.60		79	.25
1945	4.35		167	.40
1950	4.50	.69	129	.75
1955		.82	138	.75
1960		.97	148	1.00
1965		1.14	163	1.25
1970		1.65	191	1.60
1975		2.45	207	2.10
1980		3.67	211	3.10
1985		4.31	193	3.35
1989		5.17	199	3.35

[a] Dollars per day 1890-1950; dollars per hour 1950-1989.
[b] Deflated by the Consumer Price Index, 1890-1960, and the Implicit Price Deflator for Personal Consumption Expenditure, thereafter (Source: Economic Report of the President, 1991).
Source: U.S. Dept. of Commerce and Economic Report of the President, 1991.

contractors, the growers have a high quality, relatively low-wage, reliable labor force. During the late 1980s, the wage rate for hired farm workers in south Texas has been approximately equal to the U.S. legal minimum wage, and in California, it is slightly above the California minimum wage which is higher, $4.15 per hour. The linkage of labor markets (covered versus uncovered, regional, and international) as shown by Torok and Huffman and Taylor have kept the wage rate of hired farm workers on a very different path than for U.S. factory workers and one that is only consistent with low average human capital levels. With Mexicans replacing domestic workers in seasonal work, the average amount of human capital in the hired farm work force has probably fallen, at least since the Immigration Reform and Control Act of 1986.

Concluding Comments

This chapter has attempted to present new insights about conceptual issues and practical details for estimating the quantity and cost of human capital services or labor in agriculture. The conceptual framework for thinking about these issues applies to all agricultural outputs and inputs and to goods and services purchased by farm households for consumption (or home production). Also, good procedures for measuring labor input and costs are not independent of good measures for other inputs (and outputs). The procedures that I have suggested require large sample surveys and careful econometric analysis. Clearly, my attempts are just a beginning to what hopefully will be a fruitful examination and improvement of research methods in the general arena of economic costs and returns.

Notes

Phil Martin, Robert Emerson, Peter Orazem, and Mary Ahearn provided helpful insights on some of the issues presented in this chapter. Journal paper No. J/14473 of the Iowa Agricultural and Home Economics Experiment Station, Ames, Iowa, Project 2738.

1. Households are assumed to make plans that are of a long term nature. This model could be viewed as one that ignores discounting but otherwise covers a long period of time. Alternatively, the model could be modified to be multiperiod in nature, and production made to be dynamic.

2. Time allocated to commuting to off-farm jobs is likely to be important in the decision-making of farm households. It functions as a tax on time, and it is ignored here. Huffman (1976a) presents models that incorporate time to off-farm work. As commuting distance to off-farm work increases, the probability of off-farm work decreases. However, for those who chose to work off-farm, desired hours of work may be relatively large (per day), given commuting costs.

3. The variable cost of C_F can be avoided only by going out of production. It includes, for example, the cost of many information technologies of hiring and training workers (Triplett), and contractual costs associated with rights to water use.

4. A likely possibility is that specialized skills in the farm operator/manager to a particular farming business creates a quasi-fixed factor of production that is one important source of heterogeneity of farm firms (Oi; Huffman 1985; 1991b). A comparatively small number of farm firms grow to extraordinarily large sizes probably because they are managed by exceptionally talented and able entrepreneurs.

5. Of course, alternative valuation procedures exist. One is the replacement cost method where the cost of replacing each type of labor is employed. In applying the opportunity cost method, all homogeneous inputs have the same opportunity cost. Its application to the valuation of an operator's time seems to accord more with reality than the replacement cost method. What "busy" farm operator allocates his time in a farm business even to field work in the same way that a hired worker's time is allocated.

6. The first estimates of factor shares for the national economy were prepared by King. Later estimates were prepared by Martin (1939), Kuznets (1941), and Johnson (1954).

7. The first estimates of labor's share for the agricultural sector were presented by King. Later estimates by Budd substantially revised King's estimates to include unpaid family labor. Other estimates were by Johnson (1948) and MacEachern and Ruttan.

8. Labor's share in agriculture would decrease more slowly or perhaps not at all if the operator's farm labor was valued at its opportunity cost rather than at the wage rate for hired farm labor. The data on labor use in agricultural production and value per unit by Jorgenson, Gollop, and Fraumeni (Appendix B) seem to have both reasonably good estimates of the hours used and the value of labor. Their factor share for agricultural labor is approximately unchanged during 1957-1979 and is about 30 percent.

9. Part of the reason for the increased use of farm labor contractors seems to be an attempt by farmers (growers) to shift liability for using illegal workers to individuals or businesses that have few real resources on average. This is especially true after passage of the 1986 Immigration Reform and control Act which imposed fines on employers for hiring illegal immigrant workers.

10. Phil Martin has expressed a concern to me that piece-rate wages for farm workers are "too high" relative to wage rates for hired farm labor in general. Farmers/growers have an incentive to count as farm work only very narrowly defined activities, e.g., time actually spent in the field picking oranges.

11. The problem is magnified because the survey week is generally only one or four weeks of a year, but much of agricultural labor is seasonal work. Thus, sampling one of 52 or 4 of 52 weeks in a year is unlikely to provide a very accurate estimate of hours of work by seasonal labor. A better procedure is to ask a question about labor use that ends up covering the whole year.

12. If the livestock input is netted out of production to obtain an approximate measure of value added from the sector, the estimate of labor's share is even larger, 43.5% vs. 35.2% for Iowa, 54.3% vs. 40.9% for N. Carolina, and 54.0% vs. 43.8% for Oklahoma.

References

Ball, Eldon. 1985. "Output, Input, and Productivity Measurement in U.S. Agriculture, 1948-79." *Amer. J. of Agr. Econ.* 67: 475-486.

Barkley, Andrew P. 1990. "The Determinants of the Migration of Labor Out of Agriculture in the United States: 1940-85." *Amer. J. Agr. Econ.* 72: 567-573.

Becker, Gary S. 1981. *A Treatise on the Family.* Cambridge, MA: Harvard University Press.

Budd, E. C. 1960. "Factor Shares, 1850-1910," in *Trends in the American Economy in the 19th Century.* NBER Studies in Income and Wealth 24. Princeton, NJ: Princeton University Press, pp. 365-406.

Capalbo, Susan M. and J. Antle. 1988. "Agricultural Productivity: Measurement and Explanation." Washington, D.C.: Resources for the Future.

Coltrane, Robert. 1984. "Immigration Reform and Agricultural Labor." USDA, ERS. Ag. Econ. Report No. 570.

Emerson, Robert D. 1989. "Migratory Labor and Agriculture." *Amer. J. Agr. Econ.* 71: 617-29.

Goldin, Claudia. 1990. *Understanding the Gender Gap.* New York: Oxford University Press.

Hauver, James H. 1989. "Major Statistical Series of the U.S. Department of Agriculture, vol 2. Agricultural Production and Efficiency." Agricultural Handbook No. 671, U.S. Department of Agriculture, Economic Research Service.

Huffman, Wallace E. 1991a. "Agricultural Household Models: Survey and Critique" in M. Hallberg, J. Findeis, and D. Lass, eds. *Multiple Job-Holding Among Farm Families in North America.* Ames, IA: Iowa State University Press. pp. 79-111.

Huffman, Wallace E. 1976a. "A Cross-Sectional Analysis of Nonfarm Work of Farm Family Members." Springfield, VA: National Technical Information Service Report No. DLMA 91-19-75-81-1.

Huffman, Wallace E. 1988. "An Econometric Methodology for Multiple-Output Agricultural Technology: An Application of Endogenous Switching Models" in Capalbo, S. and J. Antle, eds. *Agricultural Productivity: Measurement and Explanation.* Washington, D.C.: Resources for the Future, Inc., pp. 229-244.

Huffman, Wallace E. 1985. "Human Capital, Adaptive Ability, and the Distributional Implications of Agricultural Policy." *Amer. J. Agr. Econ.* 67: 429-34.

Huffman, Wallace E. 1991b. "Human Capital for Future Economic Growth" in Glenn Johnson and James Bonnen, eds., *Social Science Agricultural Agendas and Strategies.* East Lansing, MI: Michigan State University Press. Part III, pp. 61-67.

Huffman, Wallace E. 1977. "Interactions Between Farm and Nonfarm Labor Markets." *Amer. J. Agr. Econ.* 49: 1054-1061.

Huffman, Wallace E. 1976b. "The Productive Value of Human Time in U.S. Agriculture." *Amer. J. Agr. Econ* 58: 672-83.

Huffman, Wallace E. 1984. "Some Analytical Approaches for Human Resource Issues of Seasonal Farm Labor" in Robert D. Emerson, ed., *Seasonal Agricultural Labor Markets in the United States.* Ames, IA: Iowa State University Press, pp. 35-63.

Huffman, Wallace E. 1976c. "The Value of the Productive Time of Farm Wives: Iowa, North Carolina, and Oklahoma." *Amer. J. Agr. Econ.* 58: 836-841.

Huffman, Wallace E. and R. E. Evenson. 1990. "The Development of U.S. Agricultural Research and Education: An Economic Perspective." Iowa State Univ., Department of Economics.

Huffman, Wallace E. and R. E. Evenson. 1989. "Supply and Demand Functions for Multiproduct U.S. Cash Grain Farms: Biases Caused by Research and Other Policies." *Amer. J. Agr. Econ.* 71: 761-773.

Huffman, Wallace E. and M. D. Lange. 1989. "Off-Farm Work Decisions of Husbands and Wives: Joint Decision Making." *Rev. Econ Stat.* 71: 471-480.

Johnson, D. Gale. 1948. "Allocation of Agricultural Income." *J. Farm Econ.* 30: 724-745.

Johnson, D. Gale. 1954. "The Functional Distribution of Income in the U.S., 1850-1952." *Rev. Econ. and Stat.* 36: 175-182.

Jorgenson, Dale W., F. M. Gollop, and B. M. Fraumeni. 1987. *Productivity and U.S. Economic Growth.* Cambridge, MA: Harvard University Press.

King, W. I. 1915. *Wealth and Income of the People of the U.S.* New York: Macmillian, pp. 154-57.

Kuznets, Simon. 1955. "Economic Growth and Income Inequality." *Amer. Econ. Rev.* 45: 1-28.

Kuznets, Simon. 1966. *Modern Economic Growth.* New Haven, CT: Yale University Press.

Kuznets, Simon. 1941. *National Income and Its Composition, 1919-1930.* New York, NY: NBER

Lianos, Theodore. 1971. "The Relative Share of Labor in United States Agriculture, 1949-1968." *Amer. J. Agr. Econ.* 53: 411-422.

Loomis, R. A. and G. T. Barton. 1961. "Productivity of Agriculture, United States, 1870-1959." Technical Bulletin 1238. U.S. Dept. of Agr., Econ. Res. Serv.

Lopez, Ramon. 1984. "Estimating Labor Supply and Production Decisions of Self-Employed Producers." *Eur. Econ. Rev.* 24: 61-82.

MacEachern, Gordon and Vernon Ruttan. 1964. "Declining Labor Shares" in *Farmers in a Market Economy.* Ames, IA: Iowa State University Press, pp. 190-213.

Maddala, G. S. 1981. "Identification and Estimation of Limited Dependent Variable Models" in A. S. Blinder and P. Friedman, eds., *Natural Resources, Uncertainty and General Equilibrium Systems: Essays in Memory of Rafael Lusky.* New York: Academic Press.

Martin, Philip L. 1987. "California's Farm Labor Market." University of California-Davis, AIC Issues Paper No. 87-1.

Martin, Philip L. 1986. "A Profile of California Farm Workers." Giannini Information Series No. 86-2, University of California.

Martin, Philip L. 1985. "Seasonal Workers in American Agriculture: Background and Issues." Research Report Series RR-85-04, National Commission on Employment Policy, Washington, D.C.

Martin, Philip L. and J. S. Holt. 1987. "Migrant Farmworkers: Number and Distribution." University of California-Davis, Dept. of Agr. Econ.

Martin, Robert F. 1939. *National Income in the United States, 1799-1938.* New York, NY: National Industrial Conference Board, Inc.

Oi, Walter Y. 1983. "The Fixed Employment Cost of Specialized Labor" in J.E. Triplett, ed., *The Measurement of Labor Cost.* Chicago: The University of Chicago Press for the NBER, Studies in Income and Wealth, Vol. 48, pp. 1-60.

Orazem, Peter, and J. P. Mattila. 1986. "Occupational Entry and Uncertainty: Males Leaving High School." *Rev. Econ. and Stat.* 68: 265-273.

Polopolus, Leo C., S. Moon, and N. Chunkasut. 1989. "Farm Labor in the Florida Fruit, Vegetable, and Ornamental Industries." University of Florida, Department of Food and Resource Economics, SS-FRE-901.

Rees, A. 1973. *Long Term Economic Growth, 1860-1970.* Washington, D.C.: U.S. Dept. of Commerce, Bureau of Economic Analysis.

Rosen, Sherwin. 1986. "The Theory of Equalizing Differences" in O. Ashenfelter and R. Layard, eds., *Handbook of Labor Economics*, vol. I. New York: North Holland, pp. 641-692.

Saupe, William E. 1990. "Status of Wisconsin Farming, 1990." Department of Agricultural Economics, University of Wisconsin-Madison.

Schultz, Theodore W. 1977. "The Economic Value of Human Time Over Time" in *Lectures in Agricultural Economics.* U.S. Dept. of Agriculture, ERS, pp. 1-24.

Schultz, Theodore W. 1972. "The Increasing Economic Value of Human Time" *Amer. J. Agr. Econ.* 54: 843-850.

Shumway, Richard C. 1983. "Supply, Demand, and Technology in a Multiproduct Industry: Texas Field Crops." *Amer. J. Agr. Econ.* 65: 748-60.

Singh, I. J., L. Squire, and J. Strauss. 1986. *Agricultural Household Models.* Baltimore, MD: The Johns Hopkins University Press.

Smith, J. P. and M. Ward. 1985. "Time Series Growth in the Female Labor Force." *J. Labor Econ* 3: S59-S90.

Sumner, Daniel A. 1982. "The Off-Farm Labor Supply of Farmers." *Amer. J. Agr. Econ.* 64: 499-509.

Taylor, J. E. 1986. "Differential Migration, Networks, Information and Risk" in Stark, O., ed., *Migration, Human Capital and Development.* Greenwich, CT: JAI Press.

Tokle, J. G. and W. E. Huffman. 1991. "Local Economic Conditions and Wage Labor Decisions of Farm and Rural Nonfarm Couples." *Amer. J. Agr. Econ.* 73: 652-670.

Torok, S. J. and W. E. Huffman. 1986. "U.S.-Mexican Trade in Winter Vegetables and Illegal Immigration." *Amer. J. Agr. Econ.* 68: 246-260.

Triplett, Jack E. 1983. "Introduction: An Essay on Labor Cost" in J. E. Triplett, ed., *The Measurement of Labor Cost.* Chicago: The University of Chicago Press for the NBER, Studies in Income and Wealth, Vol. 48, pp. 1-60.

Tweeten, Luther. 1989. *Farm Policy Analysis.* Boulder, CO: Westview Press.

U.S. Department of Agriculture. 1990. "Economic Indicators of Farm Sector--Costs of Production, Livestock and Dairy, 1989." ERS, ECIFS9-1.

U.S. Department of Agriculture. 1943. "Labor Requirements for Crops and Livestock." Bureau of Agricultural Analysis.

U.S. Department of Agriculture. 1939. "Labor Requirements in the United States, 1939." Bureau of Economic Analysis.

U.S. Department of Agriculture. 1944. "Labor Requirements in the United States, 1944." Bureau of Economic Analysis.

U.S. Department of Agriculture. 1954. "Labor Use for Field Crops." Statistical Bulletin No. 144.

U.S. Department of Agriculture. 1955. "Labor Use for Livestock." Statistical Bulletin No. 161.

U.S. Department of Agriculture. 1980. "Measurement of U.S. Agricultural Productivity: A Review of Current Statistics and Proposals for Change." Technical Bulletin 1614. Econ. Stat. Coop. Serv.

U.S. Department of Commerce. 1975. "Historical Statistics of the United States: Colonial Times to 1970, Part 1." Wash., DC: U.S. Government Printing Office.

U.S. Department of Labor. 1989. "Employment Cost Indexes and Levels, 1975-1989." Bureau of Labor Statistics. Bulletin 2339.

U.S. Department of Labor. 1983. *Handbook of Labor Statistics.* Washington, D.C.: U.S. Government Printing Office.

U.S. President. 1991. "Economic Report of the President." Washington, D.C.: U.S. Government Printing Office.

Vaupel, Suzanne and P.L. Martin. 1986. "Activity and Regulation of Farm Labor Contractors." University of California, Gianini Information Series No. 86-3.

Welch, Finis. 1979. "Effects of Cohort Size on Earnings, the Baby Boom Babies' Financial Bust." *J. Pol. Econ.* 87: 565-598.

Zepp, G. A. and Simmons, R. L. 1979. "Producing Fresh Winter Vegetables in Florida and Mexico: Costs and Competition." Washington, DC: U.S. Dept. Agr., Econ. Statist. Coop. Serv.

Estimating Costs of Human Capital Services: Discussion

Carlyle Ross

The main objective of the Huffman chapter is to provide some concepts and practical suggestions for improving the estimates of the quantity and cost/value of operator and family labor used in agriculture. The main hypothesis of the chapter is that USDA greatly underestimates the hours of farm work as well as the unit cost or value of the hours worked. Consequently, labor's share of total farm costs is also grossly underestimated.

Huffman gives a number of reasons why USDA underestimates the number of hours worked on farms. These include: (1) The exclusion of management functions, e.g. marketing, buying, etc. (2) Farm operators working through Farm Labor Contractors (FLC) may not know the actual number of hours worked by hired labor, since the logistics are handled by the FLC. (3) Farm labor surveys miss labor performed by family members, especially homemakers. Unpaid family labor of 15 hours or less per week are not recorded in USDA quarterly surveys, Bureau of Census and Bureau of Labor Statistics.

On the cost or value side, Huffman claims that the hired farm labor rate is an inappropriate proxy for unpaid operator and homemaker farm wage rates. As the operators and homemakers are generally endowed with greater human capital stocks, they will command higher wage rates than hired farm labor. He also discounts paid operator and family wages as being suspect. Instead, Huffman argues in support of the use of opportunity cost of off-farm work as the appropriate value/cost of operator and family labor.

Huffman presents an economic model of an owner operator farm household, where the time of the operator, homemaker and non-family hired labor is fixed. This "farm household" maximizes utility subject to an expected cash income/technology constraint. The constraint is derived by substituting a production function (equation 4) into a cash income constraint (equation 6). The constraint (equation 6) consists of wage income (off-farm), household non-farm assets, net farm income, variable costs and household expenditures. By solving the constrained maximization problem, deriving first order conditions and rearranging variables, Huffman arrives at an inequality, whereby, the marginal value of operator labor hours (MVOL) exceeds or is equal to the expected off-farm wage rate. It therefore follows that when the MVOL exceeds off farm wages, the operator would choose off-farm work. In effect the off-farm wage was a floor price. If the MVOL fell through the floor, the operator would conceivably cease farming.

Where operator and spouse choose not to take off farm work, it implies that the MVOL on the farm exceeds the expected off-farm wage rates, and in these instances, the opportunity costs underestimate the value/cost of their labor.

Despite the increasing commercialization of farming in North America, I would suggest that the vast majority of farmers see farming as a business as well as a way of life. Therefore, the annual gross benefits of farming consist of an income component obtained from the sale of produce and a non-monetary component emanating from the farming life-style. We are also reminded that traditionally farmers live poor and die rich. In this latter situation, the operator accepts lower annual returns for his labor in exchange for a larger contribution to equity or real wealth. Should the life style and/or postponed consumption be included in Huffman's model, the inequality relating MVOL to the expected off-farm wage rate or opportunity cost no longer holds.

Consider the following inequality:

$$\text{Expected off-farm wage rate} \geq \text{MVOL} + B,$$

where B is the life-style/postponed consumption variable. Depending on the size of B, the MVOL could fall significantly below the opportunity cost or expected off-farm wage rate. If this factor B is important, and it is excluded from the model, opportunity cost as measured by the expected off-farm wage rate would greatly over-estimate the value/cost of unpaid farm labor.

There are several other reasons why we should seriously question the opportunity cost approach. Farm management skills are not easily transferable to the non-farm sector. Employment opportunities in rural America and rural Canada are very limited. Transportation costs can be a factor negating the attractiveness of a higher wage rate. Off-farm work may necessitate moving. Indeed non-participation in off-farm work may be more indicative of the absence of work as opposed to a higher MVOL. Conceivably, the opportunity cost could be lower than the hired wage rate.

We need to ask the proponents of the opportunity cost approach, to clarify the relevance of this expected off-farm wage rate at precisely the time when the operator is needed on the farm. If an individual is contemplating entering or exiting the industry I can see where the off-farm wage is relevant.

However, for a bona fide farmer, his opportunity cost is related to alternative enterprises on the farm not off the farm.

As I see it, USDA is attempting to measure the value/cost of unpaid labor. Huffman is attempting to measure the value/cost of human capital service of the operator and spouse. This human capital comprises of a labor component. It seems reasonable and/or appropriate to value the strictly labor component at the market wage rate which is the hired farm wage rate. If indeed the opportunity cost of the operator exceeds the farm wage rate, the operator would be better off hiring labor for the chores in question, and working off the farm.

Management valuation requires a different approach. It is easier said than done, given the diversity/variability among farmers. Perhaps management should be viewed as a fixed factor on a farm and treated as a residual where the bottom line becomes the return to management and owner equity. This approach avoids arbitrary management fees still common in some jurisdictions. It also bypasses the very difficult problem of determining expected returns to management on individual farms.

To conclude, there are good reasons why the value of unpaid operator/family labor should be accurately measured, e.g., efficiency, equity, distribution. There are practical difficulties associated with measurement. Huffman challenges us to go beyond the common, indiscriminate practice of using the hired wage rate. He argues that the opportunity cost of off-farm work is a better measure of value or cost. The concept of opportunity cost does have merit, but much work is needed to accurately segment the monetary and non-monetary contributions associated with unpaid operator/family labor, and to identify the appropriate opportunities and hence opportunity costs associated with such labor. In the meantime I would advocate the continued use of the hired wage rate of the enterprise under consideration, with the management contribution being picked up in the residual.

Sumner begins his chapter by briefly identifying major conceptual as well as empirical problems associated with measuring and estimating cost or value. He notes that by using hired wages as a proxy for unpaid farm labor the USDA approach is at odds with the opportunity cost approach which is espoused by Huffman. In a note, Sumner also reports that the Australian system has many similarities to the USDA system, including valuing unpaid labor at the hired wage rate.

Sumner rightly suggests that if the work and time spent to complete the task are identical for hired and unpaid labor, the estimates using the hired wage are appropriate. The critical need is to have very detailed information about the tasks performed by unpaid labor to be able to select the appropriate wage.

Another issue raised by Sumner is the inclusion of a labor component in a number of cost items such as custom work, management, legal, and accounting services, veterinary services and equipment repairs. He suggests that consideration be given to isolating the labor component of these inputs to get a more complete picture of the contribution of labor and human capital.

In concluding, Sumner makes the valid point that each farm is different, and that outside observers do not know much about the alternative uses of the farm resources. More important, outsiders do not know the farmer's value of these resources. Consequently, identification of the appropriate opportunity costs is not easy especially in the short run where the farmer is not contemplating leaving the farm.

Estimating Costs of Human Capital Services: Discussion

M. C. Hallberg

In a report that is, in my view, one of the classics of the profession, J. Patrick Madden anticipates most of the issues confronting those making cost of production estimates. He first reminds us of a few Econ 1 principles governing firm behavior under atomistic competition (Madden, p. 4): (1) in the short run, the firm will produce so long as average revenue equals or exceeds average variable costs; (2) in the long run, the firm will remain in production in its present form so long as average revenue is great enough to cover average total costs; (3) over the long term average revenue will tend toward that level at which all profits are eliminated; (4) over the long term returns to any resource will tend toward that level at which the resource provides exactly enough return to keep it from being drawn into alternative employment but not enough to attract additional amounts of that resource into producing expanded levels of output, and (5) in equilibrium, all firms will produce a level of output corresponding to the low point on their average total cost curve.

Madden then points out several problems associated with attempts to apply these principles in agriculture. Farm labor issues attract much of his attention here. He notes that there is great variability in the supervisory and coordination functions required on different types and sizes of farms, and that these functions are easily underestimated. Labor resources in farming do not perform the same job throughout the production period--different skills are required at different times of the year--but hired labor must most typically be fully employed for the entire year. When resources such as hired labor become committed to production, their price is zero so they will be used until their marginal value becomes negative. That is, they will be substituted for other resources that have a positive acquisition price.

Most studies suggest that as farm size increases, average costs decrease at least over a considerable range. This would lead one to expect that all farms would tend steadily toward the "optimal" size. Clearly this is not the case and Madden suggests why. Small farms will not necessarily be forced out of production so long as their profit potential is sufficient to overcome their opportunity cost or reservation price of remaining in business. Opportunity cost is likely to be low for farmers who lack the skills, education, and mobility to be attracted into off-farm employment. Reservation price is likely to be low for managers of 1- or 2-person operations, but higher on larger operations that require supervision of several hired workers and coordination of a highly complex operation having a higher probability of failure. Many farms may have remained at less than the "optimal" size simply because the promise of a greater profit potential is offset by the uncertainty and difficulty of coordinating a larger operation.

Finally, Madden points out that many farms are "goods-and-service" firms producing not only farm commodities, but also various services such as custom work and off-farm jobs, and have the possibility of owning and operating durable resources which provide them with surplus capacity. A farm viewed as a "goods-and-service" firm may have a lower average total cost than would the same farm viewed as a firm producing only farm commodities.

I am struck by the extent to which much of Madden's guidance seems to have been treated so lightly by costs and returns practitioners. Sumner and Huffman set about to right this wrong at least with respect to estimating labor costs. Both authors deal head-on with several of the ideas treated by Madden and suggest ways of coping empirically. The key issue for both is how to value the hours of labor input by farm operators and other farm family members not paid a direct cash wage for services rendered on the farm. Both authors recognize the importance of off-farm work to farm families and the implications of this phenomenon to farm labor input valuation. Both authors are convinced that current USDA estimation methods significantly undervalue the labor input on farms.

Sumner lays out the conditions under which we should and should not value operator labor at the hired wage rate. He notes that it is generally erroneous to conclude that just because farm operators are frequently observed doing tasks also performed by hired labor, farm operator time should be valued at the hired wage rate. For him, it is the opportunity cost of operator time that is relevant. The latter will be determined by off-farm and/or custom work opportunities. Sumner believes the off-farm value of time of farm operators and unpaid family labor has risen faster than wages of hired labor and, presumably, this contributes to the farm labor undervaluation problem (Huffman concurs and presents evidence to support this claim). More fundamentally, there is a sizable difference between hired farm wage rates and opportunity costs of farm operator and unpaid farm family labor. Failure to recognize and account for this fact leads to a serious undervaluation problem.

Sumner would apparently seek more accurate detail on the distribution of job tasks performed by farm operators and unpaid family labor and would value this human capital according to opportunity costs as judged by off-farm earnings. He cautions, though, about the difficulties of doing so, and thus reminds us that valuing farm operator labor at an estimated opportunity value cannot be expected to be done with great precision. Sumner closes on the rather sage note that costs of production for a farm firm are fundamentally unique to that farm so that any averaging of farm cost information by size category, region, or type of farm is likely to be misleading for general comparative purposes, or for understanding the collective decisions farmers make. For him, outside observers cannot know enough about the farm and its operation, or about how an individual farmer determines relevant opportunity costs and, I presume, reservation values.

Huffman views the farm labor cost determination problem in the setting of a decision problem for a multi-product, multi-input, risk-neutral farm household characterized as a "goods-and-service" operation facing production time lags, uncertain output prices, certain input prices, and heterogenous labor supplies. He cautions that "...costs-returns methodologies that rely on a strong separability assumption should be viewed with much skepticism and applied only where long-term credibility is not an important virtue." He, like Sumner, argues that valuing the farm operator's time at the hired worker wage leads to an underestimate of farm labor costs. Indeed, he shows that under certain conditions the value of the farm operator's time may even exceed its opportunity cost at off-farm work which he expects, in turn, to exceed hired worker wage.

Huffman presents evidence to support the hypothesis that labor's share of the cost of farm production is considerably higher than USDA's data shows and has been falling less rapidly than these data show. To correct this problem, he would collect better data and implement more sophisticated estimation methods. He would, for example, expand USDA's cost of production data set by increasing the sample size. Huffman throws caution to the wind and advocates that, regardless of the costs, USDA estimate labor costs econometrically using the household modelling approach he outlines.

Huffman's conceptual framework is clearly appropriate to understanding the structure of a family farm business. I hope we see many more empirical studies of farm firms using this methodology. It is likely to yield the best guides to opportunity values of operator and family labor currently available. I am less convinced, however, that this approach will lead to the ultimate aggregate estimates of costs of production ERS is mandated to produce. First of all, I like Sumner, am skeptical of the possibility of learning enough about any one group of farm operations to precisely gauge how they will collectively respond to changes in exogenous conditions over the long term. I am even more skeptical about being able to obtain sufficient data on similar farm types to make the exercise meaningful or practicable. Finally,

USDA is still mandated to produce cost of production estimates for individual commodities. While it is no doubt true that the farm production function is not in every (if any) region strongly separable, we cannot easily escape ERS's mandate to generate estimates for individual commodities. It may well be, though, that Huffman's approach could be fruitfully used selectively rather than globally to provide realistic guides to opportunity values for farm operator and unpaid family labor.

The authors of these two chapters are to be complimented for emphasizing an issue that has for too long been ignored in cost of production studies (USDA's, as well as others). Many farm families do have alternative employment opportunities. They do take advantage of these opportunities as evidenced by the fact that off-farm income has made up 50-55 percent of total farm family income in recent years. Wages earned at these alternative opportunities are for most farm families more important in setting the value of operator and other family labor then are earnings from the farm. The evidence points clearly to the fact that we currently undervalue the labor component in agriculture. If current policy that puts a greater premium on farm "management" continues, it seems to me this issue takes on even greater significance in the future. That is, only the superior managers will survive and they will only remain superior managers if their labor returns are comparable with returns from alternative employment.

References

Madden, J. Patrick. 1967. "Economies of Size in Farming: Theory, Analytical Procedures, and a Review of Selected Studies." U.S. Dept. of Agriculture, Economic Research Service. Agricultural Economic Report No. 107.

PART EIGHT

The Future of Commodity Costs and Returns Estimation

19

How Economic Theory Should Guide What We Measure

Thomas A. Miller

Introduction

Producers need economic information to make production choices, and a major source of such information is the enterprise cost and returns budget. If such budgets are to be of help in making economic decisions, economic theory must be considered as we select procedures for estimating these budgets. Public decisions--that is, policy decisions--also require sound economic information. Economic policy decisions for agriculture require information developed and analyses guided by the same economic theory. Public economic decisions are only as good as the information on which they are based and the soundness of the procedures used in developing this information.

Nevertheless, economic theory is sometimes overlooked as we debate and discuss the estimation of commodity costs and returns. In Chapter 1 of this volume, Bob Robinson traced the beginnings of agricultural economics to men who considered themselves "farm managers" rather than economists and who approached their subject matter as empiricists. This approach has been called the "unendowed" school of farm management--unendowed with economic theory (Johnson). Much of the enterprise budgeting being done today still appears to be approached in this manner. In this light, economic theory may not be the "generally accepted conceptual framework" behind the process.

Problems with Applying Economic Theory

Even if economic theory is the recognized basis for resolving procedural questions concerning how to estimate commodity costs and returns, we face several inherent limitations in this process.

First, most economic theory is based on the idea of the purely competitive model: large numbers of buyers and sellers, a homogeneous product, perfect mobility of resources, free entry and exit of firms, and perfect knowledge. These conditions lead to a competitive equilibrium in the industry, a condition where products and prices, inputs and costs, and returns to fixed factors all have well-known relationships. Unfortunately, the agricultural sector of the economy is not always in equilibrium, for whatever reasons. This disequilibrium gives rise to differences between the real world and theoretical relationships; it hinders our ability to link theory and procedures.

A major problem arises in interpreting economic theory as we try to estimate "total economic costs." Theoretically all resources are variable in the long run--and all production costs are covered by returns. In agriculture (and actually in most other industries), there is little opportunity to observe these long run resource markets in action, and no objective, unambiguous method exists for estimating these theoretical economic costs. Farmers may make allocations of net returns to the fixed factors of production, but this process is arbitrary and not observable to economic analysts.

While many other limitations exist, only one more will be mentioned here. Externalities (simply, cases where one's actions affect the choices available to others) distort observed prices and restrain our ability to connect economic theory, what we observe, and what procedures should be used to generate information. From the viewpoint of society as a whole, market prices are insufficient criteria for efficient resource allocation when externalities are present. This is because the social costs of actions may not be reflected in the prices which guide resource allocation--e.g., ground water pollution from excessive nitrogen fertilizer. As a result the costs of production do not include any measure of the costs of pollutants that enter water courses. Policies based on market prices therefore may not be optimal from society's standpoint.

The Suggestions of Economic Theory

The body of economic theory contains some guidelines in developing methods for estimating commodity costs and returns. The accepted or common practices actually being used generally follow these guidelines, but there are a few important exceptions. In other cases economic theory has little to offer as we search for defensible procedures. The following points can be made, in order of their appearance in an Economic Research Service (ERS) enterprise cost and returns statement.[1]

Expected Returns vs. Actual Returns

Several observations need to be made about the returns section of the commodity budget--the yield or level of production of the enterprise and the price received for that production.

The first point is important, and relates to all components of the budget. If the estimates are to be used for planning purposes, economic theory would suggest that they be based on projected or year ahead price and yield data. (The exception of course would be when past yields and prices are the best available predictor of the future.) Consistent with this observation, Karen Klonsky has reported that 68.3 percent of the universities publish projected commodity cost and returns estimates (Chapter 10). Odell Walker mentioned that developing "forward looking budgets" is a strength of agricultural economists. In contrast, ERS is largely focused on estimating past (actual previous year) data.[2]

Farm management and farm record analysis textbooks generally recognize the distinction. Farm records, whole farm income statements and balance sheets chart the historical record. Enterprise cost and return budgets can be developed from this historical data base--these budgets are for decision making, and represent expected returns for the next production period. At an earlier conference, we argued that "All users need year ahead projections. This is especially important in times of rapid inflation and technical change. Decisions which use enterprise budgets require expected rather than experienced costs and returns" (Miller and Skold, p. 15). This statement is as important today as it was in 1980.

Nevertheless, enterprise budgeters often use actual yields, based on the argument that publication of actual historical data allows users to form their own expectations of future returns. These actual yields should be consistent with other generally accepted USDA data. As observed by Libbin and Torell, "Substantial differences exist between NASS-reported yields and prices and ERS-budgeted yields and prices, especially for years in which a crop-specific survey was not taken. USDA-ERS does not appear to use or conform to state-level data collected by USDA-NASS when formulating estimates of prices and yields" (Libbin and Torell, p. 309).

Additional problems arise when costs are expressed on a per unit basis. ERS states that "Per-unit-of-output costs are...especially useful for examining effects of annual changes in yield on costs" (USDA, p. 4). I do not agree. Farmers to not view random yield fluctuations as affecting costs (i.e., a hail storm does not increase costs; it lowers returns). Even less justifiable is the ERS process of estimating enterprise costs for each individual farm and then arraying the per unit estimates to form a cumulative distribution function (for example, see Ahearn, *et al.*). Per unit cost estimates based on actual yields imply a distribution of costs from the stochastic yield component, even if all farmers have identical cost and production functions. In my judgement, such distributions provide no information about differences in costs among farms (Miller, p. 17).

Current ERS budgets omit direct Government payments and attempt to omit the (economic) costs associated with participation in Government commodity programs. Conceptually, the cost and return estimates thus represent a "no program" or a "non-participant" situation. However problems arise in omitting the effect of payments on land returns. I applaud the support of Bud Stanton and Otto Doering to include such program impacts in both the returns and cost sections of budgets as was done recently by ERS for rice production (Salassi, *et al.*).

There is some consensus on several procedures that appear to be consistent with both economic theory and the logic of using an enterprise budget as a planning tool. First, it is correct to use the harvest month price received by farmers, at the same time excluding storage costs, marketing costs, and any additional revenues farmers may earn by storing the crop for later sales. If some later season average product price is used to estimate returns, then storage costs relevant to this time period should also be shown in the budget. Second, the proper technical unit of production is the planted acre--all tillage and production costs are defined on a planted acre basis, and total production is divided by planted acres in computing yields. This procedure is important for internal consistency and comparability with crops like winter wheat, where harvested acreage in some Plains states averages 70 to 90 percent of planted acreage.

Variable Cash Expenses

From economic theory, we know the physical production function underlies each enterprise cost and return budget. Additionally, if producers maximize income, if the purely competitive model holds, and if there is no internal or external capital rationing, the real world will reflect a particular point on this production function--the point where all variable inputs are applied at the level which maximizes returns to fixed factors. Given input prices, P_x, each variable input is applied until the value of its marginal product equals this price:

$$MVP_x = P_x$$

Estimating variable cash costs by multiplying the input quantity times its purchase price therefore reflects a profit maximizing point on the production function. Furthermore, this cost (X times P_x) gives the rightful share of total returns to the variable cash cost factors of production. This allocation of returns to input costs is consistent with procedures discussed below for estimating economic costs by imputing returns to fixed factors.

Fixed Cash Expenses and Cash Flow Analysis for the Firm

The next section of the ERS commodity cost and returns budget shows fixed cash expenses: (a) general farm overhead, (b) taxes and insurance, (c) interest actually paid on operating and real estate loans, and (d) the remaining "net returns less cash expenses". Many, if not most, budgets prepared at universities do not show these fixed cash expenses, although they have some value in cash flow analysis. What is the place of a cash flow budget in economic planning?

ERS explains that "Net returns after cash expenses are the difference between gross value of production and total cash expenses. They are a measure of short-term returns to production and are the amount of cash that would be left over after all cash expenses, including interest payments, have been made. They indicate what would be available to cover longer term costs, such as capital replacement, or to retire debt" (USDA, p. 7). Of course, this item also contributes to the living expenses of the farm family.

This information is subject to misinterpretation--I question the advisability of displaying net cash flow at such a prominent place in the budget. Cash flow is a firm financial constraint, and not a measure of enterprise profitability. Economic and financial theory suggests that a firm must meet such cash flow needs to continue existence in the sense of maintaining long term use rights to the fixed resources. The key decisions here are investment decisions, affecting the balance sheet of the firm. They are not enterprise decisions, and generally do not affect the choice of enterprises.

These total cash expenses do not, as McElroy argues, suggest the "minimum breakeven price needed...to raise and harvest a crop" (McElroy, p. 13). The decision of whether or not to produce in the short run is not based on covering fixed cash costs. Economic theory is very clear on the fact that only variable cash expenses must be covered in the short run.

Finally, most farm families meet such cash flow needs with the help of off-farm income--more than 60 percent of the income of farm families comes from off-farm sources. Since this income is not shown in the enterprise budget, no evaluation of the cash flow generated by the budget is possible.

Issues Concerning the Allocation of
Fixed Cash Costs to Enterprises

In 1980, Miller and Skold examined what economic theory says, if anything, about the allocation of fixed cash expenses to individual enterprises. We concluded that "...such allocation is in general arbitrary[3] and is not required for decision making in farm management, research, or agricultural policy" (Miller and Skold, p. 15). Short-run decisions that are based on commodity cost and return budgets are not affected by fixed cash costs such as general farm overhead, taxes, or cash interest payments.

Each individual enterprise generates a cash flow. The sum of these cash flows, along with other cash income such as wages from off-farm jobs, is available to make interest payments, pay family living expenses, cover general farm overhead, etc. The payments vary greatly from farm to farm-- they are affected mostly by the debt structure and tenure of the individual operator. Such payments are not costs to individual enterprises.

Stated more formally, enterprise choice depends on the relative contribution of each enterprise to fixed production factors. A comparison of residual returns (gross returns over variable cash expenses and capital replacement) guides these decisions. Economic theory provides no clue that the allocation of fixed cash expense items to enterprise budgets will improve these decisions. Even worse, depending on the allocation criteria used, such allocation may lead to an inefficient allocation of resources.

The present allocation criteria used by ERS--based on each commodity's share of the total value of farm production--is an example. Theory shows that comparative residual returns over variable costs guide enterprise selection and efficient resource allocation. Therefore adding any nonproportional cost component (in this case a cost proportional to the total value of production) destroys the comparison and reduces economic efficiency. The ERS allocation criteria is incorrect when viewed from the perspective of basing enterprise selection decisions on the remaining net returns.

Capital Replacement or Depreciation

Capital replacement and depreciation charges rest on financial theory, more than economic theory. Capital budgeting and discounting procedures are often used as decision aids, and the procedures are derived from financial theory. An excellent discussion of farm machinery applications has been provided by Watts and Helmers. The chapter in this volume by Oscar Burt extends these results (Chapter 15).

For our purposes, the key point of this body of thought is the necessity to maintain consistency between the methods of computing depreciation and the method of computing the economic cost (opportunity cost) of the machinery investment. In the ERS budgets this second item is shown as the "return to other non-land capital."

Two alternatives exist and provide essentially identical estimates of the total of depreciation and opportunity cost. In one, nominal costs are used, meaning that all transactions and interest rates are expressed in dollars valued at the date of the transaction. Here, depreciation is the purchase price minus the salvage value (expressed in future inflated dollars) divided by the years of expected life. Note that salvage values increase with inflation, and high rates of inflation can reduce depreciation to zero or less. The comparable opportunity cost concept is nominal interest rates (observed) which normally include an inflation premium.

The second alternative is to use real (adjusted for inflation) values throughout. Depreciation is the current purchase price of similar machines minus the salvage value adjusted for inflation back to the present. This lower salvage value results in a higher depreciation charge than the previous method. This difference is offset by using the real (inflation adjusted) interest rate in computing the opportunity cost on investment. Both methods provide essentially the same result in total and both are mostly consistent with capital budgeting practices and financial theory.[4]

ERS follows the second alternative. Machinery prices are adjusted annually to reflect current acquisition prices. Salvage prices are computed on this same dollar index--real depreciation as defined above. The determination of the opportunity cost of machinery is made in a consistent manner--a real rate of return is charged on non-land capital. The process is consistent with theory.

The key point is the use of real discount (interest) rates in the opportunity cost calculation regardless of the rate of inflation or the nominal (observed) interest rate. I am chagrined at Karen Klonsky's finding (Chapter 10, this volume) that only 9.8 percent of universities use a real interest rate for equipment capital.

Allocated Returns to Owned Inputs--General Considerations

We now consider a group of procedural questions where economic theory is of much less help. In general, the issues concern the return to factors owned and provided by the operator: operating capital, other non-land capital, land, unpaid operator and family labor, and management and risk taking. Estimation and allocation of these so-called economic costs is so arbitrary that the returns assigned to fixed resources in an enterprise budget have little economic meaning (Miller and Skold, pp. 2-3).

Up to now in the enterprise budgeting process, variable costs have been assigned in a manner that is analogous to imputing returns to these variable factors. Under profit maximization and given input prices, P_x, each variable input is applied until the value of its marginal product equals this price:

$$MVP_x = P_x$$

Each variable resource thus earns a return equal to its marginal value product. After performing this calculation for all variable factors, we are now faced with the task of allocating the residual returns to the fixed factors of production, or to use the ERS term, estimating the "allocated returns to owned inputs."[5] In practice, there is no objective and precise way to allocate this return to the various fixed factors.

By Euler's Theorem it is known that residual imputation is accurate only under very special circumstances (Debertin, p. 162; Heady; and Harrington, Chapter 4 in this volume). The theorem holds that if each resource is allocated a share of (value of) output in proportion to its marginal (value) productivity, total output will be exactly exhausted. For this to occur, the production function must be homogeneous of degree one. Thus,

$$TVP = MVP_1X_1 + MVP_2X_2 + \ldots + MVP_nX_n$$

where TVP = total value product, MVP_i = marginal value product of each resource and X_i = the level of each resource. When the function is homogeneous to the first degree, the residual returns to the nth resource can be imputed according to the equation:

$$MVP_nX_n = TVP - [MVP_1X_1 + MVP_2X_2 + \ldots + MVP_{n-1}X_{n-1}]$$

The best guidance that economic theory offers here is that if farmers maximize returns, and if the perfectly competitive model holds in both input and product markets, and if production functions are homogeneous to the first degree, this imputation process will exactly exhaust total returns.

For durable factors that are owned and provided by the farm operator, and not purchased in the market, the problems are particularly severe. We can never measure or observe how an individual farmer allocates such returns. Imputing residual returns to these items is theoretically correct only in so far as all other inputs are being employed where $MVP_i = P_i$. Likely, this equimarginal concept is seldom achieved. Alternatively, using accounting techniques or employing the concepts of economic opportunity costs do little to provide information that is economically valid. All are inherently subjective and arbitrary (Pasour, p. 247).

The only guide economic theory provides is that if we are in a perfectly competitive economy, with all durable factor markets in perfect equilibrium, and with specific production functions, then imputing residual returns to fixed factors according to their marginal value product would work. Under a long run competitive equilibrium, this return would equal the "true cost" of each factor. Even then, theory offers no suggestion that such imputations would make enterprise cost and returns budgets more useful to any user. The sections which follow offer some observations about applying these imputed value concepts to the remaining budget items.

Return to Operating Capital

For costs and returns comparisons, accuracy requires that we always recognize the time value of money when production is not instantaneous. Both financial theory and production theory apply here. Such adjustments are especially important when inflation is high or when the production period is long. All enterprise budget items must be expressed in either beginning-of-year or end-of-year terms. The general procedure is to either (1) discount all future returns to the time when inputs are purchased, or (2) compound all input costs to the point in time represented by product prices. Nominal interest rates, which include the expected inflation premium, guide the selection of the appropriate discount rates.

A simple budgeting procedure accomplishes this compounding. Merely include a charge for operating capital for the time it is tied up in the enterprise, from Spring when inputs are purchased to the Fall when the crop could be sold and the capital repaid. For farms who borrow operating capital, this is an actual cost. For other farms, this capital charge has the effect of compounding Spring costs to later harvest period dollars--and allows comparing Spring costs and Fall harvest season prices in dollar values of the same time value.

ERS uses a risk free rate of return, the 6-month U.S. Treasury bill rate, over the period from input purchase to crop sale to reflect the imputed return to operating capital (USDA, p. 11). For 1988, this rate was 6.92 percent. This procedure appears to be a valid method of recognizing the time value of money.

Return to Other Nonland Capital

The appropriate imputed return to nonland capital has already been discussed under the category of capital replacement or depreciation. With depreciation expressed in real terms, the consistent procedure is to impute a capital return using real (inflation adjusted) interest rates. Since the value of durable goods tends to appreciate with the inflation rate, only the real interest rate represents the net cost of invested capital.

ERS analysts calculate this rate by "taking the average of the previous 10-year total return to production assets in the agricultural sector, subtracting the value of the operator's labor used each year, and dividing this figure by the value of production assets" (McElroy, p. 11). The calculated rate for 1988 was 2.81 percent and represents a real rate of return to production assets in agriculture. Gustafson, Barry, and Ali suggested in Chapter 16 that since production assets are primarily real estate, this calculation may underestimate the return to non-land capital.

Cost and return budgets prepared by university researchers follow a number of different procedures here. Colorado State University has been using a 5 percent rate, which may include a risk premium, as the factor payment to machinery capital. Karen Klonsky reported that only 4 of 41 universities used a real interest rate for this purpose (Chapter 10, this volume). Using higher rates that approach nominal interest rates on borrowed capital would appear to directly contradict financial theory.

Net Land Return

The factor payment to land is the largest of the factor payments for crop enterprises and the most difficult to estimate. Note if we were trying to make a land investment decision, we would deduct the factor payments to additional nonland capital and operator labor and compute a residual return to land that could be used in the capital budgeting analysis of the land investment. This procedure would appear consistent with the theory of land rent, as developed by David Ricardo in 1771. My review of theory suggests the budgeting process stop here.

Agricultural land prices are simultaneously determined with agricultural returns.[6] Economic theory offers little support for the idea that some exogenous land cost must be covered by agricultural returns in either the short or long run. Harrington described the uniqueness of land (Chapter 4, this volume). The major problem is that farmland in the aggregate is essentially fixed (or very inelastic) in supply. Economics textbooks generally use land as an example to introduce the concept of "economic rent" (Henderson and Quandt, pp. 98-101). For such resources, values depend on the residual returns available, and there is no basis for attempting to estimate a long run "economic cost."

Thus theory provides only one suggestion for land charges in cost of production work. The suggestion is used to conclude the earlier Krenz and Gustafson paper--"The answer's quite simple--leave it out" (Krenz and Gustafson, pp. 30-31). Twelve years later, little more can be said.

With this background two major problems are faced as budget builders attempt to impute an opportunity cost to land. First direct payments under Government commodity programs are directly and legally tied to land. Expected economic rent and land prices are both enhanced by these payments. If these payments are not included in the returns section of commodity budgets, special care must be taken to remove their influence from imputed land returns.

Another problem is that the observed (sale) price of land is always greatly distorted by both potential nonagricultural value and investor concerns for expected inflationary and real appreciation. Yet we can never measure these expectations, or know what the price of land would be based on current (real) returns.

The use of observed farmland rental rates as the opportunity cost for land (the procedure used by ERS) conceptually avoids only the second of these problems. If land rental markets are efficient and in equilibrium, the prevailing rental rate on land will equal the annual imputed value, which will also equal the return to owned land from the enterprise. However land rental markets are often imperfect. Another more important problem is that cash rental rates are distorted by special arrangements concerning direct government payments and the use of surface water rights that are legally inseparable from land.

Return to Unpaid Labor

Following the same general process as used for the other fixed factors, we would impute a return to operator labor equal to its marginal value product. Theory suggests that the relationship

$$MVP_x = P_x$$

would be approached by the farm sector in economic equilibrium. Because operator and family labor is not purchased in the market, an opportunity cost concept is usually used--ERS and (apparently) a majority of the states use the prevailing farm wage rate for this opportunity cost.

Daniel Sumner and Wallace Huffman have discussed the theoretical basis for using the farm wage rate for this opportunity cost (Chapters 17 and 18, respectively, this volume). Arguments are made for using the (higher) off-farm wage rate for the opportunity cost. Huffman suggested that the MVP of farm operator and family labor may be even higher than off-farm wage rates.

One problem with these arguments is that no single opportunity cost estimate is appropriate because the reservation price of operator and family labor likely varies by size of farm. The viability and efficiency of a home garden, with its attending low opportunity cost on labor, is an extreme but illustrative example (Miller, Rodewald, and McElroy, p. 8). Many economies of size studies can be challenged because they overlook this relationship. It also raises questions about estimating the proper opportunity cost for labor on the enterprise budget for the average sized farm.

One additional observation can be offered--that labor immobility may invalidate the competitive model. Farmers may continue working on their own farm at lower and lower returns, even though they could receive higher labor returns as hired farm laborers or in working off the farm. This practice would force the MVP of unpaid labor to be lower than the prevailing farm wage rate. These and other factors may cause the prevailing wage rate to be either above or below the actual imputed return to unpaid (operator and family) labor. Systematic errors are likely and theoretical considerations offer no objective way to resolve these problems.

Residual Returns to Management and Risk

Management and risk is used as the residual claimant in the ERS commodity cost and return budget. If the purely competitive model applies to agriculture, if the sector is in equilibrium, and if our estimates of all cost and residual imputation categories are correct, then some pervasive tendencies would appear in the various enterprise budgets over time. McElroy has argued that, "The longrun return to risk is expected to average near zero, although in any particular year it could be positive or negative, depending on weather and on supply and demand. The return to management should be positive in the long run, but the level is unknown and, therefore, speculative. Thus, the residual returns to both management and risk are expected to average above zero over time" (McElroy, p. 5).

Table 19.1 shows some of the residual returns to management and risk from a recent ERS cost of production report (USDA, pp. 78-90). Clearly for most crops since 1982, the estimated budgets do not live up to the above expectation. Each of us as economic analysts may choose a different explanation for this discrepancy--in general from the list of "if" statements in the previous paragraph. After showing this table to several colleagues, there appears to be some consensus for the third conclusion, that is, the sum total of estimates for all cost and residual imputation categories is simply incorrect. The suggestion here is that the ERS procedures are systematically either underestimating returns or overestimating total costs.[7] This conclusion is the "bottom line" of what economic and financial theory is trying to tell us about current procedures.

Balancing Conflicts Between
Theory and Practice: A New Proposal

How can the agricultural economics profession balance the conflicts between the implications of economic theory and the demand for statistical information on commodity costs and returns? The chapters in this volume

TABLE 19.1 ERS National Average Commodity Costs and Returns Residual Returns to Risk and Management

Year	Wheat	Grain Sorg.	Barley	Corn	Soybeans	Cotton	Rice	Average[a]	Sugarbeets	Peanuts
					- Dollars per acre -					
1975	19.58	8.66	17.18	34.94	21.12	17.87	27.85	21.03	N/A	151.70
1976	-1.49	-3.26	12.69	8.88	44.51	69.15	-8.08	17.49	N/A	145.03
1977	-8.94	-5.34	-5.90	-3.18	37.90	37.31	77.32	18.45	N/A	81.79
1978	-10.37	41.46	-15.21	25.59	41.66	-14.49	11.07	11.39	N/A	82.97
1979	11.53	8.38	-12.47	37.90	35.05	37.47	63.78	25.95	N/A	38.68
1980	-5.77	14.10	-18.51	16.86	19.82	-24.40	33.08	5.03	N/A	-141.51
1981	-13.89	-21.32	-21.77	-18.43	0.69	-34.18	26.81	-11.73	68.54	187.16
1982	-17.56	-33.38	-30.56	-27.23	-11.38	-39.14	-79.34	-34.08	111.54	162.21
1983	-7.81	-25.42	-25.73	-5.55	20.65	-9.79	-43.30	-13.85	115.94	81.68
1984	-15.91	-25.35	-24.53	-14.73	-8.25	-23.95	-45.58	-22.61	92.45	205.10
1985	-29.26	-22.99	-41.86	-24.39	-4.11	-40.64	-41.50	-29.25	100.08	147.67
1986	-44.06	-42.09	-64.00	-75.52	-10.57	-132.40	-202.95	-81.66	157.03	92.13
1987	-36.61	-40.36	-49.81	-58.55	1.65	30.70	-170.17	-46.16	260.73	37.92
1988	-26.43	-6.62	-53.10	-46.19	14.39	-72.75	-104.81	-42.22	53.12	29.35
Average	-13.36	-10.97	-23.83	-10.69	14.51	-14.23	-32.56	-13.02	119.93	117.90

[a] Average of grains, soybeans, cotton, and rice.
N/A = Not Available.
Source: USDA, 1989.

have suggested that the major conflict is between the arbitrariness of our processes for estimating the "economic costs" of fixed factors of production, as opposed to our client's demands for defensible and accurate data. A recent example of this conflict is presented in the Food, Agriculture, and Trade Act of 1990, which requires the Secretary of Agriculture to annually publish "...a report analyzing the return on assets resulting from the production of upland cotton, rice, wheat, corn, oats, barley, grain sorghum, soybeans, peanuts, sugar from sugar beets, and raw sugar from sugarcane." USDA is instructed to consider "...returns from agricultural price support programs, the effects of agricultural price support programs on costs of production, the factors currently used in Department of Agriculture cost of production data, current value of land, and any other information necessary to reflect accurately the return on the production of such crops."

This type of legislative request along with David Harrington's call for cost of production estimates that "mimic the factor markets" provide the background for the following proposal, consisting of three conceptual parts: (1) the definition of what could be called "planned returns," (2) the definition and estimation of "market returns," and (3) special consideration for management and risk. The proposed procedure has the potential to reduce much of the current controversy concerning our commodity cost and return estimates.

The Concept of Planned Returns--Estimation and Allocation

Viewed as a decision tool, enterprise cost and return estimates represent expectations of the decision maker. Yet we do not know what level of total returns to fixed factors a farmer expects. More importantly, we can never observe or measure how an individual farmer allocates these returns to the different fixed factors of production. That process is arbitrary and not observable.

Economic theory, as discussed earlier, suggests that under specified conditions each factor of production will receive a return according to the relationship

$$MVP_i = OC_i = \text{Opportunity cost.}$$

Given quantities of fixed factors (x_i), Total Planned Returns can now be defined as:

$$\text{Total Planned Returns} = \Sigma \, MVP_i \, x_i.$$

This quantity represents what total factor returns would be if farmer-owned fixed factors of production all received their opportunity costs. When we estimate these components using opportunity cost concepts, we are estimating what can be viewed as planned returns to the fixed factors of production.

The Concept of Market Returns

This definition of total planned returns may be contrasted to "total market returns." Total market returns are actual production returns imputed from the marketplace, with actual crop yields and actual prices received for production. Colorado State University uses the terms "Net Receipts" or "Total Factor Payments" for this item. ERS uses the label "Allocated Returns to Owned Inputs" (USDA). The quantity can be measured by survey techniques or from actual farm records. Except for some allocation issues, total market returns are observable and unambiguous.

In the real world, total market returns do not equal total planned returns because of disequilibrium in product and factor markets and because of estimation errors in all categories of the cost and return budget. These include errors in yield and price expectations, errors in estimating variable costs and cash fixed costs, and errors in selecting the proper opportunity cost measures.

What does economic theory say about market returns for the individual fixed factors of production? Under constrained profit maximization, it shows that:

$$MVP_{x_i} / P_{x_i} = \delta$$

for all factors of production (x_i), where the value imputed to a marginal dollar of capital is the Lagrangian multiplier (δ) (Debertin, pp. 134-138). The important point here is that the ratios of the MVP to the price are equal for all factors.

This profit maximizing criteria suggests a process for estimating market returns to individual fixed factors, given the observed total market returns to the enterprise. The best estimate of how farmers allocate actual market returns would be:

$$MVP_i / OC_i = m$$

for the fixed factors of production, where (OC_i) is the opportunity cost of factor (i) as discussed in the previous section, and the allocation factor (m) is the ratio:

$$m = (Total\ Market\ Returns) / (Total\ Planned\ Returns)$$

TABLE 19.2 Example for Northern Colorado Dryland Winter Wheat, 1987

Factor	Amount x_i (1)	Opportunity Cost (2)	Planned Returns (3)	Market Returns (4)	Return Rate (5)
Capital	$35.00	5.00%	$1.75	$1.20	3.44%
Operator labor	1.386 hr.	$5.00	6.93	4.76	$3.44
Land	$250.00	4.00%	10.00	6.87	2.75%
Management and risk			0	0	
Total			$18.68	$12.84	

Source: Colorado State University.

The return to management and risk is assumed to be zero at this point; these factors will be considered later. Table 19.2 shows an example computed for Northern Colorado Dryland Winter Wheat. The computations proceed as follows. Planned returns are first computed for each fixed factor based on the quantity used and its opportunity cost. The procedures follow those outlined earlier in this chapter--generally they are the current procedures used by both ERS and Colorado State University. Total Planned Returns to capital, operator labor, and land equal $18.68 per acre.

However in Colorado in 1987, the actual observed residual returns to fixed factors for the wheat enterprise--Total Market Returns--were $12.84 per acre. The allocation factor (m) therefore equals $12.84 / $18.68 or 0.6874. Market returns for each factor in column (4) of Table 19.2 are computed by multiplying each item in column (3) by 0.6874. The return rate to each factor, column (5), is the market return divided by the amount of the factor and is equivalent to the MVP_i for each factor as defined earlier. In this table, columns (2) and (3) are for computational proposes only. The estimated enterprise budget would show columns (1), (4), and (5).

Management and Risk Returns

The procedure can be extended to include a nonzero return for management and risk. The planned return to management and risk can be defined as the risk premium or certainty equivalent (CE) (Newberry and

Stiglitz, pp. 70-75). This certainty equivalent would increase Total Planned Returns in column (3) of the computation process.

The market return allocated to management (entrepreneurship) and risk includes two components. The first relates to the planned risk (actually the expected cost of risk). This part of the market return to risk would be the product of the allocation factor and the certainty equivalent:

$$m(CE)$$

A second market return component accrues to management (entrepreneurship) and is the value of any stochastic element in the budget causing variation in actual returns from the expected value or trend. Using an example of yield variability, this quantity would be

$$P_y(Y_t - Y_{trend})$$

where (P_y) is the market price of the product and $(Y_t - Y_{trend})$ is the deviation of the actual yield from the trend or expected yield.

Table 19.3 shows the same Colorado enterprise budget with a return to management and risk computed using this procedure. The planned return to risk in this example is based on an assumed certainty equivalent of 2 percent of total crop receipts of $81.00, or $1.62. The literature on risk management contains numerous estimates of risk premiums or certainty equivalents for different agricultural situations (see Anderson, Dillon, and Hardaker and Newberry and Stiglitz for summaries). Total Planned Returns now equal $20.30, as shown in column (3).

Next in Table 19.3, the stochastic return component for management is estimated. Assuming a trend yield for 1987 of 27.5 bushels per acre, an actual yield of 27 bushels per acre, and a product price of $3.00 per bushel, the return to management becomes $3.00 times (27 - 27.5) or $-1.50. Total Market Returns with this trend yield would be the sum of observed residual returns ($12.84) and this adjustment ($1.50) or $14.34. The allocation factor (m) now equals $14.34 / $20.30 or 0.7064. Market returns are column (3) times 0.7064, with the allocated market return to management and risk equal to $m(CE) + P_y(Y_t - Y_{trend})$, or 0.7064($1.62)-$1.50 = $-.36. Note that the return rates are slightly higher in this example, since they are based on the higher trend yield of 27.5 bushels per acre.

This example shows 1987 actual market returns based on actual 1987 yields. A projected budget of 1987 market returns, based on the expected (trend) yield of 27.5 bushels per acre, would show a total market return of $14.34, and a management and risk return in column (4) of $1.14. Both are increased by $1.50 in the projected budget.

The published budgets would again show columns (1), (4), and (5). These estimates are consistent with economic theory in the positivist tradition; they

TABLE 19.3 Example for Northern Colorado Dryland Winter Wheat, Including a Management and Risk Return, 1987

Factor	Amount x_i (1)	Opportunity Cost (2)	Planned Returns (3)	Market Returns (4)	Return Rate (5)
Capital	$35.00	5.00%	$1.75	$1.23	3.53%
Operator labor	1.386 hr.	$5.00	6.93	4.89	3.53
Land	$250.00	4.00%	10.00	7.06	2.83%
Management and risk			1.62	-.36	
Total			$20.30	$12.84	

Source: Colorado State University.

are estimates of actual resource returns. Such commodity cost and return estimates are the proper basis for economic decisions by either individuals or society. They inform us about what is actually happening and how the market is working (or is expected to work in theprojected budget). These estimates avoid the normative connotation often expressed that the term 'cost' represents the compensation which must be received by the owners of capital and the units of factors of production used by the firm if these owners are to continue to supply productive factors to the firm.

In this manner, the proposed procedure avoids many of the procedural issues debated within the profession. It provides estimates of commodity costs and returns that are the proper basis for economic decisions by both farmers and policy makers. It also represents a method that could be used to meet the 1990 Farm Bill mandate to estimate resource returns, subject to the firm-enterprise allocation problems discussed earlier.

Notes

1. The terms budget and statement are used interchangeable in this chapter, and are not used to distinguish between forecasts or historical estimates.

2. ERS does publish forecasts annually, but only at the U.S. level.

3. "Arbitrary" has a dictionary meaning of "subject to individual judgement; contingent solely upon one's discretion." I use it in this chapter to describe procedures where economic theory offers no guidance, or where economic theory does not suggest a "correct" procedure.

4. Oscar Burt's chapter suggests there is still a small upward bias in such estimates.

5. We usually take it for granted that variable inputs are employed at the level where $MVP_x = P_x$. There is empirical evidence that $MVP_x > P_x$ for most variable inputs. Thus subtracting variable input costs from total returns tends to leave a residual which is larger than what may really be due to the fixed factors.

6. A concise discussion of the economics of land prices is provided by (Boxley and Walker, pp. 90-93).

7. After checking the Colorado State University crop budgets, we find the same tendencies. It would be interesting to see how many other universities would find similar results.

References

Ahearn, Mary, Mir Ali, Robert Dismukes, Dargan Glaze, Ken Mathews, William McBride, Robert Pelly, and Michael Salassi. 1990. "How Production Costs Vary." Economic Research Service, U.S. Department of Agriculture. Agr. Infor. Bull. No. 599.

Anderson, J.R., J.L. Dillon, and J.B. Hardaker. 1977. *Agricultural Decision Analysis.* Ames: The Iowa State University Press.

Boxley, Robert F. and Larry Walker. 1979. "Impact of Rising Land Values On Agricultural Structure," in Economics, Statistics, and Cooperatives Service, *Structure Issues of American Agriculture.* AER 438. U.S. Department of Agriculture.

Colorado State University. 1988. "Selected 1987 Crop Enterprise Budgets for Colorado." Dept. of Agricultural and Resource Economics. DARE Information Report IR:88-7.

Debertin, David L. 1986. *Agricultural Production Economics.* New York: Macmillan.

Heady, Earl O. 1952. *Economics of Agricultural Production and Resource Use.* Englewood Cliffs, NJ: Prentice-Hall, pp. 402-414.

Henderson, J.M. and R.E. Quandt. 1958. *Microeconomic Theory--A Mathematical Approach.* New York: McGraw-Hill.

Krenz R.D. and R.A. Gustafson. 1979. "How to Price Land in Cost of Production Studies," in Economics, Statistics and Cooperatives Service, *Estimating Agricultural Costs of Production--Workshop Proceedings.* ESCS-56. U.S. Department of Agriculture, pp. 30-31.

Johnson, Glenn L. 1955. "Results from Production Economic Analysis." *Journal of Farm Economics* 37, No. 2: 206-222.

Libbin, James D. and L. Allen Torell. 1990. "A Comparison of State and USDA Cost and Return Estimates." *Western Journal of Agricultural Economics* 15, No. 2: 300-310.

McElroy, Robert G. 1987. "Major Statistical Series of the U.S. Department of Agriculture, Volume 12: Costs of Production." Agricultural Handbook No. 671, Economic Research Service, U.S. Department of Agriculture.

Miller, Thomas A. and Melvin D. Skold. 1980. "Uses and Users of Costs and Returns Data: A Need Analysis." *Developing and Using Farm and Ranch Cost of Production and Return Data: An Appraisal.* Proceedings of a GPC-10 Symposium, GPAC Publication No. 95, Dept. of Agr. Econ. Report No. 104, University of Nebraska. pp. 1-16.

Miller, Thomas A. 1979. "Cost of Production Variability--Issues and Problems," in Economics, Statistics, and Cooperatives Service, *Estimating Agricultural Costs of Production--Workshop Proceedings.* ESCS-56. U.S. Department of Agriculture. pp. 12-21.

Miller, Thomas A., Gordon E. Rodewald, and Robert G. McElroy. 1981. "Economies of Size in U.S. Field Crop Farming." Economics and Statistics Service, U.S. Department of Agriculture. AER-472.

Newberry, David M.G. and Joseph E. Stiglitz. 1981. *The Theory of Commodity Price Stabilization: A Study in the Economics of Risk.* Oxford: Clarendon Press.

Pasour, E.C. 1980. "Cost of Production: A Defensible Basis for Agricultural Price Supports?" *American Journal of Agricultural Economics* 62, No. 2: 244-48.

Salassi, Michael, Mary Ahearn, Mir Ali, and Robert Dismukes. 1990. "Effects of Government Programs on Rice Costs and Returns, 1988." U.S. Dept. of Agriculture, Economic Research Service. AIB No. 597.

U.S. Department of Agriculture. 1990. "Economic Indicators of the Farm Sector: Costs of Production--Major Field Crops, 1988." Economic Research Service, Agriculture and Rural Economy Division. ECIFS 8-4.

Watts, Myles J. and Glenn A. Helmers. 1981. "Machinery Costs and Inflation." *Western Journal of Agricultural Economics* 6, No. 2: 129-137.

20

Similarities and Differences in the Data Needs for Farmer Planning, Economic Research, and Policy Analysis

Arne Hallam

Introduction

Economists, like other social scientists, seem inherently interested in measurement. Each time a scientist collects any data about a particular phenomenon, he is implicitly making a measurement. Measurement can be defined as ascertaining the quantitative and qualitative properties of something in relation to a standard. Meaningful measurement can be defined as measurement in relation to a standard that is clearly and unambiguously defined. What is a meaningful representation of observed phenomena in relation to one standard may be useless or nonsensical in relation to another standard. The usefulness of any measurement, then, is in its relation to a specific standard, or paraphrasing from ancient wisdom, is in the eye of the beholder. Whether a particular number has any meaning is determined by its accuracy in relation to a standard, the usefulness of that standard for scientific inquiry, and the acceptance of that standard by a broad group of users. Information that does not meet these criteria is not meaningful measurement and will tend to cloud rather than illuminate scientific discussion. For example, consider measuring the price of hired farm labor. For an individual producer who is hiring, the meaningful price of labor is the price that he must pay to use another unit in the production process. For an analyst attempting to understand the effect of labor costs on the price of food, meaningful is defined in terms of the response of producers to a change in the average farm wage rate. For a policy maker considering changes in immigration laws, meaningful is associated with the effects of changes in labor supply on individual laborers, labor markets, agricultural producers, and consumers. The producer is interested in a very specific price, the analyst in an aggregate price, and the policy maker in several different individual and aggregate prices. While each is concerned with accuracy, the standard for comparison is very different. An accurate description of the wage that the marginal worker in Yolo County, California will accept for field work between August 20 and August 27 in the summer of 1991 may be very relevant to a tomato grower there, but not very useful to someone trying to estimate the demand for labor in U.S. crop farming. Meaningful is clearly defined only in relation to the user.

Meaningful Measurement of Commodity Costs and Returns

This book has been organized to address the issue of appropriate measurement of commodity costs and returns in recognition of the different purposes of the estimates. To understand what is meaningful measurement in an economic context consider three examples.

1. A profit maximizing producer is interested in information that will allow him to combine resources in a fashion to obtain the greatest net income. The unit of measurement of net income is dollars and so he is interested in the dollar effects of various production choices. A normal procedure is to separate the economic and technical aspects of production, so in addition he is interested in the effects of inputs on outputs, which will be translated into dollar terms. A meaningful measure of the quantity of an input is one such that if the same amount of the input is applied at two different times, the output response will be the same. If this response is not the same due to "quality" differences, the simple quantity measurement is not meaningful. A meaningful measurement of the cost of an input would be the actual price paid for a given unit of the input as opposed to the list price or the county average price, for example. A meaningful measurement is one which accurately reflects the effect of an action on profits.

2. A economic analyst may be interested in the aggregate response of corn producers to an increase in the "price" of corn. A meaningful measurement of aggregate corn supply is the total quantity of a given class of corn supplied to the market during a given time period. The analyst must choose a way to represent or measure the "market price" of corn. Every market transaction could in principle have its own price, therefore some way of aggregating these prices is necessary. This aggregate price is meaningful if the response of aggregate production to it does not change when changes in the component prices do not change the level of the aggregate price. A meaningful measurement for the analyst is one which reflects the actual response of economic agents or an economic variable to some clearly specified stimulus.

3. A policy person may be interested in evaluating the effects of a given policy on individual economic agents or specified groups. A meaningful measurement of individual welfare is one which accurately reflects the effect of the policy on the utility of the individual. Since utility is normally considered to be an ordinal measure, most analysts have chosen a money measure of utility such as the money metric or compensating variation. Such a measure is meaningful if it allows the analyst to rank alternative bundles in the same order as the decision maker. Policies are often evaluated in terms of their effects on different groups as opposed to individuals. Net farm income to various groups is a measure often used. This measure is meaningful if the policy induced changes in individual's welfare leads to changes in net farm income of the groups that is consistent with changes in the money measure of utility change for the group. In particular, the net farm income measure and the money metric measure should rank alternative policies in a similar order.

The examples above point out three distinct users of commodity cost and return data. The first group can be characterized as optimizers in a broad sense. This would include individual producers in both a planning and evaluation mode, extension personnel, farm management consultants and organizations, and researchers interested in decision processes. The second group consists of analysts carrying out traditional positive economic research. This group is typically interested in the response of economic variables to one another and external stimuli. The last group is made up of individuals interested in evaluating government and business policy. This group is most interested in the effects of specific policy actions on different economic agents and clientele groups. The appropriate measurement of a given economic variable may well differ between the groups as the problems they attempt to solve vary.

Macro Measurement and Micro Relations

An important issue in all economic measurement is aggregation. One group, commodity analysts for example, may be interested in the effects of weather in Brazil on U.S. soybean exports. The relevant soybean price for this group is very different than the relevant one for a farmer in Grundy County, Iowa. The relevant price for the group is the one which accurately reflects the changes that will occur in world soybean market due to the weather effect. Our economic theory would postulate that the market effect of the weather event is the result of the actions of many optimizing agents and their interaction in the market. The appropriate price to use is then one which reflects all the relevant information available in the various individual prices over time and space. This leads to the standard problems of aggregation. If changes in the prices faced by individuals cause changes in their behavior, but these changes in individual prices do not change the market aggregate price, then only if all these individual changes cancel out will the aggregate price be useful for determining aggregate changes in behavior. The conditions for correct aggregation are fairly stringent and rarely satisfied in practice. As a result, the only way to perform accurate aggregate analysis may be to aggregate up micro relations and data. This means that the data that is meaningful for the optimizers may also be meaningful and necessary for the other groups since they will need it to construct meaningful aggregate data.

Aggregation and specification are also issues as related to opportunity costs. Consider an agent who must decide the value to place on his labor in determining the costs of producing feeder pigs. The correct opportunity cost is the wage offered by the best alternative employment available to him. The analyst, however, rarely knows what this opportunity is and so must construct a wage rate to represent this cost. This representative wage may be some aggregate of various market wages. A similar story could be told for determining the value of a given stock of machinery of various capabilities and vintages. The essential information is that which represents the decision environment of the optimizer.

The first hypothesis and conclusion of this chapter is that micro-level data is the key missing ingredient in much of current economic analysis. Given quality micro-level data, the next step is to agree on standard ways to aggregate this data for use by various groups. The standard for aggregation is that the data accurately reflect the decision process and environment of optimizing agents. Inevitable errors in aggregation should be chosen so as to minimize the discrepancies between aggregate and micro-level response subject to the budget constraints of data collection and processing. Given a clear market failure in the provision of even tentative standards of data aggregation due to differing objectives, a command approach by organizations such as USDA, the American Agricultural Economics Association, and the Land Grant universities seems essential. Constant updating and revision of standards is inevitable, but standards must be developed and used if measurement is to have any meaning at all. The alternative is to continue to collect and report data in the way that best meets the needs of the clientele groups with little regard for economic accuracy or content.

This chapter will discuss the data needed for each of the three groups: optimizers, analysts, and evaluators. Data problems for each group will be presented. The first section on optimizers will be somewhat more detailed than the others since many of the issues discussed there carry over to the other groups. Following this discussion, a set of key needs for integrating these three user groups will be proposed. Some remaining issues close the chapter.

Essential Data for Economic Optimizers

Types of Data Needed

The major class of economic optimizers using cost and return data are firm level agricultural producers. The second most numerous group are extension agents and farm management consultants who attempt to assist such decision makers. Another important group are researchers attempting to evaluate new or alternative technologies through the use of conditional normative models of the farm firm. The data needed to make these decisions are of three basic types: technology data, data used to determine the economic cost of inputs, and data used to value outputs. Each type of data and some of the problems involved in its collection and use will be discussed in turn.

Technology Data. Optimizers need data that accurately reflects the effects of inputs on outputs. While most data of this type is simple input- output coefficients reflecting Leontief technologies with constant returns to scale, some information reflecting variable input elasticities is important. While micro level data are often collected, they are seldom reported in a way that would allow the estimation of variable marginal products. For the individual considering input choices this is essential. One alternative is to provide crop budgets for several alternative levels of important inputs. Another alternative is to report actual response equations for relevant ranges of the production surface. At a minimum, production elasticities should be estimated and made available in some form for researchers.

Economic Costs of Inputs. One of the difficulties in valuing inputs is that some inputs are not regularly traded in the market. When no regular markets exist, economic costs must be imputed. An obvious example is off-farm labor. Each individual farmer or spouse may compete in a different off-farm labor market. The market where they would have the highest opportunity cost if they worked there, is the appropriate market for determining their implicit wage rate. For the individual optimizer, this information is inherently personal. Prepared budgets should allow this opportunity cost to be variable, should provide some suggested values, and should assist the decision maker in determining this wage correctly. In order to provide suggested values, standard estimates for off-farm wage opportunities at a somewhat disaggregated level such as crop reporting district seem necessary.

Another difficulty for the optimizer is allocating fixed costs over time and space. The allocation over alternative enterprises is particular troublesome. To the extent that rental markets exist, they should be used to determine appropriate costs for equipment and machines. More active data collection in this area is a priority. Information on average use life, second hand markets, opportunity costs of funds in agriculture would be helpful in using direct valuation methods allocating costs over time. Data on the actual vintage makeup of the current capital stock would also be valuable.

A difficulty in allocating fixed costs over time is determining use versus ownership costs for durable equipment. Again rental market data would be very useful. Research studies which attempt to reconcile rental and ownership cost data would improve the future accuracy of both methods.

Input quality and homogeneity is important to the extent that input variation will affect profits. While the individual knows the actual input used, most cost and return data is reported using average costs for the aggregate basket of input qualities and brands. To the extent that the individuals price and quantity differ from this average, errors in optimization may occur. Of particular concern is that correcting one of the price-quantity pair is not necessarily better than doing nothing. Careful description of exactly what is being described in a given budget is critical to correct interpretation.

To the extent that individuals have non-neutral risk preferences, they may opt for input combinations that do not minimize the cost of a given output level. This means that the cost of production reported for a specific individual or group may not be the least cost combination from economic theory. Disaggregate budget procedures can help minimize the impact of this effect since individuals can adjust necessary items.

Valuing Output. A major difficulty in valuing agricultural output for planning purposes is production lag. The production lag requires that all output prices be computed ex ante. Few cost and returns budgets actually forecast output prices. In order to be relevant for decision makers some effort in this area is needed. Simple time series forecasts would be a start. A difficulty in analyzing returns ex post is the effect of government programs on returns. Anyone who has struggled with the difference between the season and average market price series for major commodities is acutely aware of this issue. The individual will clearly take government hand-outs into account in farm planning. One of the important issues is valuing the entitlements to these payments. While few researchers would suggest treating the entitlement to base acreage as a cost in determining how to set the loan rate, the implicit program rents contained in land or other input prices are a less clear issue. Clearly such effects must be modelled in analyzing the individual decision.

High Priority Data Needs

While many types of data are important as discussed in the previous subsections, a few particular needs are especially important. These are listed below followed by brief discussion.

Labor Valuation. A data series on off-farm labor opportunity costs is essential at least at the state level. More disaggregate data at the sub-state but not county level is very important. Such a regular data series could significantly improve the ability of decision makers to value off-farm opportunities. The use of the hired farm wage to reflect the cost of the owner's time is biased at the very least, and an insult to management at the worst.

Depreciation and Use Guidelines. Some general guidelines to use for depreciating equipment of various types would be helpful. Simple tables of expected life that could be used by all universities and government agencies would dramatically improve consistency.

Rental Markets and User Costs. Agreement on when to use rental data and when to use capitalization methods could help reduce discrepancies in commodity costs. This would require development and maintenance of better rental series and agreement on when to use such series.

Government Program Information. Significant economies of scale exist in preparing decision aids and in analyzing the effects of government payments. Rather than have every state and university develop their own method of presenting this information to the producer, a central method of tabulation and calculation would be beneficial.

Essential Data for Economic Analysts

General Types of Data Needed

Economic analysis for the purposes of this chapter is traditional positive tests of hypotheses and prediction of economic response. Most market level supply and demand analysis fits this category as does work on the structure of agriculture and the organization of firms. The data needed to test economic hypotheses and describe behavior are of four basic types: data on individual preferences, technology data, data used to determine the economic cost of inputs, and data used to value outputs. Each data type and associated issues involved in its collection and use are mentioned below.

Preference Data. Analysts must have information about decision makers preferences in order to predict their response. The decision makers risk preferences will have a major impact on their response. There is little consistent long term work attempting to measure and evaluate risk preferences on a large scale. Decision makers may have multi-attribute utility functions. For example, farmers may value lifestyle as much as profit. Data on these types of tradeoffs could improve understanding of farm level choices. There is a need for more work on how decisions are made in the farm household. The assumption of a single decision maker is not only incorrect, but may be highly inaccurate when used as an assumption for prediction. Analysts are ignoring important factors when they assume unqualified profit maximization and many of the errors commonly made in prediction may be due to the misspecification of preferences. Preference data should become as much a part of the standard data sets as prices and quantities.

Technology Data. Economic analysts need the same technology data as optimizers. The best form of data for the analyst is micro-level data that can be aggregated in an appropriate fashion to obtain the effects of total input levels for each input on aggregate output. If only aggregated data is available, the researcher is forced to assume the same technology across firms. This will lead to bias in estimation of input-output elasticities. While it is not feasible to collect data on each individual producer, aggregate data on inputs and outputs should be supplemented with regular micro-level studies that allow investigation of the effects of aggregation and validation of aggregate estimation procedures. Again, aggregate data must be continually supplemented with and validated in comparison to micro-level data. The ways in which aggregate data are developed should be made clear in all publications where they appear so that correct interpretations are possible. Since much aggregate analysis is conducted using time series data, information on changes in technology is also important. The changing composition of aggregate input categories over time should be factored in to all positive economic analyses.

Economic Costs of Inputs. A major difficulty with market level analysis is that individuals are non-homogeneous and thus will not purchase exactly the same inputs (either type or quality) or pay the same prices. Accurate analysis requires a way to aggregate the prices paid by various individuals into a meaningful representative price. The use of a market price is not correct since individuals may not respond to that price. To the extent that price differences relate to the same product and the only difference is due to spatial or transportation issues this is not a serious matter. Unfortunately, almost all economic analyses use rather aggregated input categories so that errors occur not just in price and quantity aggregation for individual products but in determining the composition of the aggregate. Clearly a farmer who regularly uses urea, anhydrous ammonia, and animal manure as nitrogen sources will respond differently to an increase in the cost of anhydrous ammonia than an individual who exclusively uses this liquid form of fertilizer. Or consider an increase in the minimum wage which will have very different effects on different types of farms. An aggregate model estimated during a regime of one minimum wage may not predict well the demand for farm labor due to changes in that minimum wage. Difficulties in valuing non-market inputs are increased when the analyst must do so across many individuals, markets, and/or states. While the individual may have some idea of his opportunity cost off the farm, this is particularly difficult to impute for broad groups.

Difficulties for the optimizer is allocating fixed costs over time and space are only more prevalent for the analyst since he or she must determine these without usually having good micro-level data. While the individual has good information on the type, vintage and condition of his durable equipment, the analyst usually must make due with a jelly-like aggregate measure expressed in dollar terms. This often reduces his analysis to allocation of an average bundle that does not represent the problem of any one producer. The allocation problem is also apparent for the case of fixed inputs and multiple enterprises. Recent work that has attempted to allocate fixed inputs across enterprises using cross section data is commendable and should be continued. Some detailed survey work on actual allocations would also be useful.

Information on use and ownership costs that was suggested for the optimizers would also be useful for the analyst. Regular comparison and validation of cost and return data and aggregate capital cost data would improve both efforts.

Valuing Output. Analysts need to predict prices both in the long and short run in order to predict economic response. Issues related to expectations formation and policy regimes should be better taken care of in supply and demand studies. Simply reporting past net income measures is not enough for understanding behavior or expected response. Analysts need to collect data on the entitlements associated with government programs. The actual disaggregated nature of acreage allotments, historical yields, and set aside areas may be necessary to understand the aggregate nature of response to changes in conservation or price support laws. To the extent that there is a market in such entitlements, data collection and analysis would be useful. Of particular interest is the capitalization of government entitlements into input costs. All rents may be captured by initial owners with no benefits to current users or there my be little capitalization. The effect of expected entitlements on input prices in a market equilibrium is critical in determining actual long run response.

High Priority Data Needs

The most pressing data needs and issues for the analyst are similar to those for the optimizer.

Labor Valuation. A regular data series on the value of off-farm labor should be developed at least at the state level. Such a series would reduce the standard bias associated with using farm wage rates to reflect the cost of agricultural labor. This series could be coordinated with the aggregate data prepared on the cost of labor in agriculture.

Depreciation and Use Guidelines. General guidelines for depreciating equipment of various types are needed. Simple life tables would be a start. Consistency between micro-level and aggregate series on capital costs is a desired goal. Data collection procedures at both the micro and aggregate levels should be undertaken in such a way that micro studies can validate aggregation procedures, and suggest improvements in collection of aggregate data.

Rental Markets and User Costs. Agreement on when to use rental data and when to use capitalization methods is as important at the market level as at the firm level. Again specific micro-level studies can help to check the validity of aggregate collection methods. Differences between the prices reported by dealers and those paid by farmers should be carefully studied.

Government Program Information. More work needs to be done on how to capitalize government programs into input costs and how to separate such effects from market prices. To the extent that current input costs reflect past subsidies, they are not a fair basis for determining "fair" rates of return on assets.

Disaggregate Price Information. A critical need is for disaggregate price and quantity information. The price paid by an individual farmer is not necessarily the price in a regional market. This may be due to spatial, quality or market power considerations. To the extent that the market prices do not reflect the prices firms pay, aggregate analysis may error. In cases where the aggregate price is constructed, both the aggregate and the firm level information should be reported. Depending on functional form chosen for analysis, different types of aggregation rules are needed (Pope and Chambers). With disaggregate information, statistical ways to relate individual and market prices could be developed. Such methods might allow representative firm level models to be built based on market prices with actual validation only occasionally.

Disaggregate Capital Values. Data on capital values at less than the national level would be useful. The relation of these values to data from cost and returns surveys would also validate both micro and aggregate analysis. Of particular concern is the fact that data organized for cost of production estimates is not necessarily the same as that used for the aggregate USDA productivity series. Coordination between firm level and aggregate data analyses is valuable.

Individual Firm Flexibility. The supply elasticity of an individual firm and that of the market may be very different depending on firm structure, input market responses and other external factors. Information about the ability of individual firms to respond to price changes and their traditional responses could help analysts sort supply response into changes by existing firms and entry and exit of firms.

Essential Data for Policy Evaluators

General Types of Data Needed

Those who evaluate policy must do so on an aggregate and individual level. The effects of an import quota must be evaluated in terms of farm income, income to the group protected as a whole and the effects on individual firms of different types. The most straightforward way to carry out such analyses is to aggregate micro-level effects. This is costly, however, and so various less accurate and less expensive aggregate methods are often used. Regular studies that investigate the biases associated with aggregate analysis and provide accurate individual level effects for samples of individuals are an important ingredient in improving analysis in the policy area. The same types of data needed by analysts are also needed here. Almost all data relevant to analysts is important for policy analysts interested in anything but very simple aggregate analysis. In order not to duplicate previous discussion, issues in this section will be mentioned as they are new or differ from the previous two sections.

Preference Data. Data on preferences is critical because with non-linear preferences, welfare cannot be measured using simple income measures. This is particularly important in studying the effects of programs that reduce risk, attempt to change agricultural structure or preserve the family farm. For example, farmers may be better off with less income and more conservation acres if they have strong environmental preferences.

The tendency to report the effects of all farm programs in terms of income ignores many other aspects of preference and the program provisions. Individuals may prefer lower income with a particular farm structure than higher income with an alternative structure. If such effects are ignored, estimates of producer welfare may be too low and there may be a tendency to increase program payments that are not in the best interests of society.

Technology Data. Policy evaluators are often interested in the effects of new technologies on producers' income and economic welfare. Better data on new technologies is important to evaluate their effectiveness. Models which are built on primal foundations and explicitly include technologies should probably be used for more policy work.

Economic Costs of Inputs. The most critical need to policy researchers is to determine the ways in which program benefits are capitalized into market prices and economic rents. Only as the actual welfare effects of policies on disparate groups such as landowners, machinery dealers, rural communities and individual farmers are determined, can sensible policy recommendations be made. To the extent that farm program benefits are expected or capitalized, all the critiques of Lucas about response analysis are valid.

Valuing Output. The market effects of government payments are important in determining their welfare significance. Payments accrue to individuals but over time these benefits are dispersed throughout the economy depending on the competitiveness of markets, restrictions on sale of entitlements and individual risk preferences. The extent to which a target price represents a rate of return on various resources depends on the value of those resources and the portion of that value that is due to entitlement.

High Priority Data Needs

If policy analysis is to have meaning it must be based on disaggregate measures. Therefore the data needs of the true policy analyst are no different than those of the good analyst or the individual optimizer. Simple reporting of effects in an aggregate or even cross section manner, with little attention to the underlying economic models and incentives, may satisfy simple questions but will not allow discussion of alternative policies or the effects of changes in policy on individual agents. The most pressing need in the policy area is not data, other than that needed for good analysis, but rather a commitment to do true welfare analysis of programs based on good micro-level models and economic concepts of welfare such as the money metric and compensating variation. Net farm income is not a welfare measure in the way it is usually computed and should not continue to be used as such.

Data Needs for Integrating Optimization, Analysis and Evaluation

Each of the previous three sections has identified some key data needs for different groups of users. This section will attempt to summarize those needs as they relate to all the groups.

Valuing Farm Labor

Standard ways of valuing farm labor need to be determined. There can be alternative acceptable methods but these should also have agreement. An active promotion of these standards is important. A regular series on farm labor costs at a disaggregate level is the most pressing need in this area. Hired labor wage rates are not an appropriate proxy for all operator labor. Opportunity cost principles must be included.

Valuing Durable Machinery and Equipment

Accepted methods for valuing durables is needed at both the aggregate and firm levels. Published tables on durable values would be a first step. Vintage and productivity information needs to be standardized and made generally available. More and better data on rental markets is a priority. To the extent possible, rental market data should be used to value durables.

Valuing Land

More work needs to be done on the relative merits of using rental values versus capitalization rates in determining the user cost of land. There are strong arguments for using rental rates for valuing land in cost of production estimates. The separation between input owners and producers should be maintained for reasonable welfare analyses. Work also needs to be done on the effects of crop share arrangements on land costs.

Input Costs and Government Program Effects

Government program payments may be capitalized into the cost of various agricultural inputs, the primary example being land. The actual effects of such phenomenon is critical if costs of production are to be used in any fashion to determine future government payments. Data on any transactions in entitlements could help shed light on this issue.

Disaggregate Data on a Wide Variety of Economic Variables

Disaggregate data are useful for many purposes. Such data can be used for firm and area analysis. It can be used to understand individual behavior and test economic theories. It is essential in determining the welfare effects of policies at the micro level. Disaggregate data is also essential for preforming good aggregate analysis. The use of aggregate data in an *ad hoc* fashion leads to biased estimates of everything from supply elasticities to economies of size. Disaggregate data cannot be economically collected for all or even most studies. The alternative is to continue with aggregate data collection and reporting of aggregated micro data, but to perform enough micro data collection to allow for testing of aggregation rules, some aggregate analysis based on micro data, and for checking the validity of current aggregation procedures. One alternative would be to use statistical aggregation and disaggregation techniques which are validated every few years.

Data Validation and Data Confidentiality

There must be ways to better validate current data collection and analysis procedures while maintaining data confidentiality. Wider use of current data is critical to maintain the credibility of the system and the unbiasedness of the estimates. One possibility to loosen up confidentiality is to require some disclosure in order to obtain government payments. One cannot evaluate the effect of a program without data on its impacts.

USDA and the Land Grant System

The Land Grant System and its interactions with USDA have few economies of scale or scope at present. One obvious way to use the system effectively is to adopt standards for use in data collection and analysis. This is particularly important for cost and return data. The current system promotes competition and rent seeking. A paper by Pope and Hallam postulated that differences in economists evaluations of different questions may be due to rent seeking behavior. As long as methods for valuing costs and returns are not uniform, any analyst can find a ready market for his method if it meets the needs of a particular clientele group. Such blatant self-serving behavior can only be controlled by the use of reasonable published standards, large and representative data sets, and open access to data and methods.

The way to integrate data needs is to integrate analysis and insist that consistent methods of collecting and aggregating information are used. Measurement is only meaningful in relation to standards and the current system of a standard for every agent is not of any value. Standards should be adopted for both economic modelling and data aggregation. In this way measurement can be useful to a variety of groups with disparate values and needs.

Some Remaining Issues

This chapter has discussed the data needs of alternative groups of data users. An integrative economic approach to analysis resolves many of the current data conflicts. Two important issues remain to be addressed. The first has to do with the form in which analysis is presented. Cost of production estimates are usually prepared as point estimates. They simply report the cost of production of a given entity for a given time period. Cost of production forecasts are usually of the same point estimate form. Such estimates are useful to capture a snapshot of an industry or firm but must be used cautiously for analyzing industry structure or direction. Furthermore, causal analyses are perilous with such point estimates. Supply, demand and production analysis is concerned with the determination of functional relationships between important economic variables. Such analysis is compatible with welfare analysis, policy projections and structural studies. While the same data can often be used for both types of analysis, the implications that can be drawn are often different. Special care must be taken to not use point estimates to discuss structure.

The final issue has to do with what can be called "defensibility" in relation to cost and revenue estimation. While non-sample based methods of data collection may be useful for some purposes, such as determining practices among a select group, extreme care should be used in extrapolating the costs or returns of sub-samples or 'representative' samples to entire populations. Unfortunately, non-random samples often tell a variety of stories about the population, each which may be endearing to a particular interest group. Unless economists insist on defensible procedures, they will find themselves in high demand to justify whatever policy issue becomes important. In this case, let the public beware.

References

Lucas, R.E. 1981. "Econometric Policy Evaluation: A Critique," Lucas, R.E., ed., *Studies in Business Cycle Theory*. Cambridge: MIT Press.

Pope, R.D. and R.G. Chambers. 1989. "Price Aggregation When Price Taking Firms Vary." *Review of Economic Studies* 56 (2): 297-310.

R. D. Pope and A. Hallam. 1986. "A Confusion of Agricultural Economists?--A Professional Interest Survey and Essay." *American Journal of Agricultural Economics* 68: 572-594.

21

How the Profession Could Coordinate Itself on This Issue

R. J. Hildreth

The agricultural economics profession has had many discussions, indeed, controversy over accounting for costs and returns. Henry C. Taylor, at the December, 1922 meeting of the American Farm Economics Association, presented a paper, "The Objectives in Agricultural Cost Accounting." Taylor discussed a number of then current issues in cost accounting. They included methods of assembling cost data, which he gave only passing attention. His major emphasis was on methods of analyzing and interpreting cost data. Taylor stated the primary purpose of agriculture accounting is to furnish a manager the basis for making a choice between two or more methods of carrying out a given line of production, including the most profitable method of disposing of crops. He also said agricultural accounting is the "...basis for making comparisons between returns secured in the different lines of agricultural production in the various parts of the United States" (p.77).

Taylor concluded his presentation with the following statement: "The accounting procedure must depend upon the objective, the questions to which we are seeking an answer. And if the results are to be of assistance in answering the questions which will confront us in the future, they must be expressed in terms which will admit a modification from time to time to conform to the changes in cost rates and prices of products" (p.78).

Taylor and Taylor, in their book, *The Story of Agricultural Economics*, devote Chapter 16 to the efforts made prior to 1933 to apply the techniques of cost accounting to finding the facts essential to planning and operating a farm for maximum profit. They review much of the work at the various universities and in the Bureau of Agricultural Economics, as well as the joint efforts between these groups.

They report on a conference called by the Bureau of Agricultural Economics held in Chicago in May, 1923 with the title, "Conference of Research Workers in the Field of Farm Organizations and Cost of Production." The topics were: (1) Objectives in Farm Organization Research; (2) Use of Complete Farm Accounts in Farm Organization Studies; (3) Use of Enterprise Costs in Farm Organization Studies; (4) Use of Farm Survey Method in Studying Farm Organization Problems; (5) How to Make Results Most Useful; and (6) Cooperative Relations. It is interesting that in 1991, almost seven decades later, professional controversy continues about economic accounting for commodity cost and returns on many of these issues.

Currently, concern is expressed about the lack of comparability of cost-of-production estimates and returns. Libbin and Torell, as well as others, demonstrate the divergence between various state and United States Department of Agriculture estimates for the same commodity. The North Central Department Chairs Committee requested the North Central Farm Management Extension Committee consider developing strategies for attaining greater efficiency, effectiveness, and consistency in collecting and utilizing data on cost-of-production. Consistency in methods among states' cost-of-production estimates may be lower than between the Economic Research Service's methods and some individual state's methods.

Why has the profession not resolved these issues or at least made more progress over at least seven decades? It is my hypothesis that the attention of a large part of the profession became concerned with "production economics" rather than "farm management" in the 1950s and 1960s. Linear programming and other decision models followed. The power and potential usefulness of these tools and the difficulty of dealing with cost-of-production and returns led to only modest attention to the issues. In addition, reduction in Economic Research Service (ERS) field staff resulted in the limited coordination of ERS efforts with economists at land grant universities. The lack of an organizational structure for coordination and interaction between ERS and state economists also contributed.

Much of the recent concern with the issues comes from increased use of cost-of-production estimates by policy actors: Congress (both members and staff); commodity and general farm organizations; agencies such as the Congressional Budget Office and U.S. General Accounting Office; and others. The lack of consistent estimates of cost-of-production has been noted by these actors.

How can the profession achieve coordination on this important and controversial set of issues? In some ways it is not too much different than the issues faced by the accounting profession. The Financial Accounting Standards Board (FASB) is an organizational device which has significant impact on accounting procedures. Accounting standards were basically set by the American Institute of Certified Public Accountants (AICPA) through the Accounting Principles Board until 1973, though supplemented by certain requirements imposed by the Securities and Exchange Commission (SEC) on publicly-held companies (Chazen). As the result of criticism, adverse publicity, and litigation, the AICPA recommended the demise of the Accounting Principles Board and the birth of the FASB which is an organization independent of the AICPA.

The FASB has established standards of financial accounting and reporting since 1973 as a private sector organization. The Generally Accepted Accounting Principles (GAAP), standards developed by FASB, govern the preparation of all financial reports for both public and private companies. Thus, the FASB is a voluntary group but accepted by the accounting profession and has the approval of the SEC and other government agencies. The FASB is not without criticism and concern exists about its standards. For example, a recent *Wall Street Journal* article discusses a controversial new rule requiring companies beginning in 1993 to accrue, or set up a reserve for, future medical benefits of retirees rather than deduct them from reported profits each year when the premiums are paid as is the current practice (Berton). It is estimated this rule may reduce corporate profits by at least $200 billion and perhaps as much a $1 trillion over a period of years.

I would raise the issue of whether or not the profession needs to develop, not an FASB, but some organizational means of examining the continuing controversies about economic accounting for farm costs and returns in order to achieve coordination among states and the ERS.

Two standards entities now exist. The Farm Financial Standards Task Force, led by the American Bankers Association, developed standard farm financial statement formats, identified certain financial measures that are common in all areas of the country, identified standard methods for calculating these measures and developed content standards and calculation standards for farm financial software. The members for the Task Force were from many of the facets of the agricultural financial industry: lenders, regulators of financial institutions, research and extension personnel, accountants, representatives of farm groups and the ERS. The other entity, described in chapters 1 and 8, is the National Agricultural Cost of Production Standards Review Board.

But the need to improve consistency and achieve coordination remains. Could the AAEA's Economic Statistics Committee develop a plan for the implementation of task forces to examine and develop accounting procedures and, perhaps in the next seven decades, agree upon a few standards? Since the 1920s the issue of coordination and cooperation between universities, farmers, lenders and government has arisen during the dialogue on the difficult issues of the economic cost accounting. Why not formalize the cooperation and develop a means whereby different points of view can be expressed and judgments made about accounting procedures issue by issue? As a first step, the Economic Statistics Committee, or a subgroup, would have to structure the issues. The Cost of Production

Standards Review Board could contribute to the discussion as well as consider judgments reached.

Considerable dialogue and debate on accounting procedures for a given issue would have to take place, as well as careful consideration of alternative methods to achieve the goal. All of the parties would have to be willing to sit down, talk to each other and consider alternative points of view. But it seems possible to me that, within the framework of the American Agricultural Economics Association, the profession could arrive at a more useful method for continuing, into the next seven decades, discussion about economic accounting for commodity costs and returns.

References

Berton, Lee. 1990. "FASB Issues Rule Changes on Benefits." *The Wall Street Journal.* Dec. 20, p. A3.

Chazen, Charles. 1979. "Accounting and Auditing Pronouncements: Their Source and Authority." *The Practical Accountant.* pp. 54-60. September.

Libbin, James D. and L. Allen Torell. 1990. "A Comparison of State and USDA Cost and Return Estimates." *Western Journal of Agricultural Economics* 15: 300-310.

Taylor, H.C. 1923. "The Objectives in Agricultural Cost Accounting." *Journal of Farm Economics* 5: 65-78.

Taylor, H.C., and Anne DeWees Taylor. 1952. *The Story of Agricultural Economics in the United States, 1840-1932.* Ames, IA: Iowa State College Press.

APPENDIX

Computing Considerations

Appendix

Computing on Personal Computers: Three Case Studies

Lawrence A. Lippke, Jerry Crews, and James C. Wade

Introduction

The type of calculator one uses for computing does not affect the quality of cost and return budgets. In effect, the shortcomings or concerns when using personal computers (PCs) for computing costs and returns are the same as for any method. The concern about accuracy and representation of the data, together with the economic assumptions underlying the calculations, are no different when using PCs.

What is different, and critical in some situations, is the ability of analysts computing on PCs to put in the hands of end-users the same data, calculation procedures, and reporting capabilities used by those publishing the costs and returns estimates. Depending on institutional restrictions, this delivery to end-users may or may not occur, but technologically it is possible.

To better understand the types of PC budgeting software that is available and how these PCs are being used, we describe three different types of software currently in use to develop enterprise budgets. The three types do not fully represent the software in use across the U.S., but simply serve as examples. The first type to be described is a full budget generator. This is a coded program which closely emulates the features of the Oklahoma State Budget Generator which has been in use in many states for over 20 years. The second type of budget software is a template which has been written for use with an electronic spreadsheet. These spreadsheet programs are widely used by both university faculty and agricultural producers as both decision aids and record keeping programs, and they allow for easy transfer from budget developer to budget user. Finally, the third type of budget software which will be described is written using the data storage and programming capability of a database manager.

Microcomputer Budget Management System

The Microcomputer Budget Management System (MBMS) was developed in the mid 1980s as a cooperative effort between Texas, California, and Oklahoma and was supported financially through a grant from the Kellogg Foundation. It is currently used in Texas to develop, annually update, and publish about 230 crop and livestock budgets for 14 Extension districts.

MBMS provides for systematic storage of all the factors affecting the costs and returns of enterprise production, retrieves information about those factors whenever a budget is calculated, and prints the budget in any of several formats available to the user. MBMS also has the capability of merging several budgets to create a whole farm income statement and a cash flow projection. In developing the whole farm budget, however, the user may also enter costs and returns which have not been allocated to the individual enterprises. This feature accounts for whole farm overhead, non-enterprise income, and other costs and returns not directly associated with any given enterprise.

Four types of information are stored by MBMS. First are the resources used by the farm. Resource information for machinery, breeding livestock, land, operating inputs, custom operations, labor, and management are stored for later retrieval. The information stored with each resource generally consists of the resource price, unit, weight per unit, and cash flow type. In addition, for machinery, irrigation equipment, and other depreciable assets, factors for estimating repair cost and depreciation are stored. The second type of information stored is the parameters affecting the use of resources. Such items as interest rates, fuel prices, wage rates, property tax rates and insurance rates are kept for use in the budget. Third, information about the commodities produced on the farm, specifically prices and weights, are stored. Finally, the budget itself is stored, but the budget consists at this point of a calendar of production operations and operating inputs used by an enterprise. In operations, the "times over" are specified; with operating inputs and products produced, the number of units are entered.

With these four sets of information in place, MBMS then calculates the budget by combining the appropriate number of resources and products, calculates their values as affected by the parameters, and prints the results.

Budget reports can take any of several formats and can be printed for different resource owner situations. When specifying the budget, the user can indicate for each income and expense item what percent a landlord receives or pays. In addition, each item can be specified as a cash or non-cash item and whether it is a fixed or variable factor. The user can then print owner operator, landlord, and tenant budgets, and can print them with both cash/noncash and economic cost assumptions.

While MBMS is a powerful budgeting tool, its major drawback is that it is complex and not likely not be used by producers. Other concerns about MBMS are its extensive use of agricultural engineering models for estimating machinery cost and its inability to explicitly allocate overhead expenses to individuals, but these are not peculiar to MBMS. Its main advantage, especially for those publishing annual budgets, is its ease of update. For example, if all that changes from one year to the next was the price of fuel, that change need only be made to the parameter file. All budgets could then be recomputed and the new fuel price would be reflected in the budgets.

Electronic Spreadsheets

Extension economists at Auburn University and several other universities across the U.S. use electronic spreadsheets for enterprise budget development. In doing so, they are providing information to agricultural producers regarding profit potential of various enterprises, breakeven prices to use in developing marketing plans, farm program alternatives, cash flow, and the sensitivity of net returns to price and yield variability.

The reasons for using electronic spreadsheets vary from state to state, but likely have certain considerations in common. First, budgets developed on electronic spreadsheets can easily be delivered to end users, and those end users can adapt them to their individual operations. Second, developing budgets on electronic spreadsheets requires less programming skill than with a coded program, thereby cutting the cost and time of development. Spreadsheets also have built in features to allow for statistical and financial calculations, for database backup, and for graphics. Finally, spreadsheets can be made easier to use though the use of keyboard and programming "macros" in which certain features, such as printing, can be invoked with a single keystroke.

Factors affecting the design of a spreadsheet template concern both the use and the user. In the use, one must consider whether the template will be used primarily as a budget generator, calculating and printing a large number of budgets, or whether it will be single application oriented. These considerations may impact the degree to which data linking or lookup is used.

Concerning the user, one must consider the level of computer literacy held by those users. The designer must decide whether to make the template plain, possibly putting the user at a disadvantage, or to give it a lot of "bells and whistles", a disadvantage to the developer.

The spreadsheets developed at Auburn are totally integrated, combining a machinery complement (and machinery costs) and government program options with the prices and quantities of other inputs and production to develop the enterprise budget. These budgets can then be merged to whole farm analysis and cash flow projection.

Once the budgets are calculated, they can be printed in a fairly standard format. Along with variable and fixed costs, the budgets also reflect breakeven prices and yield to cover both variable and total costs. The budget templates also create a sensitivity table, varying prices, yields or other factors peculiar to the enterprise to determine the effects on net returns as stochastic variables change.

Another feature of the Auburn spreadsheets is a section to assist in determining pesticide costs. This section specifies chemical or trade names of those pesticides approved for the crop being budgeted along with a default set of prices and application rates. By specifying the number of applications of each chemical being applied, the spreadsheet is able to estimate those pesticide costs. The costs are then transferred to the budget itself.

Electronic spreadsheets have been found both to enhance the educational efforts of Extension agents and specialists and to enhance technology and delivery. Spreadsheets also allow for site specific modification, making the budgets much more useful for producers. Electronic spreadsheets also decrease the distance between developer and end user, as the same set of data, formulas, and economic assumptions used by the developer can be placed on the desk of the end user. This carries some risk of unwarranted modification by the end user, but spreadsheet enhancements which allow for protecting critical sections of spreadsheets can be used to minimize this risk. All in all, the ease of development and use, the allowance for end users to directly use the templates, and the compatibility of spreadsheets with other microcomputer software, makes them highly useful for those concerned with producer decisions.

Database Management Software

The data requirements for cost estimates can be extensive if several enterprises and/or locations are considered. As previously pointed out, such considerations are not new. The MBMS system is a prime example of a cost estimation system that utilizes extensive data. Thus, the database orientation described here is simply a new twist on an old theme. Database management techniques offer the opportunity to handle large amounts of data in an efficient and systematic manner with a minimum of programming skills. They also provide a basis for extensive summary reporting of the data through precoded (supplied by the database software developer) and user coded techniques.

Database management systems (DBMS) rely on extensive organization of data to provide a basis for operations and analysis. To illustrate the use of DBMS in facilitating the estimation of production costs on microcomputers, the Arizona Crop Budgeting System (ACBS) is briefly described. This system is an outgrowth of about 20 years of work on Arizona's version of a coded budget generator much like MBMS, that considers some of the unique characteristics of diversified irrigated agriculture, but operated on a main frame computer. New computer systems required a new approach to the crop budgeting that used PC's. Building on the concepts previously developed, a DBMS approach was designed and implemented. ACBS is used, in combination with an extensive data collection effort, to create annual cost estimates on about 125 crop-area combinations including field, vegetable, and tree and vine crops.

The basic organization for ACBS is the "farm" and much of the data organization is similar to that of MBMS. A "farm" can be generalized to reflect a single business or to represent a geographical area such as a county, irrigation district, or state. Using a business definition, a farm is composed of the resources available to the business to carry out farm production enterprises and produce an output. A farm level database contains relevant data about the farm organization and resources. At the next level of abstraction, enterprises are defined in terms of production activities and inputs and products produced. Input requirements are defined in terms of the individual operations of the enterprise. The operations require specific inputs such as labor, machinery, chemicals, and other materials. A database contains all operations of a farm's enterprises. Additional databases for labor, machinery, material (fertilizer, pesticides, and others), and custom services costs are maintained on a farm or regional basis. While all of this organization is not substantially different from other techniques, several differences in data management increase the efficiency through DBMS.

Resource constraints played an important role in the choice of DBMS for cost analysis. The lack of funds to provide large blocks of time for data collection and analysis required an approach that could be controlled and operated by a few individuals. Since the budget analyses are in high demand by a large cross section of users from farmers to consultants, information provided had to be detailed and easily traceable to provide the accountability required by the users. Finally, the product had to provide information on a large and growing number of enterprises. DBMS provided characteristics that aided in meeting the goals and constraints of Arizona's cost estimation process. Eight of these characteristics are outlined below:

1. Many databases can be made available at any one time. For example, data on labor, fertilizer, and pesticides are available to the user without

bringing all of the data into computer memory. Organization of data can be as detailed as the program requires.

2. The advantages of DBMS's are in their abilities to facilitate data organization. All files are indexed. Indexing is a type of data sorting using key variables noted in separate files. These keys, or indexed variables, point to the desired data in the prespecified order. Any data file can have many indexes. However, none of the indices actually modify or physically rearrange the data file. If the data file is changed, the indices are automatically updated.

3. Relational characteristics of DBMS provide the user with the ability to describe a variable in one data file which automatically advances a number of additional data files to provide data that is related to the defined variable. A simple example is selecting a fertilizer for an operation and instantly having available from other data files the physical characteristics, purchase units (e.g., tons), application units (e.g., pounds), purchase price and recommended application rate. The relational characteristic creates efficiencies in memory requirements and processing time.

4. DBMS also provides for evaluation and updating of data with preprogrammed editing and reporting features reducing programming requirements and providing for easy checks on large quantities of data. The ease of data update increases the amount of input data and the number of cropping alternatives that can be considered using the limited resources available.

5. Modularity of coding allows expansion to consider other aspects of cost estimation without substantial change in existing functions. This modularity is available in the large family of dBase languages and through linkages to several expert systems and the "C" family of languages. The dBase family is available on both IBM-compatible and Macintosh personal computers. ACBS is programmed in FOXBASE+,and extremely fast dialect of dBase which is compilable. The speed and compilability of the software was chosen to facilitate the ultimate distribution of the system.

6. Data compatibility allows linkages to other software and to other data sets. For example, large databases such as farm record keeping services could be linked to systems like ACBS to information on input usage or overhead costs.

7. Annual updates of data such as machinery or chemical prices require only the distribution of new databases.

8. Context sensitive HELP is available in ACBS. This feature allows the user to examine the feasible responses to data entry inquiries. For example, if an operation requires a tractor, pressing the HELP key will bring to the user's attention all tractors available to the farm. The user selects from this list. This type of help is available for all inputs available to the farm. This feature, while not unique to DBMS, is an innovative use of the full screen

features of PC's to lead the user to provide the appropriate response without extensive training or the use of long lists of equipment or material codes.

In summary, the efficiencies of data handling available using DBMS and the innovative features of PC's allow more farms and enterprises to be analyzed more completely and effectively with limited resources. The development of ACBS is an on-going activity of the University of Arizona's Cooperative Extension with additions planned for whole farm, perennial crops, water pumping, and irrigation applications costs.

Conclusion

PCs provide new and improved opportunities to create much better produces, not just for the distribution to end users, but to provide better data handling and more innovation in the data collection and processing. PCs provide these features at low prices, and we, and a profession, must learn to use these features effectively to our advantage. Our concerns for data quality and calculation procedures are the same, but is PCs offer anything, the offer is low case, high quality software to go with the hardware. These have been major revolutions in software that will affect everything we do.